W9-CBI-596

Analog and Digital Electronics for Scientists

A typical electronic component, shown both as a complete unit and with its cover removed to expose the actual silicon chip. The background is a portion of the artwork for the same device. (Photo courtesy of Analog Devices, Inc.)

ANALOG AND DIGITAL ELECTRONICS FOR SCIENTISTS

Basil H. Vassos
Professor of Chemistry
The University of Puerto Rico

Galen W. Ewing
Professor Emeritus of Chemistry
Seton Hall University

Third Edition

A WILEY-INTERSCIENCE PUBLICATION
JOHN WILEY & SONS
New York • Chichester • Brisbane • Toronto • Singapore

Library of Congress Cataloging in Publication Data:

Vassos, Basil H.
Analog and digital electronics for scientists.

"A Wiley-Interscience publication."
Includes index.
1. Electronic instruments. 2. Analog electronic systems. 3. Digital electronics. I. Ewing, Galen Wood, 1914– . II. Title.

TK7878.4.V37 1985 621.381′028 85-625
ISBN 0-471-81138-6

Printed in the United States of America

10 9 8 7 6 5 4 3 2

PREFACE

The five years since the publication of the previous edition have seen great strides in connection with microprocessors. They are now designed into almost every type of laboratory instrument as well as a large variety of consumer and industrial products. This has been made possible principally by expanding horizons in the design and fabrication of large-scale integrated circuits. It has also required the development of fast and flexible interfacing devices and circuits. Any textbook on basic electronics written today must take these developments into account, even though the actual design of microprocessor circuitry is not within their purview.

This book is intended, as were previous editions, for the use of scientists who need to understand the circuitry employed in instruments found in their laboratories and who may be called upon to design and construct some instruments of their own. A brief introduction to microcomputers is included, with the objective of indicating how they can best be utilized. This treatment should serve to bridge the gap between electronics and computer design.

The third edition is somewhat more rigorous than its predecessors, with more extensive use of mathematical techniques. It is designed as a first book in electronics for use in college and university classes at both undergraduate and graduate levels. In terms of background, calculus and a year of physics are assumed.

The experimental section of the book has been expanded, including material suitable for a half-day laboratory per week for a semester.

It may be of interest to note that the text has been written with the aid of a word processor consisting of an Osborne-1 computer using the WordStar software (MicroPro, Inc.) and prepared in the form of camera-ready copy on a Texas Instruments Model 855 printer.

BASIL H. VASSOS
GALEN W. EWING

Rio Piedras, Puerto Rico
Las Vegas, New Mexico
May 1985

CONTENTS

PART ONE

PART TWO

PART ONE

I

INTRODUCTION AND OVERVIEW

Modern laboratory instruments contain such extensive electronic circuitry, including microcomputers, that scientists of all disciplines must have at least a rudimentary acquaintance with the principles of electronics.

A working knowledge of basic electronics is important to any scientist who needs to design a special-purpose device to meet specific requirements. Trouble-shooting or repair, or even intelligent conversation with a service technician, is greatly facilitated by a knowledge of electronics.

ANALOG AND DIGITAL SIGNALS

Electronics shows a marked dichotomy between an analog and a digital domain, as suggested by the title of this book. This is reflected in both hardware and circuitry, and it is essential that the distinction be stressed at the very beginning of our study.

The majority of laboratory instruments used for measuring physical properties accept information in a form that may vary continuously over a wide range of values, an *analog* signal. In some cases, however, information is more readily available in a *digital* form, in which the measuring instrument need only respond numerically to its input.

This distinction can be made clear by an example. The temperature of water in a container is a continuously variable quantity that can have any value between the freezing and boiling points. Hence an electrical thermometer used to observe this temperature must be capable of producing a voltage (or current) that follows exactly any temperature changes; we say that it yields an analog signal. If the information is to be processed by a computer, however, it must be provided in digital form, because that is the only type of signal that a computer can understand. Some form of analog-to-digital conversion, or "digitization," is required.

The distinction between digital and analog properties of signals can be carried over into the internal structure of instruments. Lift the cover of any digital computer, and you will see a myriad of conductors, every one of which can be electrically in one of only *two* voltage states. For a particular computer these might be +5 and 0 V. In contrast, an analog instrument contains conductors that can have *any* potentials within a continuous range.

Both analog and digital techniques have their areas of special competence. For some applications, either of them can be used, but generally the two are not easy to combine. Nevertheless, a great many laboratory instruments utilize both, and the technique of conversion from one domain to the other requires serious attention.

One major advantage of digital over analog circuitry is its low vulnerability to noise and interference. This is because in any well designed digital circuit, moderate amounts of noise can almost never produce a false bit of information. Devices operating in the analog domain lack such a safety margin against interference, and so their ultimate ability to respond to small signals is limited.

ELECTRONIC MEMORY

Frequently in electronics, especially in connection with computers, it is necessary to store information either temporarily or permanently. Electronic memories are available in both analog and digital forms. To retain an analog signal, the corresponding voltage is impressed on a capacitor, which can hold it for a time. This has its important applications, as is seen in a subsequent chapter. For long-term storage or for large files of data, however, it is more appropriate to use digital memory. It can be implemented with a variety of solid-state devices, which will retain the data reliably unless the power is interrupted. For storage on a permanent basis, magnetic materials are used; these include rigid disks, diskettes, and tapes.

DISCRETE AND INTEGRATED CIRCUITS

Most of the circuits treated in this book could be assembled from an assortment of individual transistors, resistors, capacitors, and a few other components. However, much time and effort can be saved through the use of *integrated circuits* (ICs). These are modules, each containing the equivalent of many discrete components, all fabricated simultaneously on a substrate of crystalline silicon. The internal components of an IC are interconnected to form a complete circuit, optimized for a specific application. A large variety of ICs are available commercially.

The advantages of using ICs rather than discrete assemblies are many: (1) most of the circuit design and optimization has already been done by experts; (2) less space is required in the final product; and (3) ICs are considerably less expensive. There are still some vital applications that require discrete components, and these will be mentioned as they arise, but our emphasis is on the use of ICs.

COMPUTERS

A few decades ago analog computers were quite important in military and industrial applications. These were built up from an array of units called *operational amplifiers* (op amps), interconnected in such a way that elaborate mathematical tasks could be performed. Analog computers are now seldom seen and are not discussed in detail here; however, operational amplifiers are discussed.

Most computers today are digital. They all follow roughly similar designs, from the large mainframe computers, through minicomputers to the popular microcomputers. They are built around a central processing unit (CPU), to which are connected other items, including various types of memory and interfaces to the outside world. Microcomputers are frequently included in modern laboratory instruments, but it is also possible to connect an instrument to an external computer. So-called personal computers are often appropriate for this purpose. The implications of computers in laboratory instrumentation are extremely significant and are considered in a later chapter.

SIGNAL-TO-NOISE RATIO (S/N)

By "noise" we mean spurious voltages of the same sort as the information-carrying signals. The ratio of signal voltage to noise voltage (the S/N ratio) is a widely used figure of merit.

Improvement of the S/N ratio can be approached by either of two different routes: (1) increasing the signal while maintaining a constant noise level and (2) diminishing the noise at its source or during subsequent data processing steps.

II

SIGNALS AND MEASUREMENTS

Since nearly all aspects of instrumentation are concerned with the generation, processing, and measurement of signals, it is important that the reader understands thoroughly what is meant by the term "signal."

For our purposes a signal is best defined as a flow of energy that carries information. The *carrier* can be an electric current, a radio wave, or a light beam; the concept can even be extended to a piece of paper. A steady current or a steady light beam itself carries no information, any more than does a blank sheet of paper. Just as the paper must be written on to provide the information content, so must the current or beam of light be varied in some way from its normal condition if it is to carry information. This process of endowing the carrier with information is called *modulation*. Its recovery from the carrier is *demodulation*.

The process of modulation of electrical carriers can be performed in various ways. The simplest is to make the current or voltage take a value that represents directly the desired information. For example, a thermocouple gives a voltage output that is a direct measure of the temperature difference between its two junctions. We can refer to this type of signal as an analog-modulated signal, or in short, an analog signal.

The carrier itself can be a DC current, as in the thermocouple, but analog modulation can also be accomplished with an AC flow of energy. Examples of this latter type are found in radio transmission, where the sound pressure at the microphone is first transformed into an electrical current that is then used to modulate an electromagnetic wave.

It is also possible to modulate a carrier by digital techniques. This process requires two distinct states of the carrier, usually designated as "logic 0" and "logic 1." The carrier of digital modulation can be a DC current that may have two values, such as 4 mA for 0 and 20 mA for 1, or it can be an AC current that assumes different amplitudes or different frequencies for the two logic states. For example, consider a digital signal consisting of the sequence of logic states designated as 1,0,0,1. This could be represented by either a current that takes the successive values of 20, 4, 4, and 20 mA or by an AC current that follows the sequence 120, 100, 100, and 120 kHz. Equally well, the information could be carried by a DC voltage of 5, 0, 0, and 5 V.

ELECTRICAL CIRCUITS

Information is carried within instruments in closed paths called *circuits*. A circuit consists typically of an energy source and various electronic components connected by conductors to close a loop (Figure 2-1). The connecting wires can be considered to be perfect conductors, so that the potential is identical throughout a given wire. The wires as a rule present no impediment to the passage of the current. The devices designated as *A*, *B*, and *C*, in contrast, do interact with the current flow. These may include resistors, capacitors, inductors, or transistors. Conventionally the return path of the current to the power supply is not shown explicitly in circuit diagrams (Figure 2-1*b*).

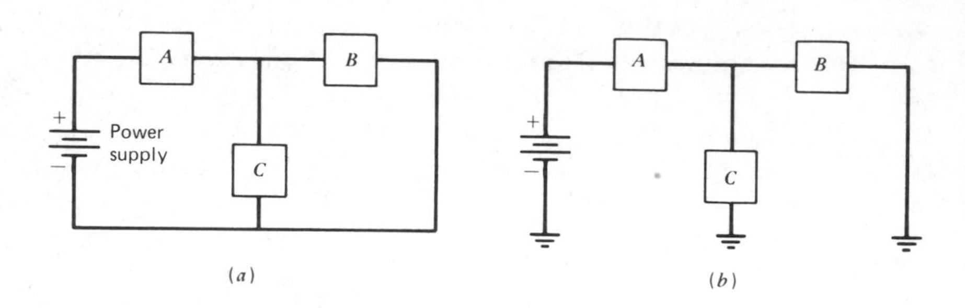

Figure 2-1. (*a*) An example of a circuit consisting of two loops. (*b*) The same circuit, shown with the ground convention.

Instead, a "ground" is established for the entire circuit, to serve not only as the return or "common" lead for the power supply, but also as the zero or reference level for all potentials. The "ground" of a circuit may or may not be connected to the third wire of the power line.

ELECTRICAL QUANTITIES

The electrical quantities of primary concern to us are listed in Table 2-1, together with the names of the basic units and commonly used symbols. A distinction is sometimes made between instantaneous voltages and currents (with lowercase symbols e and i) and the steady-state quantities (E and I). In this book the symbol V is used mostly for power supply voltages, where it is sometimes given a double subscript, such as V_{CC}. Vectorial quantities are shown in boldface.

The basic form of alternating current (AC)* is the sinusoidal wave shown in Figure 2-2. Such waves need

* The term "AC" is used loosely to describe both alternating currents and alternating voltages. There is no "AV."

TABLE 2-1

Electrical Quantities[a]

Quantity	Unit	Symbol
Potential	Volt (V)	E, e, V
Charge	Coulomb (C)	Q
Current	Ampere (A)	I, i
Power	Watt (W)	P
Resistance	Ohm (Ω)	R
Conductance	Siemens (S)[b]	g
Impedance	Ohm (Ω)	Z
Admittance	Siemens (S)[b]	G
Capacitance	Farad (F)[c]	C
Inductance	Henry (H)	L
Frequency	Hertz (Hz)	f

a The usual multipliers (k, m, M, μ, etc.) are used with the unit abbreviations.

b The siemens is often given the older name "mho" with the abbreviation Ω^{-1}.

c One is likely to encounter in the literature various nonstandard abbreviations for the microfarad (mfd, MFD, uF) and for the picofarad (μμF, mmf, uuF).

three parameters for a complete description: the amplitude A, the frequency f, and the phase ϕ. Since the signal has different amplitudes at different moments, it is convenient to define a type of average called the *root-mean-square* (RMS) voltage:

$$E_{RMS} = \left(\overline{E^2}\right)^{1/2} \tag{2-1}$$

where the bar over the E^2 indicates the time average. A similar formula is valid for the RMS current. For sine waves, the RMS value is $1/\sqrt{2}$ or 70.7% of the peak value A. The line voltage is conventionally expressed in terms of RMS quantities: thus "115 V RMS" means 115 $\times \sqrt{2}$ = 163 V amplitude. The excursions are both positive and negative with respect to ground, resulting in 326 V peak to peak.

The power in an AC circuit has a pulsating characteristic since it must be zero whenever the current passes through zero, twice in every cycle. The RMS voltage can be defined as the value of an equivalent DC voltage that would cause the same heating effect when applied to a resistor. Thus 115 V AC (RMS) and 115 V DC will provide equal total heat in a kitchen toaster. Figure 2-2*b* shows that two signals of the same frequency and amplitude can have different values at the same instant. Signal *B* reaches any particular point earlier than does *A*; *B* is said to lead and *A* to lag in *phase*.

Sinusoidal currents or voltages have a number of unique properties. For instance, any sum of sine waves *of the same frequency* reduces to a single sine wave with phase and amplitude determined by its components. Another interesting property is that all derivatives and integrals of sine waves are also sine waves, differing only in amplitude and phase from the original.

Consider a circuit that contains a single source of AC at frequency f. Because of the interaction between the sine wave and the different components of the circuit, the sinusoid may have different phases at different points. It is often convenient to assign zero phase angle to the source voltage and describe the phases of all other voltages with respect to this standard. These voltages can be described by an equation of the type

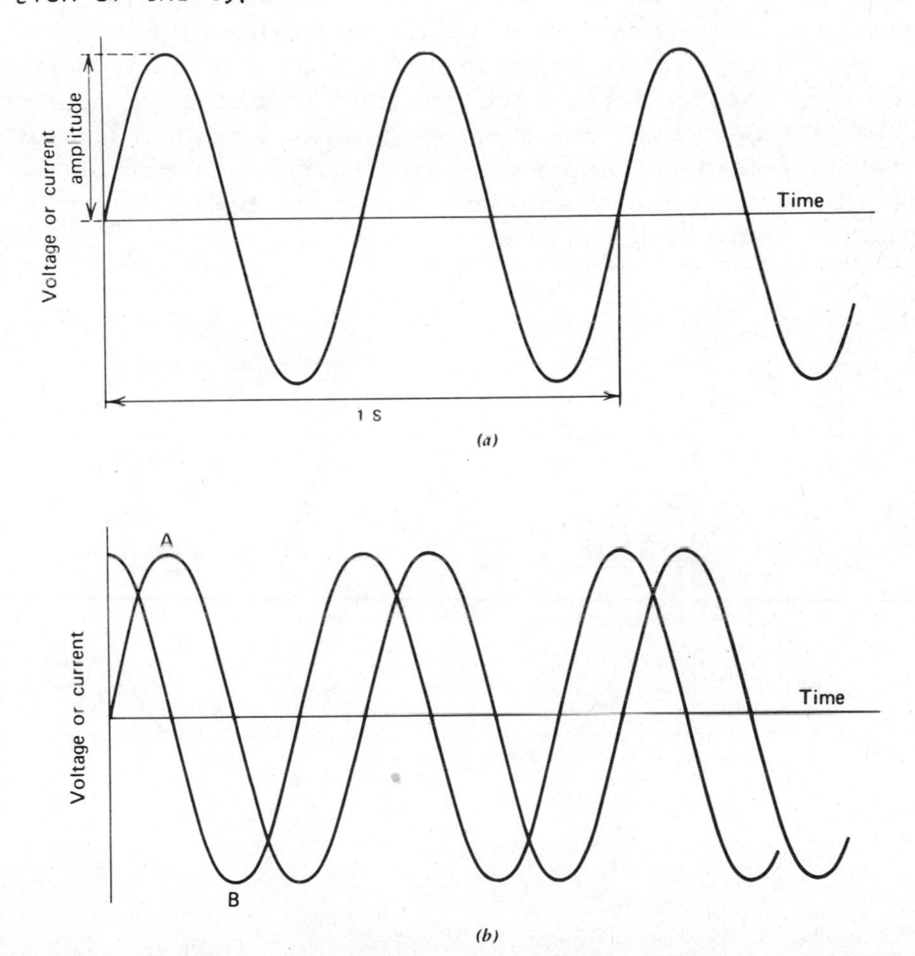

Figure 2-2. (*a*) An example of an AC signal; the frequency is 2 Hz, determined by the number of complete cycles in 1 s; (*b*) two signals of equal frequency and amplitude, exhibiting a phase difference.

$$E = A \sin (2\pi f t + \phi) = A \sin (\omega t + \phi) \qquad (2\text{-}2)$$

where t is the time, A the amplitude, and ϕ the phase difference with respect to the reference. If two signals have the same value of f but differ in ϕ, they will appear displaced with respect to each other along the time axis, as in Figure 2-2*b*. Usually this displacement is expressed in terms of angles of rotation of a vector rather than in time units (Figure 2-3). For example, a difference of 180° (π radians) means that the two waves are exactly opposite to each other (out of phase). The waves in Figure 2-2*b* differ by 90° or $\pi/2$ rad and are thus said to be 90° out of phase. Note that curve B is a cosine wave.

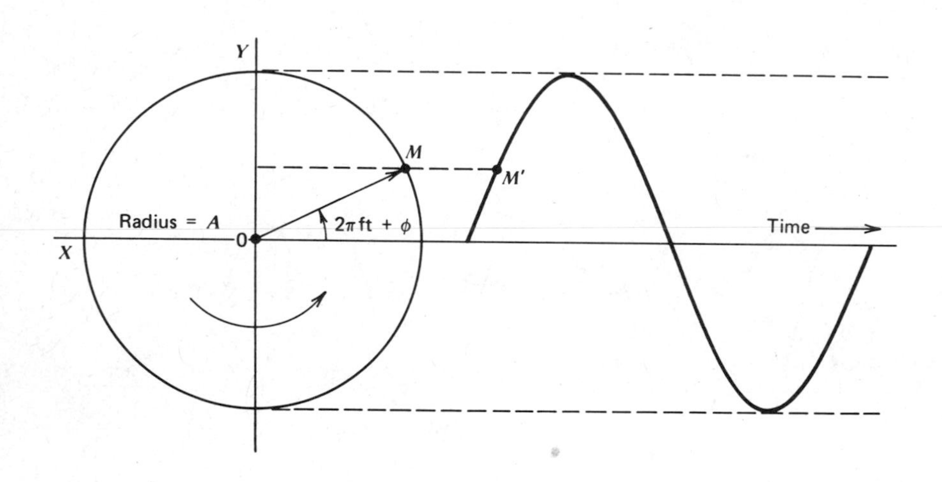

Figure 2-3. Generation of a sine wave from circular motion. The radius **OM** (magnitude A) can be considered to be a vector rotating in the counterclockwise direction. The position of M, projected on the Y axis, becomes a sine wave when represented as a function of time.

PHASORS*

The representation of sinusoids in terms of rotating vectors can be simplified if we consider the situation at time zero. The sine wave can then be described by a *stationary* vector of amplitude A and angle ϕ with respect to the x-axis. This corresponds (in Figure 2-3) to the vector **OM** standing still rather than rotating. It is customary to represent such a vector in a complex plane, that is, a plane in which the abscissa x is real and the ordinate y is imaginary, with j taken as $\sqrt{-1}$. This representation, describing a voltage or current as a vector in the complex plane, is referred to as a *phasor* (Figure 2-4). Note that only a single frequency is considered in any given vectorial representation.

Also indicated in Figure 2-4 are the mathematical relationships between the *polar coordinates* A and ϕ and the rectangular coordinates x and jy. Thus a phasor can be equally well described as a vector by the notation $A \angle \phi$ or as the complex number $(x + jy)$. Simple trigonometric relationships permit the interconversion of the two representations:

$$x = A \cos \phi \qquad A = (x^2 + y^2)^{1/2} \qquad (2\text{-}3)$$

$$y = A \sin \phi \qquad \phi = \arctan (y/x) \qquad (2\text{-}4)$$

The fact that AC currents and voltages can be represented by complex numbers implies that the voltage/current ratio in a circuit, defined as the *impe-*

* A more complete exposition of phasors is given in Chapter VI.

dance, can also be represented by a complex number. For example, if the current I is taken as phase reference and the phase angle of the voltage E is found to be $+90^\circ$, the impedance is $\mathbf{Z} = (E \angle 90)/(I \angle 0) = (E/I) \angle 90$. The impedance in this case is a purely imaginary number, as the reader can ascertain by referring to Eqs. 2-3 and 2-4.

The impedance, as defined previously, can be used to calculate the current when the voltage is known, or vice versa:

$$\mathbf{E} = \mathbf{IZ} \quad \text{or} \quad \mathbf{I} = \mathbf{E}/\mathbf{Z} \tag{2-5}$$

These relationships are valid for AC of any single frequency, including DC. (Direct current can be considered as zero frequency.) In the case of DC, the impedance reduces to resistance, and Eqs. 2-5 become the familiar forms of Ohm's law.

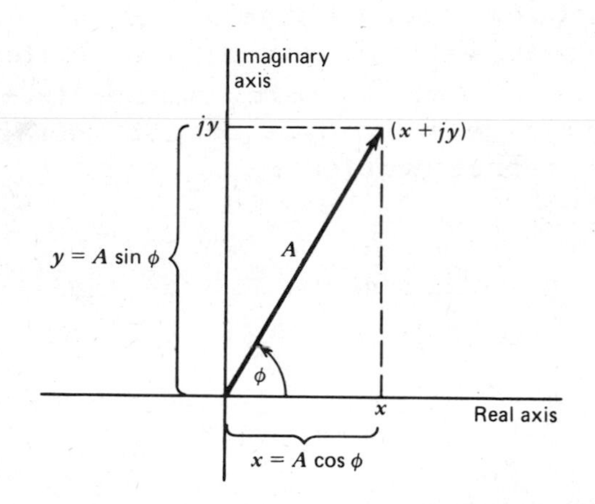

Figure 2-4. Phasor representation of an AC quantity. Note that at $t = 0$, the quantity $\sin(2\pi f t + \phi)$ becomes $\sin \phi$.

Fourier Theorem

The question remains as to how to handle AC signals that are not simple sinusoids, a common occurrence in scientific measurements. A fundamental principle called the *Fourier theorem* states that a periodic function of frequency f is equivalent to the sum of an infinite series of sine waves with frequencies that are multiples of f (harmonics). This is called the *Fourier expansion.** Ordinarily a few terms will suffice to describe the original function with sufficient accuracy for practical purposes. One seldom needs to go beyond the first 10 terms or so (the tenth harmonic) to obtain an excellent approximation of an AC signal.‡ The Fourier theorem can assist in phasor calculations, in that each component frequency can be treated separately.

POWER

Of interest in many electronic operations is the amount of energy lost to the system by transformation into heat. It can be shown that this lost energy is given by

$$P = EI \cos \phi \tag{2-6}$$

where ϕ is the difference between the phase angles of the RMS voltage and current. The cosine term is called

* An additional constant term must be added if the time average of the alternating signal is other than zero. See Chapter XIV for details.

‡ This principle is well known in connection with music. The sound quality of nearly any musical instrument can be approximately reproduced with a synthesizer that operates by adding together the fundamental and a few harmonics with suitably adjusted amplitudes.

the *power factor*. This equation is valid for any circuit or device. In the case of resistors, the phase difference is zero, and

$$P = EI \tag{2-7}$$

or

$$P = \frac{E^2}{R} = I^2R \tag{2-8}$$

An interesting case occurs when the phase difference is 90°, which causes the power factor to become zero. Thus even though both voltage and current are present, no energy is converted into heat. The power, as defined above, is in fact zero. This does not mean, however, that energy is absent, but merely that it is being transported around the circuit without loss.

AMPLIFICATION AND ATTENUATION

An amplifier is a device that serves to augment a signal. The quantity to be amplified can be voltage, current, or power, thereby defining a corresponding voltage, current, or power gain. Since the power in an electric circuit is equal to the product of current and voltage (assuming the power factor to be 1), it follows that the power gain of an amplifier is equal to the product of its voltage and current gains:

$$\frac{P_{out}}{P_{in}} = \frac{E_{out}}{E_{in}} \cdot \frac{I_{out}}{I_{in}} \tag{2-9}$$

It is quite possible for a particular amplifier to have a voltage (or current) gain of unity or even less while still having a large power gain.

An amplifier is an *active device*, meaning that it must extract power from an external source, such as a battery, to increase the power in the signal. No *passive* device (one without access to an energy source) can do this.

The Decibel Scale

An *attenuator* can be considered to be the inverse of an amplifier, diminishing the voltage supplied to it. A familiar example of attenuator is the volume control of a radio.

The gain or attenuation of a device is commonly represented on a logarithmic scale, the *decibel scale*, defined in terms of output/input ratios:

$$dB_E = 20 \log \frac{E_{\text{out}}}{E_{\text{in}}} \tag{2-10}$$

and

$$dB_I = 20 \log \frac{I_{\text{out}}}{I_{\text{in}}} \tag{2-11}$$

For example, a gain of 10^6 corresponds to 20 log (10^6) = 120 dB. The decibel scale has the advantage of using small numbers to represent large ratios. Consecutive stages of amplification or attenuation lead to a sum (rather than a product) of individual gains.

Consider as an example three cascaded amplifiers with gains of 10, 100, and 1000, which give an overall gain of 1,000,000. The corresponding individual decibel gains are 20, 40, and 60, and the total gain is the sum of these, 120 dB.

In addition to the current and voltage definitions of the decibel, a distinct formula is applicable to power amplification. Recalling that the power dissipated in a resistor is $P = E^2/R$, we can substitute for E its equal $(PR)^{1/2}$:

$$dB_E = 20 \log \left(\frac{P_2 R}{P_1 R}\right)^{1/2} = 20 \log \left(\frac{P_2}{P_1}\right)^{1/2} \qquad (2-12)$$

which gives us an equation with a coefficient of 10 rather than 20:

$$dB_P = 10 \log \left(\frac{P_2}{P_1}\right) \qquad (2-13)$$

Gain or attenuation expressed in decibels is often plotted as a function of frequency, with the latter also on a logarithmic scale, as in Figure 2-5. This is called a *Bode plot* and is used for representing the frequency response of any device. The use of logarithmic scales on both axes makes it possible to display a very wide range of frequencies and of gain ratios on a single graph.

TRANSFER CHARACTERISTICS

To increase the amplitude of a 1-mV signal to give 10 V, we can implement an amplifier in a variety of ways, all giving the same end result. The user can ignore the details of implementation and regard such a unit as a "black box" characterized only by a gain of 80 dB. This black-box concept is widely used to simplify electronic thinking. The action of any black-box amplifier can be completely described by a relationship of the form

$$S_{out} = GS_{in} \tag{2-14}$$

Here S_{out} and S_{in} represent output and input signal amplitudes and G is a frequency-dependent constant called the *transfer coefficient*, which may be a complex number. Thus if G is equal to $(-j100)$, the circuit will put out a signal that is amplified one hundred-fold and has a phase angle of 270° with respect to the input. The concept of the transfer coefficient can be extended to the case where the output/input relationship is more complicated, in which case we can write

$$S_{out} = F(S_{in}) \tag{2-15}$$

with F representing some function, such as logarithmic or quadratic. (Note that the term "transfer function" has another meaning and is not to be used in the present context.)

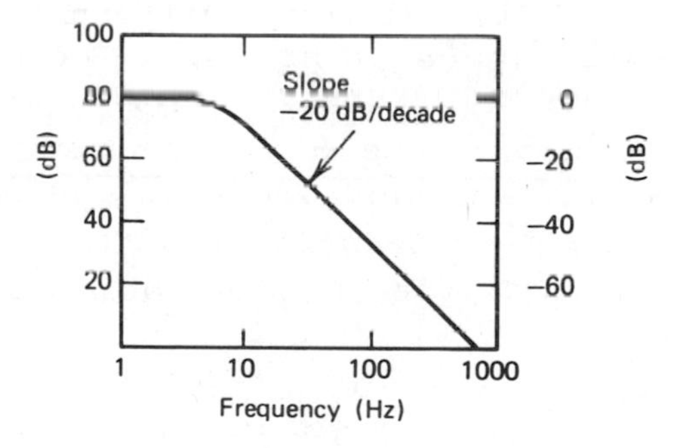

Figure 2-5. Frequency response curve (Bode plot) for an amplifier or attenuator. The same curve is applicable to either, but with different scales. The frequency-selective networks responsible for the sloping portions of this curve often produce a "roll-off" of 20 dB per decade, as shown. An advantage of such representations is that the curves usually consist of intersecting straight lines, either horizontal or at a 45° slope.

The relationship between input and output can also be represented graphically. For example, if G is a real number, the graph of E_{out} versus E_{in} becomes a straight line through the origin. In practice, G remains constant over only a finite range of the variable E. This may be made clear by considering an example. Suppose that we have a linear amplifier with a gain of 1000, hence following the relationship

$$E_{out} = 10^3 \cdot E_{in} \tag{2-16}$$

If this amplifier is an IC, E_{out} will be limited physically to some such range as −10 to +10 V. It follows from the equation that the input should be restricted to the range −10 to +10 mV. If a voltage greater than this is applied, the output will not be able to respond linearly but will level off (*saturate*) at 10 V. At the other extreme, as the input approaches zero, linearity is again lost, but for a different reason, specifically, the presence of *noise*. The noise in well-designed circuits is generally small compared to bona fide signals, but as the signal is reduced, the noise becomes significant and eventually limits the range. Figure 2-6 shows the *transfer plot* (E_{out} as a function of E_{in}) for an amplifier, illustrating the limitations at both ends.

Dynamic Range

The extent of the linear region is known as the *dynamic range* of the amplifier, in this case from approximately 0.1 to 10 mV at the input. If this were a logarithmic amplifier, the term "dynamic range" would describe the region over which the log function is accurately followed.

A major objective in circuits that handle signals is to increase the signal-to-noise ratio (S/N). If the input has an S/N ratio of less than 2, the signal is obscured by the noise to the point of being almost useless. Amplification affects both signal and noise equally without changing their ratio. Enhancement of the S/N ratio is often difficult to accomplish. Various techniques for doing this are considered in later chapters.

NOISE

Noise can arise in many different ways, a few of which can be eliminated at the source, but most are unavoidable and can merely be minimized. The basic difficulty is that the noise is observable by the same

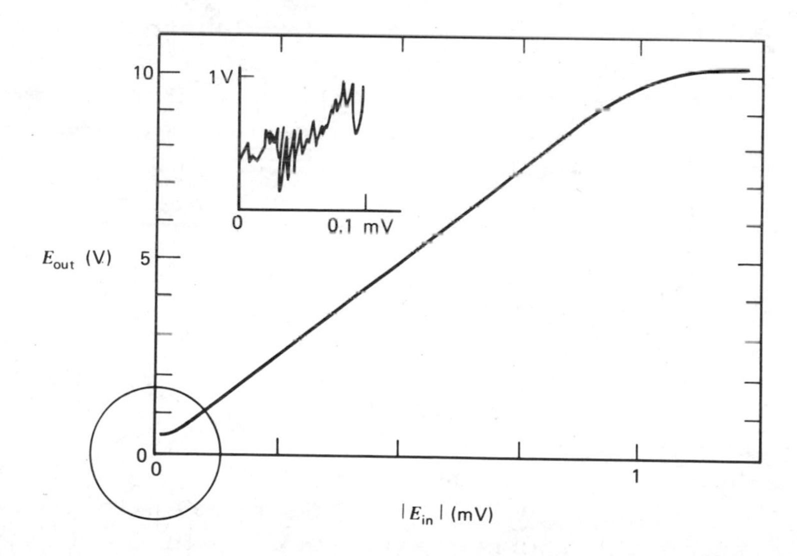

Figure 2-6. Measured transfer plot for a linear amplifier. The noise-limited region is shown magnified in the inset; the fluctuations are caused by noise rather than by the signal and consequently vary with time.

means as the desired signal.

Noise is usually observed at the output of an amplifier but for a quantitative measure is conveniently referred to the input by means of a parameter called the *noise-equivalent power* (NEP). Thus an amplifier with a power gain of 1000 and 10 mW of noise at the output is said to have an NEP of 10 μW.

There are four main categories of noise of importance to us: resistance noise, shot noise, current noise, and environmental noise. The first two are inherent in the electronic systems themselves, the third results from properties of particular components, and environmental noise is produced by external phenomena.

Resistance Noise

Resistance noise, also called *Johnson noise*, results from the random thermal motion of electrons in resistive components. The voltage that it produces, squared and averaged over a period of time (see Eq. 2-1), is given by the expression

$$\overline{e^2} = 4kTR\ \Delta f \qquad (2\text{-}17)$$

or

$$e_{\text{RMS}} = \left(\overline{e^2}\right)^{1/2} = \sqrt{4kT}\ \sqrt{R}\ \sqrt{\Delta f}$$
$$= 1.3 \times 10^{-10}\ \sqrt{R}\ \sqrt{\Delta f} \qquad (2\text{-}18)$$

in volts at 25°C. The quantity k is the Boltzmann constant (1.38×10^{-23} J/K), T is the Kelvin temperature, R is the source resistance in which the noise originates, and Δf represents the width of the band of frequencies over which the measurement is made (the *bandwidth*). This means that within a particular frequency span, say, 100 Hz, the resistance noise will be the same, regardless of what part of the spectrum is selected. (The term "white noise" is used in this case by analogy with white light, which contains all frequencies.)

Resistance noise is specified in the units $V/\sqrt{Hz}$ which come directly from Eq. 2-18 from the units of the factor $\sqrt{4kTR}$. Thus if a device shows 1.0 μV (RMS) of noise (at 25°C) over the frequency span of 150 to 250 Hz, the noise is expressed as $10^{-6}/\sqrt{100} = 10^{-5}$ $V/\sqrt{Hz}$.

For a given resistance value, the square of the voltage is proportional to power; hence we can rewrite Eq. 2-17 as

$$\overline{P} = \frac{\overline{e^2}}{R} = 4kT\ \Delta f \tag{2-19}$$

which gives the rather surprising result that the power associated with resistance noise is independent of the resistance.

It must be emphasized that since resistance noise is due to a fundamental fluctuation that occurs in all devices that dissipate power (capacitors and inductors are free of it), it cannot be eliminated. It represents the minimum amount of noise that one can hope to obtain under a given set of circumstances. It can, however, be decreased by changes in *instrumental* parameters such as by lowering the resistance, the temperature, or the bandwidth.

Shot Noise

Shot noise (*Schottky noise*) is observed in addition to resistance noise whenever a current passes through some interface (e.g, the *pn* junction in a transistor). It reflects the fact that electricity is quantized and can flow only in units of single electrons. Shot noise is usually specified as a random fluctuation i of the current I, again squared and

fluctuation i of the current I, again squared and averaged over a time interval:

$$\overline{i^2} = 2I\varepsilon\ \Delta f \tag{2-20}$$

and

$$\overline{P} = 2IR\varepsilon\ \Delta f \tag{2-21}$$

Here ε is the electronic charge, 1.602×10^{-19} C, and R refers to the load resistance through which the current $(I \pm i)$ is flowing. Shot noise is also white and dependent on the bandwidth Δf.

Current Noise

Current noise, or *flicker noise*, originates by a rather involved mechanism in those resistors and other components that are granular in structure. It is not white, but has an inverse dependence on the frequency:

$$\overline{e^2} \propto \frac{\Delta f}{f} \tag{2-22}$$

and

$$\overline{P} \propto \frac{\Delta f}{Rf} \tag{2-23}$$

Current noise is found to some extent in all components but is especially marked in transistors, carbon-paste resistors, and photocells. Metallic film and wire-wound resistors show the effect to a smaller degree.

The $1/f$ dependence suggests that this type of noise can be important at very low frequencies, including DC.* This is indeed true, and means that it is advantageous to avoid DC measurements when working with very small signals.

One of the most annoying phenomena encountered with low-level DC signals is that of *drift*. This is a continuing change in output caused by some minor variation in the circuit. It is most often the result of slow temperature changes but may also represent aging of components. It is a form of low-frequency $1/f$ noise.

Impulse Noise

There are several types of *impulse noise*. One of these, a result of switching transients, is especially noticeable if both analog and digital circuitry are located in close proximity. Digital circuits are characterized by sudden shifts in voltage from one level to another, often synchronized over a whole system. At the moment of switching, an impulse is produced that can be sensed by nearby analog components. This type of interference is best minimized by physical separation and shielding of analog and digital circuits and by the judicious use of ground connections.

Another variety of impulse noise is called "popcorn noise" because in an audio system it sounds like popping corn. This originates within the integrated circuits. Specially processed ICs may have little of this type of noise.

Environmental Noise

Most other types of noise can be lumped together under the heading of *environmental noise*. Figure 2-7

* The fact that the expressions in Eqs. 2-22 and 2-23 become infinite for $f = 0$ need not disturb us, since zero frequency implies an infinite period not realizable experimentally.

suggests qualitatively the various sources of interference to be expected in a typical university laboratory. Observe that below perhaps 10 Hz such noise appears to show an inverse frequency response, similar to that of current noise. It can be avoided by conversion to higher frequency AC.

A prevalent source of interference arises in the coupling of signals from one system to another by means of electromagnetic radiation. This is sometimes designated as electromagnetic interference (EMI). An instrument can be a source of EMI, even if it does not operate at radio frequencies, for example, if it contains very fast switching circuits such as those in computers. As a rule, EMI noise is not random and consequently does not fall under the previous statistical considerations. Often the power lines conduct

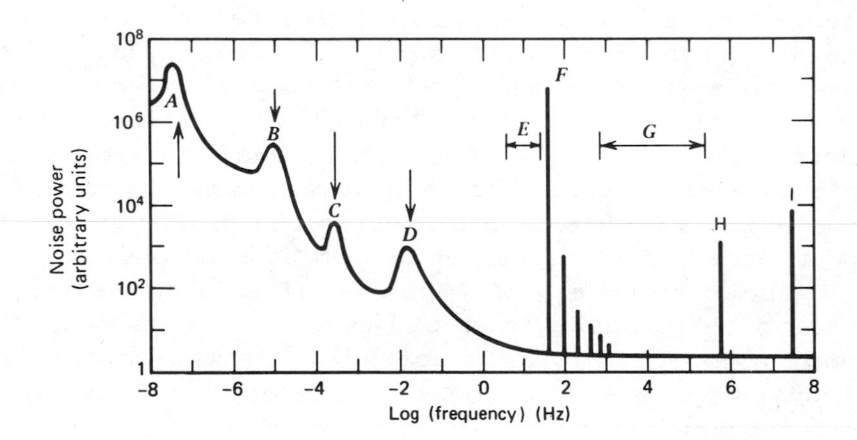

Figure 2-7. The $1/f$ dependence of environmental noise. [From T. Coor, *J. Chem. Educ.*, *45*, A583 (1968), an excellent short treatment.] The noise sources are identified as follows: *A*, temperature variations per year; *B*, temperature variations per day; *C*, change of classes per hour; *D*, elevator operation per minute; *E*, fairly noise free region; *F* power-line frequency and its harmonics (60, 120, 180 Hz, etc.); *G*, good quiet region; *H* local AM radio interference; *I*, television interference.

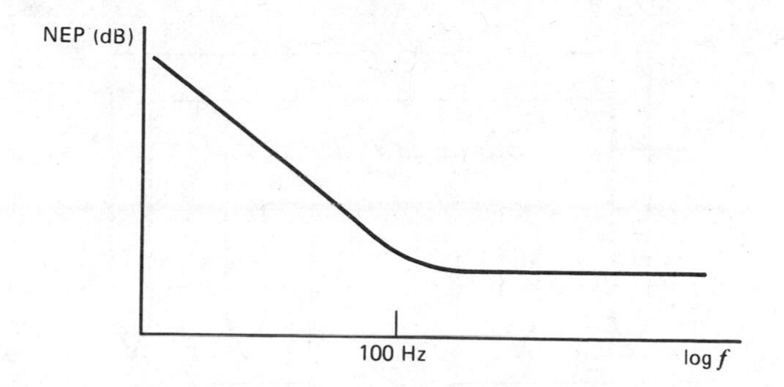

Figure 2-8. Frequency dependence of total noise in a typical instrument.

such interference from one instrument to another if plugged into the same branch line.

A special case that can be particularly troublesome is pickup of spurious signals at the power line frequency itself and its first few harmonics. This pickup can occur through a direct conductive path (e.g., leakage through a cracked insulator) or by magnetic or capacitive coupling. It can be controlled by a combination of carefully planned grounding and shielding.

In general, one particular form of noise will be dominant. Resistive sources are limited by Johnson noise, whereas certain devices, such as photomultipliers, give mostly shot noise. As a rule, most instruments show predominantly $1/f$ noise at low frequencies. White noise takes over at higher frequencies. The situation is illustrated in Figure 2-8.

TRANSDUCERS

In the majority of scientific measurements, the immediate objective is to obtain the value of some property of a physical system. This is implemented by

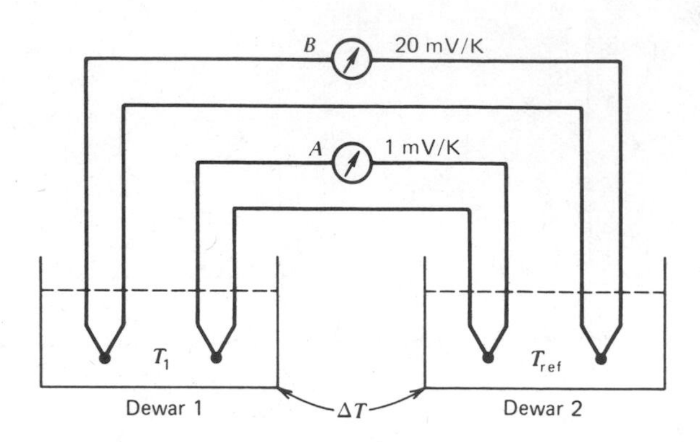

Figure 2-9. Two thermocouples of different sensitivities.

means of a sensor called a *transducer*.* Some transducers are able to monitor the desired property by themselves. The thermocouple is a good example; it produces a voltage proportional to a temperature difference without the need of any additional source of power. On the other hand, a resistance thermometer requires a current to produce a voltage across the temperature-sensitive element.

The requirements of a good transducer are stability, linearity, and sensitivity. *Stability* implies that, once calibrated, the output should continue to follow the calibration plot for an extended period of time. *Linearity* is a property of a transducer that gives an output proportional to the quantity being measured; it is by no means an indispensable condition, but is desirable. The *sensitivity* is related to the amplitude of the output for a given value of the measured property. The larger this output for a given level of noise, the greater will be the S/N ratio.

Consider as an example (Figure 2-9) two thermo-

* At this point transducers are considered solely as circuit elements; the use of transducers as information converters is discussed in Chapter XIII.

couples with sensitivities of 1 and 20 mV/K. If we assume that the noise level is 10 μV for both, and the temperature differential ΔT is 1 K, thermocouple *A* will exhibit an S/N ratio of 100, whereas thermocouple *B* will have the larger ratio of 2000. Hence in both cases the noise has practically no effect on the measurement process. If ΔT becomes small, however, the situation will be much different. For example, at ΔT = 0.01 K, thermocouple *A* shows an S/N ratio of 1, which is unacceptable, whereas *B*, which is more sensitive, has an S/N ratio of 20.

The ratio can be further improved by electronic signal processing. The fundamental requirement is the availability of some advance information about the properties of the signal to be received. Since the exact value of the quantity to be measured is unknown, such advance information can be of only an approximate nature. The most useful types of information that can improve the S/N ratio are (1) the maximum expected slope of the signal curve as a function of time and (2) the frequency range of the signal.

If the noise is largely in the form of impulses with steep rise times and short duration, it can be reduced by limiting the slew rate of the circuit. This is called *slew-rate filtration*; it must be used sparingly to avoid distortion of the signal. An inherent form of slew-rate filtering is encountered in graphic recorders, where the pen speed is limited by the inertia of the moving system.

Another approach by which the S/N ratio can be improved is by taking advantage of the frequency distribution of the signal. If it is known, for example, that the signal extends only from DC to 10 Hz, all frequencies greater than 10 Hz should be eliminated as thoroughly as possible by means of a *low-pass filter*, the action of which is illustrated in Figure 2-10. The shaded area represents the part of noise that can be eliminated by filtering without affecting the sig-

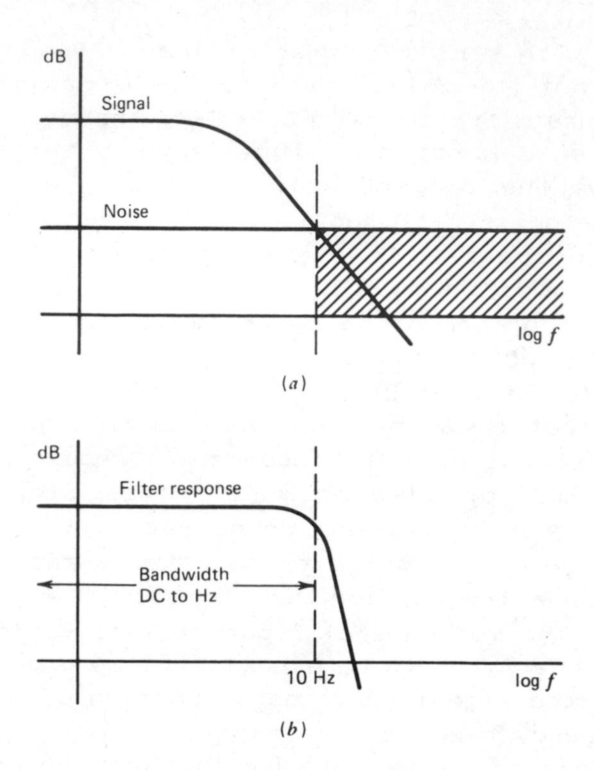

Figure 2-10. An example of filtering: (*a*) signal with noise, (*b*) response of an appropriate filter for the signal in (*a*), a "low-pass" filter.

nal. Other types of filter are available for cases in which a different frequency response is desired. Details are given in Chapter III.

The case of a DC signal requires special attention. It might appear that AC filtration would not be applicable to a DC signal, but this is not the case, as the following discussion will show.

Let us assume that a DC signal is switched on at time t = 0 (Figure 2-11*a*), and filtered through a low-pass network with cut-off frequency f_0. Irrespective of the value of f_0, the output will eventually reach a steady DC level. The sloping part of the response curve is an example of a *transient*, and its duration depends inversely on the value of f_0. Consequently, if the measurement is to be made within a reasonable period of time, it is essential that f_0 not be too small. As a rule of thumb, if a true measurement is to be made in t seconds, the cutoff frequency must be larger than about 10/t Hz.

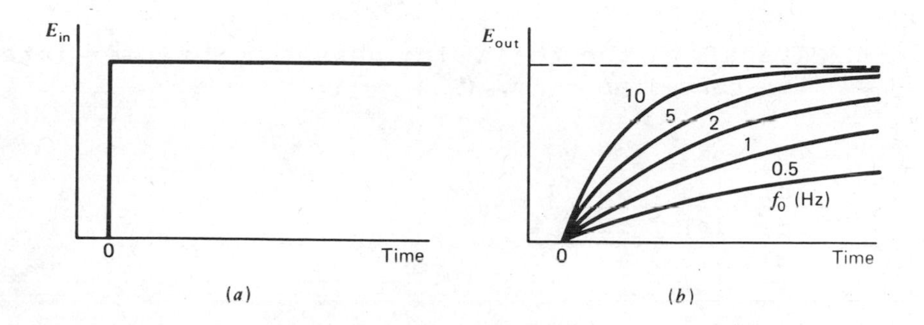

Figure 2-11. (*a*) Excitation step voltage applied to a filter; (*b*) transient response of various low-pass filters, all approaching the dashed line asymptotically.

PROBLEMS

2-1. By way of review, write all the general relationships possible between the quantities listed in Table 2-1. (For example, $E = RI$.)

2-2. For each of the following pairs of DC input and output conditions for an amplifier, determine the value of the transfer coefficient and indicate its units:

(a)	input: 1 mA	output: 0.5 V
(b)	input: 0.5 mA	output: 1 V
(c)	input: 1 mA	output: 10 mA
(d)	input: 5 V	output: 15 V
(e)	input: 15 V	output: 5 V

2-3. Transform the following phasor quantities to the polar form: $A \angle \phi$:

(a) $(5 + j5)$ V
(b) $(-3 + j0)$ A
(c) $(-4 - j4)$ V
(d) $(8 - j16)$ A
(e) $(\sqrt{2} - j\sqrt{2})$ V

2-4. Transform the following phasor quantities into the cartesian form, $(x + jy)$:

(a) $35 \angle 30^\circ$ V
(b) $115 \angle 90^\circ$ V
(c) $141 \angle 225^\circ$ A
(d) $100 \angle -30^\circ$ V

2-5. (a) Calculate the impedance in a circuit element that passes $0.30 \angle 60^\circ$ A if a voltage of $120 \angle 0$ is applied across it. (b) Calculate the power dissipation.

2-6. Consider an amplifier with a voltage gain of 32 dB, followed by a filter with an insertion loss (attenuation) of -12 dB at the frequency of interest, followed by a voltage divider formed from a 10- and a 6-kΩ resistor. What is the overall gain if the output is taken across the 6-kΩ resistor?

2-7. An amplifier is fed from a voltage source with negligible noise. Assume the input noise of the amplifier to be solely of the Johnson type. For a given bandwidth, as an average, the noise power is 10^{-14} W. Compute the RMS voltage noise and current noise for R_{in} = 100, 1000, and 10,000 Ω.

2-8. (a) A square wave with an angular frequency of 10 rad/s has peak potentials of +6 and -6 V. What is the RMS voltage? What is the average voltage? (b) What are the RMS and average voltages for a triangular wave of the same amplitude and frequency?

2-9. Prove by means of Eq. 2-1 that the RMS value of a sine wave is $E_0/\sqrt{2}$. Note that the average of any periodic function $f(\theta)$ over the interval of zero to 2π is

$$\overline{F} = \frac{1}{2\pi}\int_0^{2\pi} F\, d\theta$$

2-10. A voltage $E = 10 \cos (240t + 135^\circ)$ across a passive electronic component causes a current $I = 0.1 \cos (240t - 45^\circ)$ to flow through it. If E is in volts, I in amperes, and t in seconds:

(a) What is the frequency in hertz?
(b) What is the phase difference between E and I in degrees and in radians?
(c) Calculate the peak-to-peak and RMS values of both E and I.
(d) What is the impedance of the component at the frequency calculated in (a)?
(e) Compute the power dissipated in the component. What is the power factor?

2-11. Compute the decibel values for each of the following ratios of E_{out}/E_{in}: (a) 0.01, (b) 0.1, (c) 1.0, (d) 100, (e) 2, (f) 3.142, and (g) 90.

2-12. Two circuits, *A* and *B*, are connected in series. The voltage attentuation of *A* is -20 dB, and that of B is -3 dB. Show mathematically that the overall attenuation is -23 dB, the *sum* of the individual attenuations.

2-13. Suppose that a circuit has an attenuation of -73 dB. In the decibel table (Appendix I) we can find the ratios corresponding to 60, 10, and 3 dB. Show mathematically that the overall voltage ratio for the given circuit is obtained by taking the *product* of the three individual attenuations.

2-14. An amplifier with a power gain G_p = 100 generates an output noise of 1 mV into a load of 1000 Ω. What is its NEP figure?

2-15. An amplifier has a noise voltage, referred to the input, of 0.1 $mV/Hz^{1/2}$. Compute the signal-to-noise ratio at the output if an input signal of 100 mV is applied, and the bandwidth of the amplifier is (a) 100 Hz, (b) 10 Hz, and (c) 1 Hz.

III

ELECTRONIC COMPONENTS

The heart of any electronic circuit lies in its active components, such as transistors, but no active device can function alone. It must have a selection of passive units, including resistors and capacitors, to support it. This chapter is intended to provide insight into the properties and modes of operation of both types so that they can be used intelligently.

PASSIVE DEVICES

The chief passive components are resistors, capacitors, inductors, and diodes. The properties of capacitors and inductors are inherently dependent on the frequency of the current with which they are used. Direct current at a constant level can pass through a pure inductance without any effect being noted but, in contrast, cannot pass through capacitors at all. Resistors, on the other hand, behave identically in AC and DC circuits. The response of diodes is more complex.

Impedance and Resistance

Passive components are characterized by their

impedance, denoted by the letter Z, and measured in ohms. As seen previously, impedance is conveniently taken as a vector quantity in the complex plane. In connection with DC circuitry, however, the impedance can be replaced by a scalar quantity, the *resistance* R. Thus resistance is a special case of impedance.

Resistors, capacitors and inductors* all obey Ohm's law, $E = ZI$, where I is the current flowing through the device and E is the voltage developed across it.‡ This relationship is of fundamental importance, and is referred to repeatedly.

Resistors

A resistor is rated not only by the value of its resistance, but also by its power-dissipation ability (the maximum power in watts that it can handle without damage) and its temperature coefficient. Commercial, fixed-value resistors are of several types: carbon or composition resistors, which are the least expensive; metallic film types, which have less tendency to introduce noise into the system; and wire-wound units, some of which are capable of dissipating considerable power as heat.

Variable resistors are particularly important as a means for adjustment of a circuit. Those units that are provided with three connections, as symbolized in Figure 3-1a, are called *potentiometers*. If only two terminals are needed, it is best practice to wire one end permanently to the variable contact (the wiper), as

* These terms, which end in *-or*, refer to the physical objects or components, whereas the corresponding terms ending in *-ance* refer to the magnitude of the physical property of such a component.

‡ The original statement of Ohm's law ($E = RI$) applies only to pure resistances.

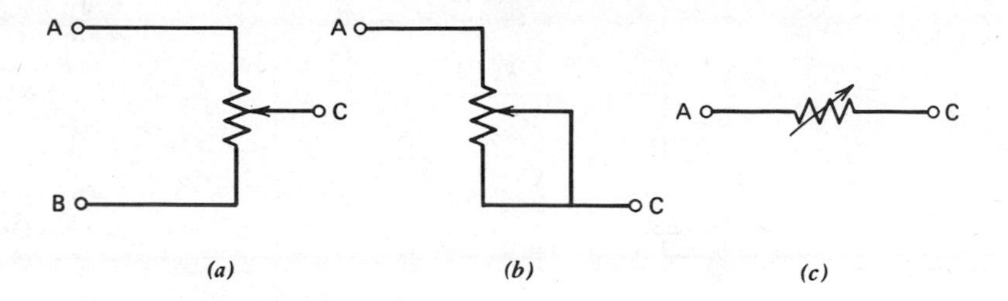

Figure 3-1. (*a*) A three-terminal variable resistor, also called a "potentiometer" or "pot." (*b*) The same unit wired for two-terminal operation, sometimes called a "rheostat". (*c*) An alternative symbol for (*b*).

shown in Figure 3-1*b*. This tends to reduce the "contact noise" produced as the wiper moves. A two-terminal variable resistor, especially one capable of carrying high currents, is called a ***rheostat***.

Actual resistors are never perfect. Wire-wound resistors are very stable but may have enough inductance to affect their operation at high frequencies. Carbon and composition units have negligible inductance but have higher temperature coefficients and also generate more noise. Metal film resistors are the preferred type for high-quality circuits but are more expensive.

Combinations of resistors are often needed, and it

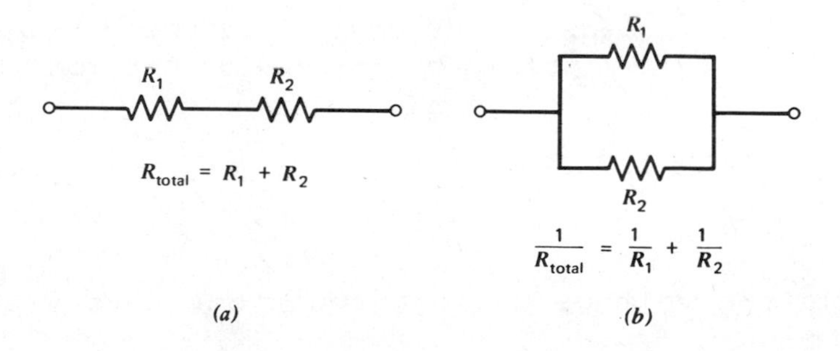

Figure 3-2. Two resistors connected (*a*) in series and (*b*) in parallel.

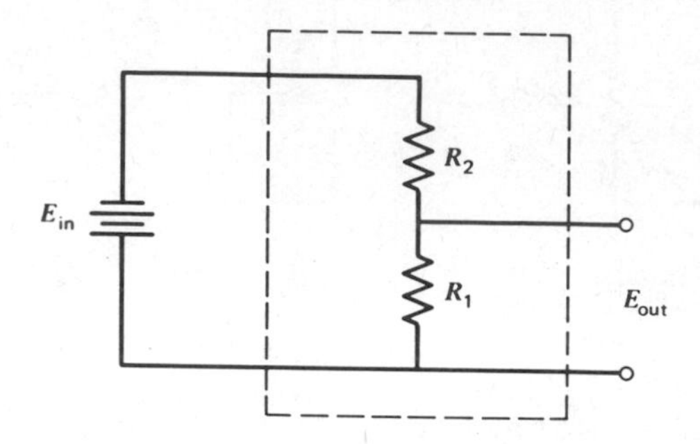

Figure 3-3. The voltage divider.

is desirable to be able to compute the associated voltages and currents. Two rules for combinations are basic: (1) resistances in series are additive, and (2) resistances in parallel follow a reciprocal relationship. These are illustrated in Figure 3-2, where it is shown that each combination is equivalent to a single resistor of value denoted by R_{total}.

An important application of these rules is seen in the *voltage divider* (Figure 3-3). In this circuit, the input voltage E_{in} (symbolized as a battery) is impressed across the two resistors, but only a fraction of it appears across R_1 to give the output voltage E_{out}. By Ohm's law, the current I_{in} is given by $E_{in}/(R_1 + R_2)$. This current produces in R_1 a drop of voltage $E_{out} = I_{in}R_1$. By combining these two relationships, one can obtain the response function:

$$\frac{E_{out}}{E_{in}} = \frac{R_1}{R_1 + R_2} \qquad (3\text{-}1)$$

This is the voltage divider equation, a very useful relationship.

Two applications warrant more detailed treatment. If the sum of R_1 and R_2 is constant, as happens when

(212) 639-7753

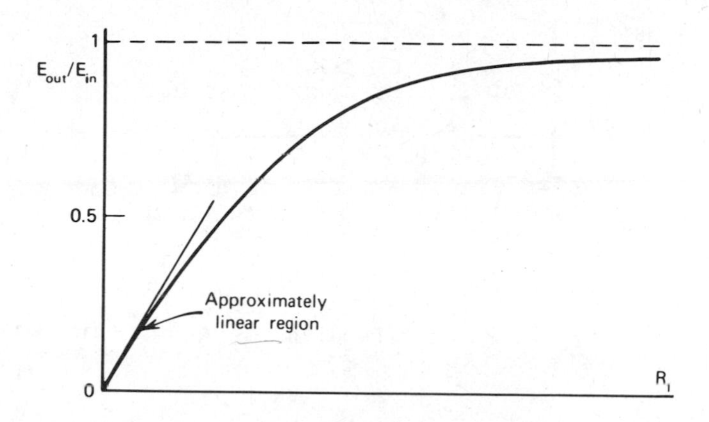

Figure 3-4. Variation in output from the voltage divider shown in Figure 3-3 when R_2 is held constant and R_1 is varied.

they correspond to the two branches of a potentiometer, the output is directly proportional to the value of R_1. Many such potentiometers have R_1 proportional to the rotation of a shaft, causing E_{out} to be proportional to the angular displacement.

If only one of the resistors is a variable, the total resistance is also variable and the response function becomes somewhat more complicated. From Figure 3-4 one can see that the relationship between E_{out} and R_1 is nearly linear for small values of R_1 as compared to R_2 but deviates for larger values.

<u>Loading:</u> The output of a voltage divider is affected by other resistances connected to it (see Figure 3-5*a*). The output is given by $E_{out} = 15(5\ k\Omega/15\ k\Omega) = 5.0$ V. This is valid only if no current is drawn from the output terminal. On the other hand, suppose that a 20-kΩ load is added (Figure 3-5*b*). The output voltage will be significantly diminished. To calculate its value, the resistance of R_1 and R_L in parallel must be substituted for R_1 in Eq. 3-1. This parallel resistance is $(5\ k\Omega)(20\ k\Omega)/(5\ k\Omega + 20\ k\Omega) = 4\ k\Omega$. Hence

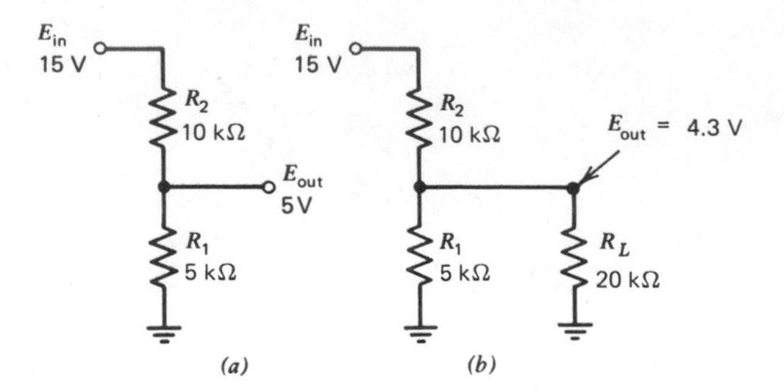

Figure 3-5. A voltage divider, with (*a*) no load and (*b*) a load resistor R_L.

E_{out} becomes 15(4 kΩ/14 kΩ) = 4.3 V, obviously much smaller than 5.0 V. This loading effect becomes negligible only if $R_L >> R_1$. Otherwise it must always be kept in mind.

Electronic circuits in general have internal input and output impedances, $\mathbf{Z}_{in}$ and $\mathbf{Z}_{out}$, defined by Ohm's law as the ratios of the respective voltages and currents:

$$\mathbf{Z}_{in} = \boldsymbol{E}_{in}/\boldsymbol{I}_{in} \qquad \mathbf{Z}_{out} = \boldsymbol{E}_{out}/\boldsymbol{I}_{out} \tag{3-2}$$

These impedances can influence the behavior of voltage dividers, as shown in the following example.

Consider a typical situation that might occur in the laboratory: the coupling of a measuring instrument to a strip-chart recorder that was not designed for it. Suppose that the instrument has an output range of 0 to 100 mV and a purely resistive output impedance R_{out} of 10 Ω. On the other hand, the recorder indicates full scale for 10 mV and its input resistance R_{in} is 10^7 Ω. Figure 3-6*a* shows the situation schematically and suggests the use of a voltage divider (Figure 3-6*b*) to make the instrument and recorder compatible. Now consider Figure 3-6*c*. The resistance R_{out} appears in series with R_2, whereas R_{in} parallels the lower leg of the divider. Let us assume for the moment that the

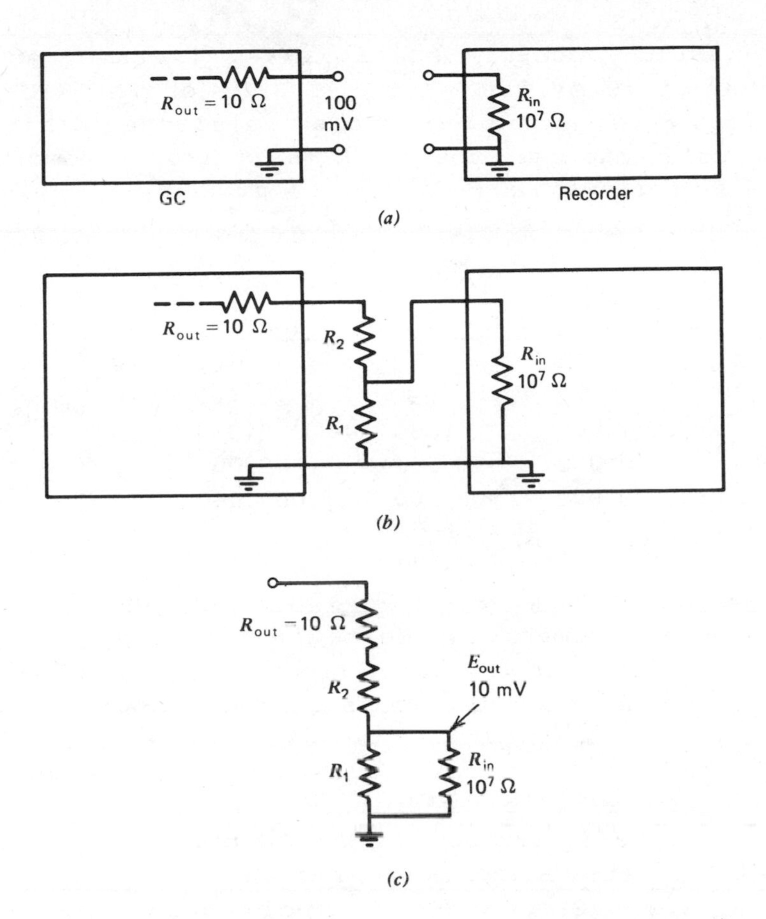

Figure 3-6. A laboratory instrument and recorder to be interfaced so that 100 mV provides 10 mV to the recorder: (*a*) the problem; (*b*) a voltage divider as a possible coupling device; (*c*) an equivalent circuit for the calculation of R_1 and R_2.

input resistance of the recorder is so large that it exerts no significant loading effect on the voltage divider and thus that Eq. 3-1 can be applied without correction:

$$\frac{E_{out}}{E_{in}} = \frac{R_1}{R_1 + R_2 + 10} \tag{3-3}$$

On the other hand, E_{out}/E_{in} is required to be equal to 10 mV/100 mV = 0.1. We are at liberty to select a value for one of the resistances arbitrarily and calculate the other. A few possible combinations are as follows, where $R_T = R_1 + R_2 + 10$:

R_1	R_2	R_T	I
10	80	100	10^{-3}
100	890	1000	10^{-4}
1,000	9,000	10,000	10^{-5}
10,000	90,000	100,000	10^{-6}
100,000	900,000	1,000,000	10^{-7}

Note that for $(R_1 + R_2) > 10^4\ \Omega$, the output impedance of the instrument can be neglected, simplifying the calculation. On the other hand, if R_1 is greater than about $10^5\ \Omega$, loading effects become evident. Hence $(R_1 + R_2) = 10^4\ \Omega$ would be the best choice, intermediate between the impedances of the instruments that we are trying to render compatible.

The merit of such a nonloaded voltage divider is that the attenuation is independent of the exact values of R_{out} and R_{in}. Thus R_{out} may change by 100 % to 20 Ω, without any noticeable effect.

An important application of voltage division occurs when an indicating meter is to be provided with a number of ranges. Suppose a moving-coil meter has an inherent sensitivity of 1 V at 100 μA (which can be expressed as 10,000 Ω/V) and ranges of 1, 5, 10, 50, and 100 V, full scale, are desired. This can be implemented by means of a five-position range switch with corresponding resistors (Figure 3-7*a*). For the 1-V range, no resistor is needed. For the 5-V range, Eq. 3-1 gives:

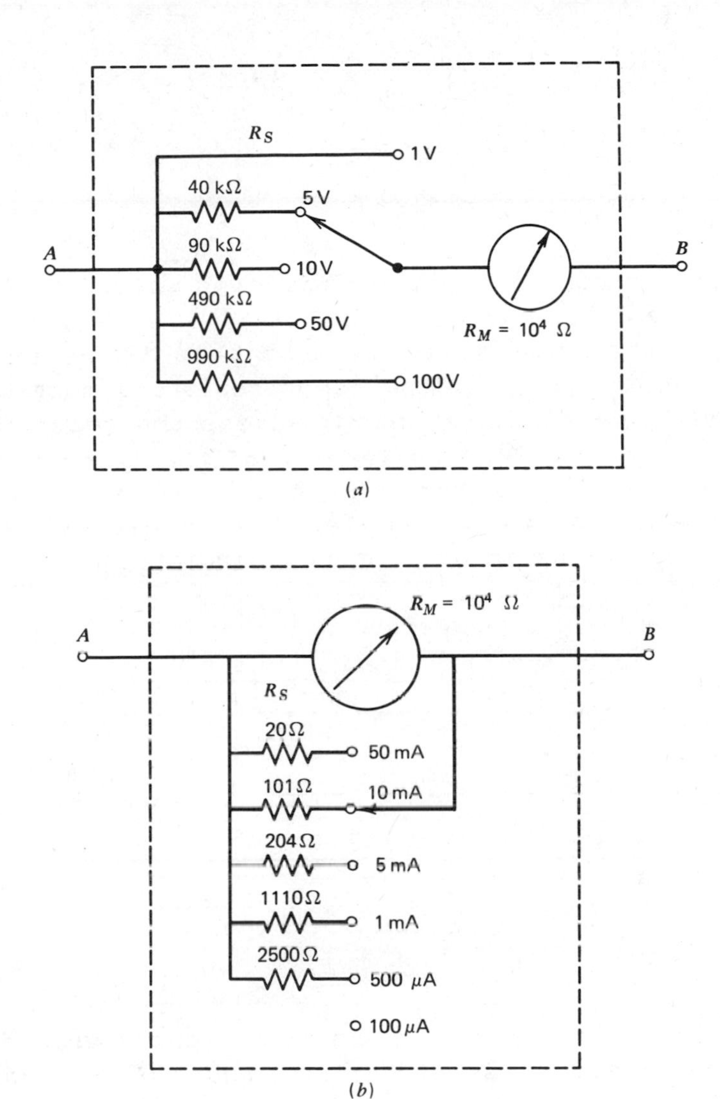

Figure 3-7. A laboratory meter with selectable ranges: (*a*) as a voltmeter, (*b*) as an ammeter; the switch must be of the "make-before-break" type to protect the meter from overload.

$$\frac{E_{out}}{E_{in}} = \frac{1}{5} = \frac{10^4}{R + 10^4} \tag{3-4}$$

$$R = 4 \times 10^4 \ \Omega$$

The reader can verify the remaining values of *R* shown in Figure 3-7*a*.

If the same meter movement is to be used for current measurements, the range resistors must be connected in parallel (shunted) across the meter (Figure 3-7*b*). The parallel resistance of the meter and shunt must in each case be such that the desired current will produce a 1-V drop, causing full-scale deflection. The value of the shunts can be calculated by the relationship

$$R_{total} = \frac{R_M R_s}{R_M + R_s} = \frac{10^4 R_s}{10^4 + R_s} \tag{3-5}$$

In the case of the 50-mA range, for example, R_{total} = 1 V/0.05 A = 20 Ω, which results in

$$R_s = \frac{20 \times 10^4}{20 + 10^4} \cong 20 \ \Omega \tag{3-6}$$

The meter shunt is an excellent example of a *current divider*, as shown in Figure 3-8. One can readily prove that

$$\frac{I_1}{I_2} = \frac{R_2}{R_1} \tag{3-7}$$

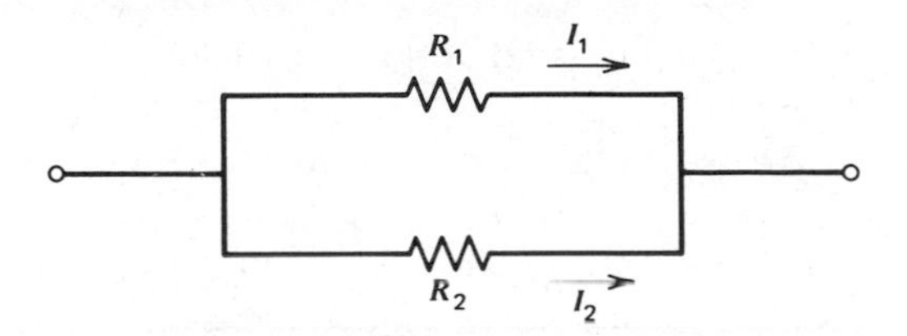

Figure 3-8. A current-dividing circuit.

The shunt values in Figure 3-7*b* can be calculated from either Eq. 3-5 or 3-7.

Capacitors

A capacitor consists of two conductors separated by a thin insulating layer. There are three principal types of capacitors: (1) *solid-dielectric*, in which the separating layer may be made of various plastic films, mica, paper, or ceramic; (2) *electrolytic*, composed of aluminum or tantalum foil covered by a thin film of oxide produced electrolytically; and (3) *air-dielectric*, which are variable capacitors used primarily in tuning radiofrequency circuits.

Capacitors are rated by their capacitance [in farads, or more commonly in microfarads (μF) or picofarads (pF)], and by the maximum permissible voltage at which they can safely be used. The several types mentioned in the preceding paragraph vary in secondary characteristics, which may be of great practical importance. For instance, an electrolytic unit (unless it is marked "NP" or "nonpolarized") must be connected in such a way as to maintain a DC voltage of proper polarity across it; wrong polarity may well result in explosive failure. These units, particularly of aluminum, can be obtained with very large capacitance at low cost and are especially useful in power-supply circuits. Electrolytic capacitors behave as though they had a resistor of a few megohms in parallel (the leakage resistance), and this limits their field of application.

Of the nonelectrolytic types, those with dielectric of polystyrene or polycarbonate are preferred over paper, since their capacitance remains constant with time and their leakage resistance is very high. Ceramic disk capacitors are often more convenient than the paper types and of similar low cost. Many types of capacitors have considerable inductance. When it is important to pass a wide band of frequencies, this inductance may be troublesome, and it is good practice to wire a low-inductance ceramic capacitor in parallel with any large paper or plastic film unit.

Inductors

A typical inductor consists of a coil of wire containing a core of iron or ferrite* or simply air and is characterized by its inductance, expressed in henrys. The iron core types have high inductance and are used principally in high-current power supplies. Ferrite inductors are important in radiofrequency equipment, for example, as the antenna coil in a radio receiver. Inductors with two or more windings can be used as *transformers* to couple one circuit to another without direct connection.

Impedance of Capacitors and Inductors

Impedance is a vectorial quantity that can be expressed in terms of three components: resistance, capacitance, and inductance. The latter two are grouped together under the term "reactance." Resistors,

* Ferrite is a ceramic-like material made by sintering together various heavy-metal oxides. It has particularly favorable magnetic properties and extremely high electrical resistance.

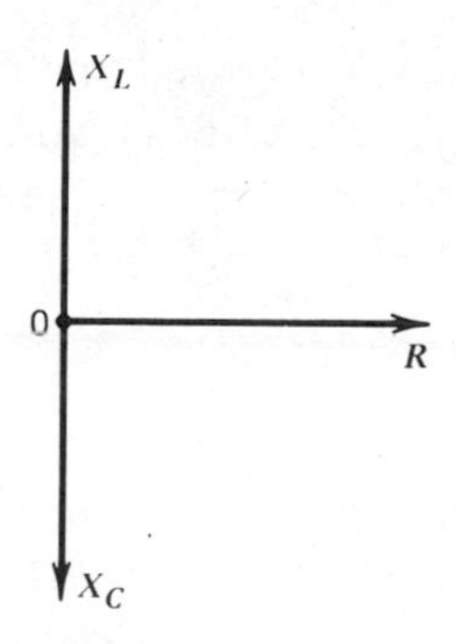

Figure 3-9. Coordinate system for the representation of complex impedances.

capacitors, and inductors all possess these three quantities in varying proportions and derive their individual characters from whichever predominates.

Reactances are frequency dependent, whereas resistance is not. Hence at extremes of frequency the dominant component may change. For example, a wire-wound resistor has sufficient inductance that at high frequency it acts as an inductor.

Capacitive reactance $\mathbf{X}_C$ and inductive reactance $\mathbf{X}_L$ are in many respects opposites. The *vector* $\mathbf{X}_C$ is plotted downward in the complex plane, whereas $\mathbf{X}_L$ points upward. The resistance vector runs from left to right (Figure 3-9). The rules for series and parallel combinations of impedances are the same as for resistances.

Reactances are given by the relationships

$$\mathbf{X}_C = 1/j\omega C \qquad \mathbf{X}_L = j\omega L \tag{3-8}$$

where C is the capacitance in farads and L the inductance in henrys. The use of these formulas can be illustrated by considering the impedance of capacitors and inductors that have significant internal resistance. In the case of the inductor, the resistance is that of the wiring and is in series with the inductive reactance,

$$\mathbf{Z} = R + j\omega L \tag{3-9}$$

The situation is more complicated with a capacitor that has appreciable parallel (leakage) resistance. Since parallel impedances combine as reciprocals, we must write

$$\frac{1}{\mathbf{Z}_C} = \frac{1}{R} + \frac{1}{\mathbf{X}_C} = \frac{1}{R} + j\omega C \tag{3-10}$$

or

$$\mathbf{Z}_C = \frac{R(1/j\omega C)}{R + (1/j\omega C)} \tag{3-11}$$

Multiplication of both numerator and denominator by $j\omega C$ and simplification gives

$$\mathbf{Z}_C = \frac{R}{1 + j\omega RC} \tag{3-12}$$

This can be brought to the standard form by multiplying both numerator and denominator by the quantity $(1 - j\omega RC)$:

$$\mathbf{Z}_C = \frac{R(1 - j\omega RC)}{(1 + j\omega RC)(1 - j\omega RC)} \tag{3-13}$$

which is equivalent to

$$\mathbf{Z}_C = \frac{R}{1 + \omega^2 R^2 C^2} - j\left(\frac{\omega R^2 C}{1 + \omega^2 R^2 C^2}\right) \qquad (3\text{-}14)$$

Both expressions (Eqs. 3-9 and 3-14) describe the impedance as the sum of two terms, one real and the other imaginary. The relative importance of the two terms depends on the frequency. For capacitors the resistive part is unimportant beyond a few hundred hertz. For example, if the leakage resistance R of a 1-µF capacitor is 10^5 Ω (a rather poor capacitor), at 100Hz, $\mathbf{Z}_C$ is 1.6 kΩ,* about 1.5% of R. Equation 3-14 indicates that $\mathbf{Z}_C$ is (25 − j1600), whereas Appendix III gives $\mathbf{X}_C = -j1600$. The real component (25 Ω) is so small compared to the imaginary part (only 1.5%), that the vector is practically coincident with the vertical axis. The deviation increases with frequency.

At higher frequencies another factor also becomes important, namely, the residual inductance. Many capacitors consist of alternate layers of insulator and conductor rolled into a cylinder; this configuration contributes to the inductance. For a 1000-µF capacitor with an inductance of 0.01 µH, as an example, the inductive impedance becomes 1% of the total impedance when $(2\pi f L) = [1/(2\pi f C)](1/100)$, a condition that occurs for this example when f = 50 kHz.

Inductors, as a rule, are further removed from ideality than capacitors. Their residual resistance is that of the wire with which they are wound, and this is seldom negligible, especially at low frequencies. At higher frequencies, the capacitance between adjacent windings becomes significant. Figure 3-10 illustrates the vector representation of the impedances of real inductors and capacitors.

* Appendix III tabulates values of $\mathbf{Z}_C$ as a function of frequency.

Resistors are affected by their inductance only in the cases of low resistance value and high frequency. For example, if $R = 10\ \Omega$ and $L = 0.02\ \mu H$ at 1 MHz, the percentage contribution of the inductance is given by $(2 \times 10^6)(2 \times 10^{-8})(100/10) = 1.2\%$ meaning that the imaginary part is about 1.2% of the real part. This inductance is to be considered in series with the resistance.

The capacitive reactance present in resistors is apparent only at high frequencies. Thus 0.1 pF of residual capacitance gives an impedance of 1.6 MΩ at 1 MHz, in parallel with the resistor. This effectively limits the usefulness of resistors greater than about 1 MΩ at high frequencies. For a 1.6-MΩ resistor, this amounts to equal real and imaginary parts of the impedance.

Finally, a rather unexpected source of error in resistors is thermoelectric in origin. The potential developed at the copper/carbon junctions at each end of

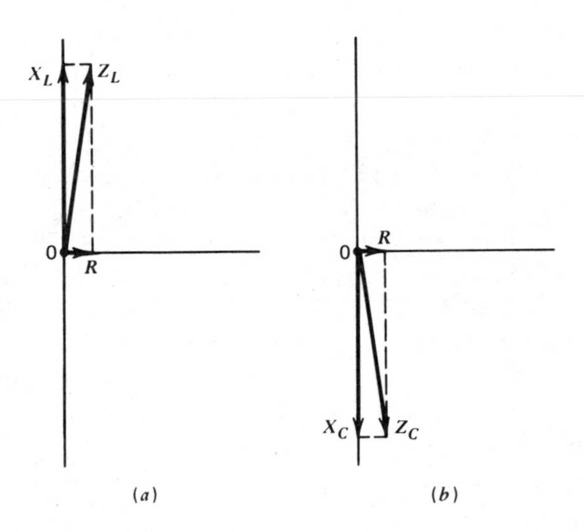

Figure 3-10. Impedance vectors for (*a*) an inductor and (*b*) a capacitor.

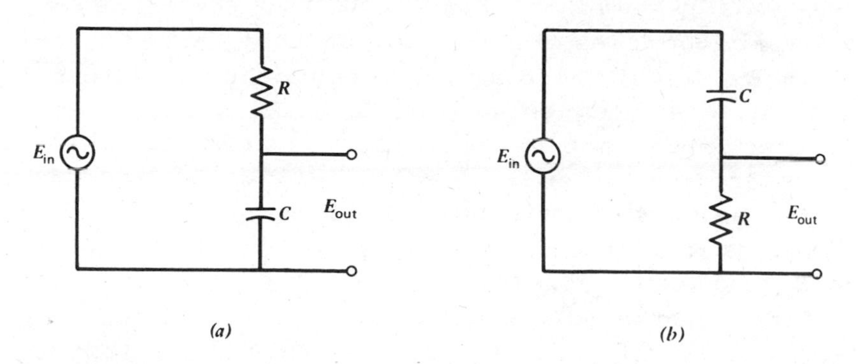

Figure 3-11. Two *RC* voltage dividers: (*a*) a low-pass filter; (*b*) a high-pass filter.

a resistor can be as great as 30 μV per degree difference of temperature.

Passive Filters

It is instructive to investigate the properties of voltage dividers composed of resistors and capacitors (Figure 3-11). In (*a*) of the figure, the output voltage is given by

$$E_{out} = E_{in}\left(\frac{\mathbf{Z}_C}{R + \mathbf{Z}_C}\right) = E_{in}\left(\frac{1/j\omega C}{R + 1/j\omega C}\right) = E_{in}\left(\frac{1}{j\omega RC + 1}\right) \tag{3-15}$$

whereas for (*b*) it is

$$E_{out} = E_{in}\left(\frac{R}{R + \mathbf{Z}_C}\right) = E_{in}\left(\frac{R}{R + 1/j\omega C}\right) = E_{in}\left(\frac{j\omega RC}{j\omega RC + 1}\right) \tag{3-16}$$

The outputs of these voltage dividers are functions of the frequency and hence they can be used to attenuate selectively (filter out) portions of the frequency spectrum. In the circuit shown in Figure 3-11a, if frequencies are low enough that $\omega RC \ll 1$, then $\boldsymbol{E}_{out} = \boldsymbol{E}_{in}$, and there is no attenuation. In contrast, at high frequencies, $\boldsymbol{E}_{out} \ll \boldsymbol{E}_{in}$, and the signals themselves are filtered out. This is called a *low-pass filter.*

In the circuit of Figure 3-11*b*, the effect is just the opposite: if the frequency is low so that $\omega RC \ll 1$, then $\boldsymbol{E}_{out} = \omega RC\boldsymbol{E}_{in}$, which means that $\boldsymbol{E}_{out} \ll \boldsymbol{E}_{in}$, thus filtering out the low frequencies. On the other hand, if $\omega RC \gg 1$, then $\boldsymbol{E}_{out} = \boldsymbol{E}_{in}$. This constitutes a *high-pass filter.*

The input-output relationships are shown in Figures 3-12*a and 3-12b* as functions of frequency. These plots consist of two straight lines, with their intersection somewhat rounded off. The intersection point lies at the frequency $f_0 = 1/(2\pi RC)$, where the impedances of R and C are equal. It is easily shown that the rounded curve should lie below the intersection of tangent lines by $20 \log \sqrt{2}$, which is very nearly the equivalent of -3 dB; this is often called the "3-dB point."

A *band-pass filter*, in which both low and high frequencies are attenuated (Figure 3-12*c*), is also useful but is not conveniently constructed with passive components alone.

Let us now consider some applications of these

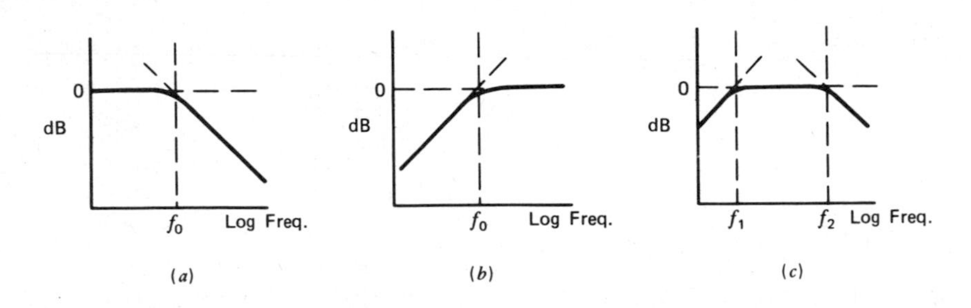

Figure 3-12. Bode response plots of passive filters: (*a*) low-pass; (*b*) high-pass; (*c*) band-pass.

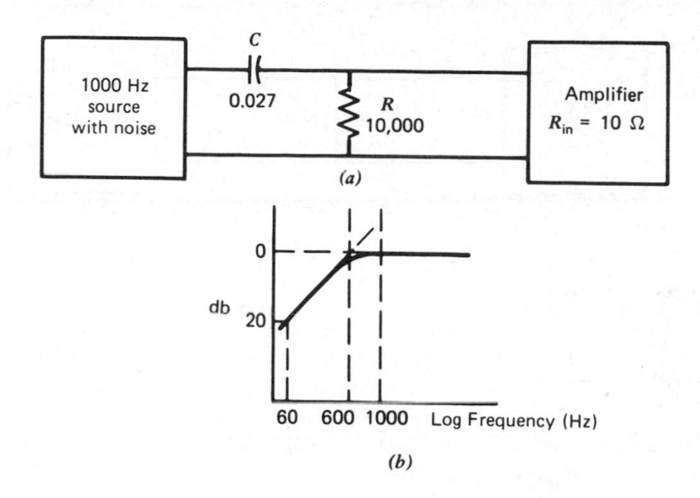

Figure 3-13. (*a*) A high-pass filter coupling a 1000-Hz noisy source to an amplifier. (*b*) Corresponding Bode diagram; at f_0 (600 Hz) the signal is attenuated to 71% (3 dB down).

several types of filters. Figure 3-13 shows a high-pass filter being used to reduce line noise in a 1000-Hz sine wave. A good choice for the cutoff frequency is 600 Hz, 10 times the line frequency. Since the slope of the Bode curve is 20 dB per decade, the attenuation at 60 Hz is 20 dB. At the same time 1000 Hz is sufficiently far removed from f_0 to be left unchanged by the filter. The net effect is an increase in S/N by a factor of 10. Noise of the $1/f$ type is also attenuated since its upward slope toward lower frequencies is offset by the downward slope of the Bode curve. The values of R and C are selected to conform with the formula $f = 1/(2\pi RC)$. Tolerances for the components are not critical.

High-pass units are useful in improving the S/N ratio when information is introduced onto an AC carrier by modulation since the $1/f$ noise and 60-Hz interference can be filtered out without appreciably attenuating the information. Low-pass filtering benefits

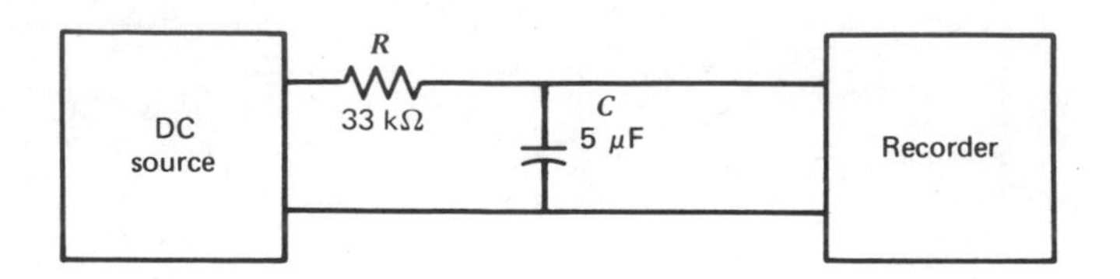

Figure 3-14. A low-pass filter with a time constant $RC = 0.16$ s, meaning that a sudden step change of input voltage will cause 63% of the complete output change to occur within that time. This may slightly round off sharp peaks in the recorded signal. The value of f_0 is about 1 Hz.

low-frequency or DC measurements as a means of reducing noise, as seen in Chapter II.

An alternate mode of description of filters is by the product RC, which has the dimensions of time and is called the *time constant*. If a DC step signal is applied to such a filter, the potential will rise, as shown in Figure 2-11. The curves attain 63% of their final voltage at the time equal to RC (one time constant).* An example is shown in Figure 3-14.

A more complex type of filter is the *notch filter*, an example of which is given in Figure 3-15. This network, called a *Twin-T filter*, can be regarded as a parallel combination of a modified low-pass filter (R, R, $2C$) with a matching high-pass unit (C, C, $R/2$). Because of the phase interaction between the two branches, the impedance of the filter is very large (many megohms) at the notch frequency. The sharpness of the notch depends on the precision of the components. This filter, tuned to the power frequency, is

* The mathematical relationship between the input voltage and the voltage on the capacitor as a function of time is given by

$$E_{cap} = E_{in} \, [1 - \exp(-t/RC)]$$

When $t = RC$, the parenthetical expression takes the value $1 - e^{-1}$, which equals 0.63.

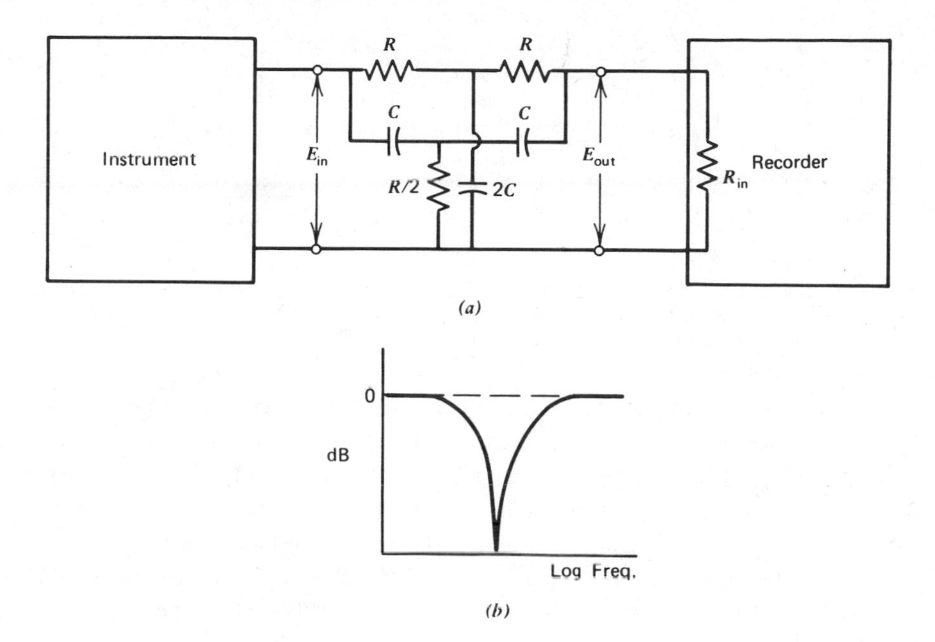

Figure 3-15. The Twin-T notch filter. (*a*) An example of its use to couple the output of an instrument to a recorder with the elimination of interference at frequency $f_0 = 1/(2\pi RC)$. The value of R_{in} must be larger than R but smaller than the impedance of the filter at f_0. (*b*) The frequency response.

widely used to remove traces of 60-Hz noise from DC circuits.

SEMICONDUCTORS

Elementary silicon is the basis of all modern semiconductor devices. Its crystal consists of a three-dimensional array of identical atoms covalently bonded to each other.* This is conveniently diagrammed in simplified form in two dimensions, as in Figure 3-

* The properties of germanium are similar to those of silicon and enable it to be used under certain circumstances in semiconductor devices.

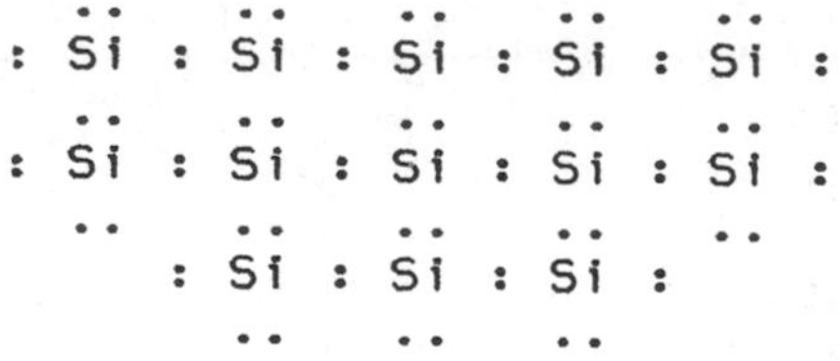

```
  ..    ..    ..    ..    ..
: Si  : Si  : Si  : Si  : Si  :
  ..    ..    ..    ..    ..
: Si  : Si  : Si  : Si  : Si  :
  ..    ..    ..    ..    ..
      : Si  : Si  : Si  :
        ..    ..    ..
```

Figure 3-16. The 2-dimensional analog of a silicon crystal. Each pair of dots represents a pair of electrons in the covalent structure.

16. The silicon atom has four valence electrons. In the crystal structure each pair of neighboring atoms are joined by a bond consisting of two electrons, one from each atom. Thus each atom of silicon is surrounded by a stable complement of eight electrons. The crystal is essentially a nonconductor of electricity, as the electrons are tightly held in the bonding structure and are not free to move about in the crystal.

Silicon becomes very useful in electronic devices when certain impurities are present in extremely small amounts, to render the crystal conducting. One appropriate type of additive is an element that has five valence electrons, such as arsenic, antimony, or phosphorus. Figure 3-17 shows the result of the addition of a small amount of arsenic to the silicon crystal. Each added arsenic atom is accompanied by one available

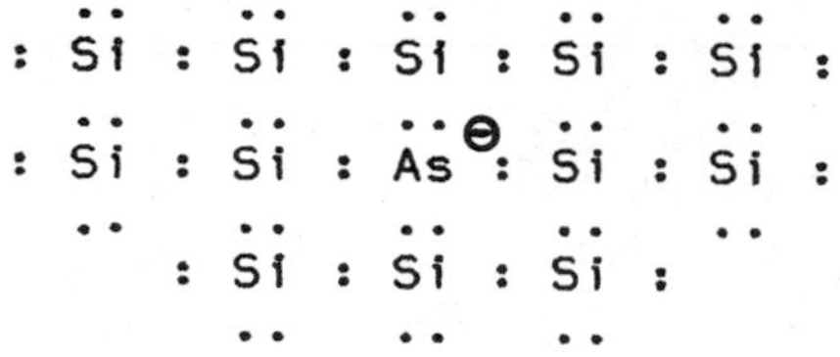

```
  ..    ..    ..    ..    ..
: Si  : Si  : Si  : Si  : Si  :
  ..    ..    ..⊖   ..    ..
: Si  : Si  : As  : Si  : Si  :
  ..    ..    ..    ..    ..
      : Si  : Si  : Si  :
        ..    ..    ..
```

Figure 3-17. A silicon crystal with an atom of arsenic substituted into the lattice. The encircled electron is mobile.

```
    ..     ..     ..     ..     ..
  : Si  : Si  : Si  : Si  : Si  :
    ..     ..     .. o ..     ..
  : Si  : Si  : Al  . Si  : Si  :
    ..     ..     ..     ..     ..
         : Si  : Si  : Si  :
           ..     ..     ..
```

Figure 3-18. A silicon crystal with an atom of aluminum replacing one silicon atom. The circle represents a "hole" where one electron is missing.

electron in excess of the number needed for covalent binding. These extra electrons are relatively free to move around through the crystal, while leaving positive As^+ ions immobilized in the lattice.

On the other hand, impurity atoms with only three valence electrons (aluminum, boron, or gallium), when present in the silicon, give structures deficient in electrons, as shown in Figure 3-18. In this case, for each added atom, a "hole" results that could accept an electron to complete a pair. It is always possible, as a result of thermal energy, that an electron from a neighboring Si-Si bond will free itself and jump into the hole, leaving a new hole behind. By this means the hole can effectively move step by step through the crystal.

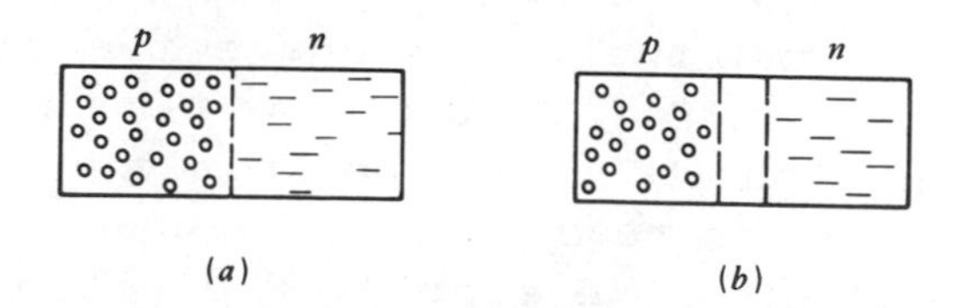

Figure 3-19. A *pn* junction: (*a*) before charge diffusion occurs and (*b*) following diffusion, showing the depletion region. Circles represent holes and minus signs, electrons.

Materials that conduct electricity by either of the two mechanisms just discussed are called *semiconductors*. The addition of impurities to silicon is called *doping*. Additives with excess electrons give rise to *n*-doped silicon, and those deficient in electrons produce *p*-doped material. The concentrations of impurities are very low, much less than a part per million.

It is instructive to consider what will happen when *n*- and *p*-doped silicon come in contact with each other. The situation is diagrammed in Figure 3-19, in which holes are represented by circles, and electrons by minus signs. There will be a tendency for electrons to cross the boundary from right to left and holes from left to right, as a result of coulombic attraction between their negative and positive charges. This quickly produces a narrow region, called the *depletion region*, around the boundary, free of both holes and mobile electrons. The result of this charge diffusion is that the *p* region becomes charged negatively and the *n* region positively, with about 0.7 V between them, the *barrier potential*.

Now suppose we provide the crystal with metallic electrical connections. Once these connections are made, the unit becomes the circuit component called a *diode*. If a potential larger than 0.7 V is applied, with the *p* material positive, as shown in Figure 3-20*a*, the added field will drive electrons toward the depletion region from the right where there are excess electrons available. At the same time it will drive holes in the opposite direction from the left. These holes and electrons readily combine where they meet at the *pn* junction. The supply of electrons in the *n* region is maintained by additional electrons injected from the external circuit at *B*, while electrons are drawn away on the opposite side, creating additional holes. The result is that current flows readily and the net voltage across the unit is essentially the

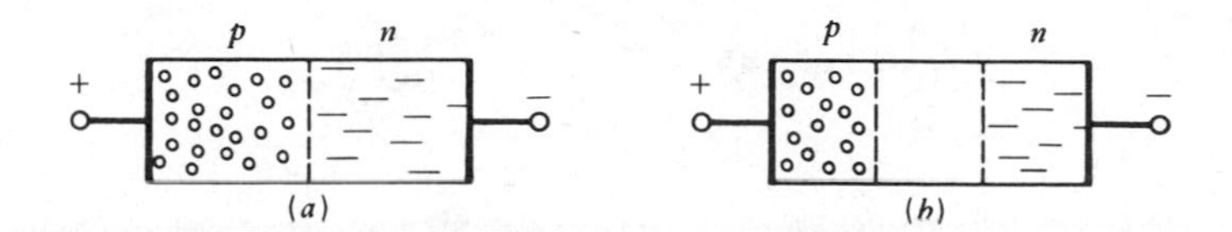

Figure 3-20. A *pn*-junction diode with electrical connections: (*a*) forward biased, with both electrons and holes driven toward the depletion region, and (*b*) reverse biased, with both electrons and holes withdrawn from the depletion region. The metal-to-silicon junctions (called "ohmic" junctions) do not form depletion regions.

barrier potential, 0.7 V. In this condition the diode is said to be *forward biased.*

If the applied potential is reversed (Figure 3-20*b*), the situation will be quite different. Now both electrons in the *n* region and holes in the *p* material are pulled *away* from the depletion region, thus effectively preventing any current from flowing (*reverse-biased* condition). Thus the chief area of application of diodes becomes evident, namely, the ability to conduct current in one direction while preventing its reverse flow.

If the voltage is increased in the reverse-bias condition, a point will be reached at which the diode will break into reverse conduction. This occurs when the electric field becomes large enough to produce new ion-hole pairs in (or close to) the depletion region. For heavily doped diodes the depletion zone is very narrow, and breakdown occurs in the range of 2 to 6 V. If the diode is less heavily doped, breakdown takes place only at higher voltages and is due to the secondary ejection of electrons from covalent bonds as the result of collision with fast-moving electrons accelerated by the field. Diodes optimized to make use of this breakdown effect are called *Zener diodes.*

Diodes as Circuit Elements

Whereas resistors, capacitors, and inductors are linear circuit elements, diodes show a nonlinear current-voltage characteristic, as depicted in Figure 3-21, where also is shown the conventional symbol for a diode. A typical unit, the 1N3605, has a resistance of 10 Ω in the forward direction and 3 GΩ (3×10^9 Ω) in reverse.

The residual current that flows when a diode is reverse biased (the *leakage current*) is very small, but not zero; it can be measured in micro- or nanoamperes. The potential drop across the diode during forward conduction (*forward voltage drop*) for a germanium diode

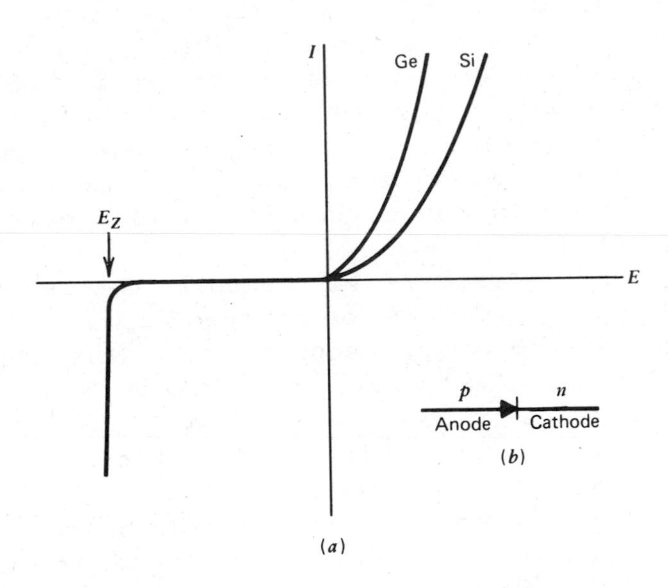

Figure 3-21. (*a*) The current-voltage curves characteristic of silicon and germanium diodes (E_Z is the Zener potential). (*b*) The symbol for a diode; the arrowhead shows the direction of flow of conventional (positive) current.

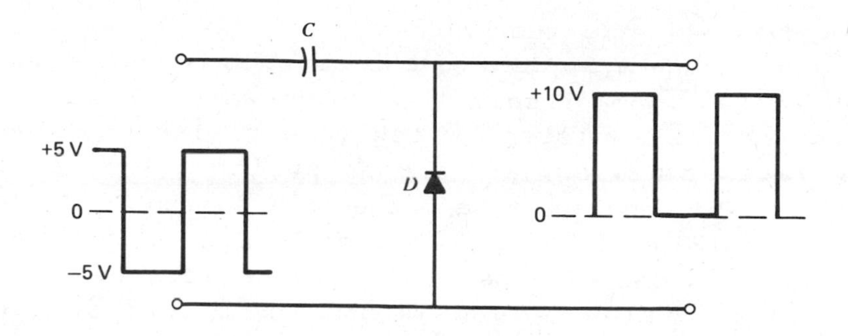

Figure 3-22. A diode clamp.

is approximately half as large as that for a silicon diode. At higher current values, the forward drop is fairly constant, about 0.3 to 0.4 V for germanium and 0.6 to 0.8 V for silicon, but in the examples that follow, we shall assume the value of 0.7 V. The 1N3605, at its rated current of 80 mA, gives a forward drop of 0.8 V.

The primary function of a diode is to allow current to flow in one direction only, and hence one of its major uses is in *rectification*, the conversion of alternating to direct current, dealt with in Chapter VIII.

Another application of a diode is as a *clamp* or "DC restorer." Suppose that a square wave with 10-V peak-to-peak amplitude is passed through a capacitor (Figure 3-22). Unless the *RC* product is too small, the shape of the wave will not be impaired appreciably, but in the absence of the diode, it would no longer have a predictable relationship to ground. This relationship can be reestablished by the diode clamp, which prevents the signal from becoming negative. The entire 10-V swing is then forced to be positive with respect to ground. Allowance must be made for the forward voltage drop of the diode, so that (for silicon) the square wave would actually swing between about -0.7 and +9.3 V. Hence this kind of clamp is not satisfactory for very small amplitude signals.

Diodes are also used as automatic switches, activated by the signal itself. When forward biased, the diode acts as a closed switch (low resistance) and, when reverse biased, as an open switch (high resistance). This analogy to a mechanical switch is demonstrated in Figure 3-23. As long as E_{in} is less than +4.3 V, the diode will be in its low-resistance state. Current will flow through R and the diode, and the whole voltage drop (except 0.7 V) will appear across R, so that $E_{out} = E_{in} + 0.7$. When E_{in} becomes greater than +4.3 V, the diode becomes nonconducting, no current flows through R, and E_{out} = +5 V. The plot in Figure 3-24*a* depicts this relationship.

Note that the switching action resides in the diode alone; thus, in Figure 3-24, E_{out} is either connected directly to E_{in} or to the 5-V source. An-

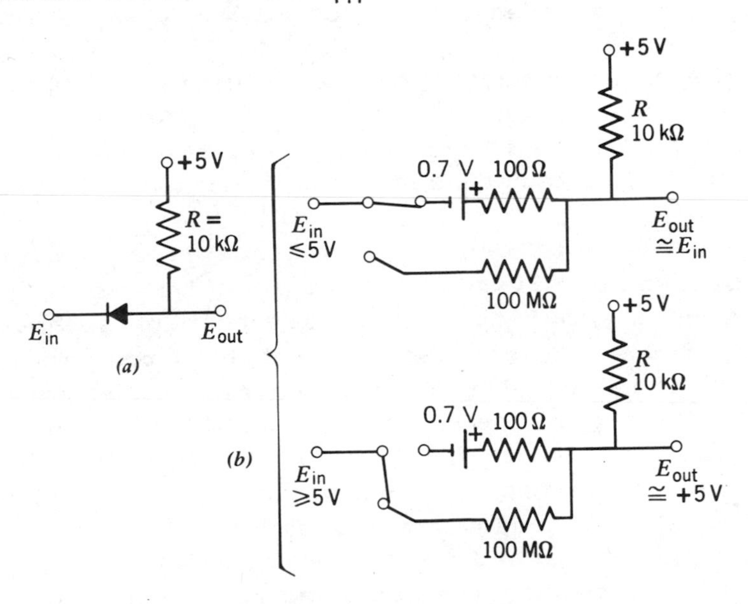

Figure 3-23. (*a*) A diode switch and (*b*) its equivalent circuits. By applying the principle of the voltage divider, one can see that the output must be either +5 V or (E_{in} + 0.7) V.

other form of diode switch with its corresponding plot is shown in Figure 3-24*b*. Other configurations of diode and resistor can be utilized in switching circuits.

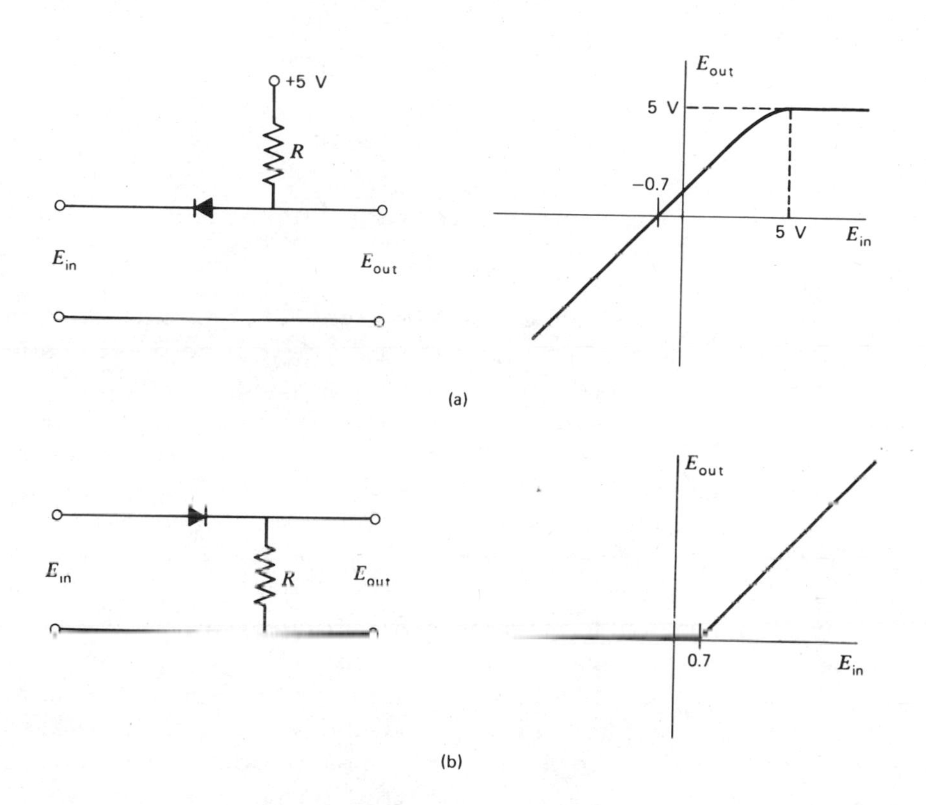

Figure 3-24. Two diode switches and their response plots. In either case, zero input gives nearly zero output and +5 V *in* gives +5 V *out*. Circuit (*a*) acts as a leveler for logic "1" since any higher voltage *in* gives +5 V *out*; (*b*) is a leveler for logic "0," as any negative voltage *in* produces zero *out*. In both cases it is assumed that the impedance of the load is large compared to *R*, so that the current drawn from the output is negligible.

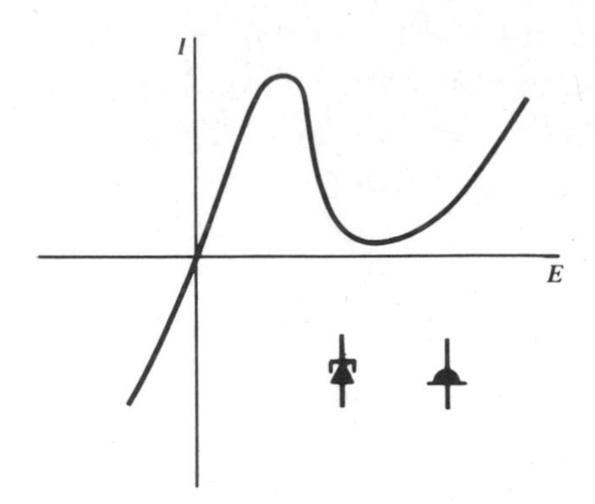

Figure 3-25. The current-voltage characteristic of a tunnel diode and two alternative symbols.

Another class of diode application depends on the very nearly logarithmic shape of the current-voltage curve of a silicon diode when forward biased. This permits multiplication and division of electrical quantities through logarithmic transformation, considered in a later chapter.

Diodes with an unusally high level of doping and an abrupt transition from *n* to *p* show the current-voltage characteristic illustrated in Figure 3-25. For reverse as well as very small forward voltages, the diode acts as a pure ohmic resistance. With increasing forward bias, the current passes through a maximum, followed by a marked valley, beyond which it increases exponentially as in a conventional diode. Such units are called *Esaki diodes*, from their inventor, or *tunnel diodes*, from the theoretical explanation, which involves quantum-mechanical "tunneling" of electrons through an energy barrier. Tunnel diodes are inherently low-power devices (usually less than 100-mW dissipation), which find important applications in high-speed switching for computers and in microwave electronics.

Zener Diodes

Zener diodes operate in the reverse breakdown mode with the voltage E_Z (Figure 3-21). They are widely applied to voltage regulation, for which an elementary circuit is shown in Figure 3-26. (Note that the symbol for a Zener diode is that of a normal diode with wings added, reminiscent of the letter Z.) The object of the circuit is to maintain a constant voltage across the load resistor R_L, in this case 10 V, regardless of variations in either E_{in} or R_L. In the absence of the Zener diode, the circuit would act as a voltage divider, and the objective would not be attained. With the Zener in place, the system still acts as a voltage divider, as long as the voltage at A is below 10. As the input voltage is raised, that at point A also rises, but when it reaches the Zener voltage, current begins to flow through the diode. Reference to Figure 3-21 shows that the output voltage becomes clamped (at exactly 10 V). Further increase in E_{in} increases the current through the diode but causes no appreciable change in the voltage to the load. The extra voltage appears across the resistor R_S. Variations of the load resistance will likewise have no effect provided the voltage at A as calculated by the voltage-divider

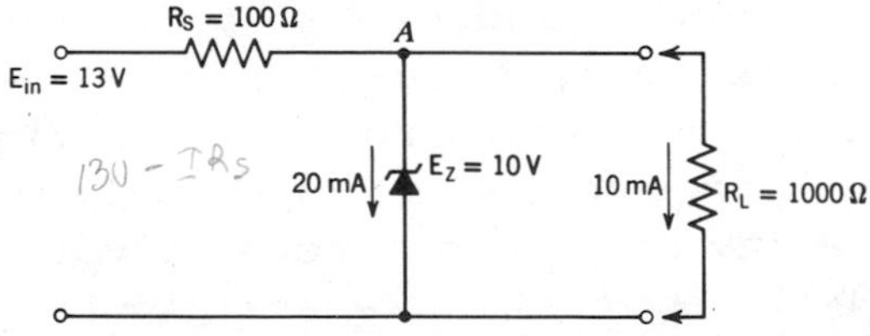

Figure 3-26. Circuit illustrating the use of a Zener diode.

equation is not less than 10 V. For the values indicated in Figure 3-26, the current is given by $(E_{in} - 10)/R_s = 3/100 = 30$ mA, of which 10 mA will flow through the load and 20 mA through the Zener.

As remarked previously, components are never perfect, and the Zener diode is no exception. The Zener voltage is dependent on the current to the extent that the descending portion of the curve in Figure 3-21 is not exactly vertical. If the input voltage and load resistance are not expected to vary by more than a few percent, this source of error may be insignificant.

The Zener voltage has a temperature coefficient, and for high precision some kind of thermostatic protection may be needed. The coefficient is nearly zero for a Zener voltage of about 5 or 6 V, increasingly negative below this and positive above, corresponding to the two different mechanisms previously described producing low or high Zener voltages. For a 10-V Zener, the coefficient is about +6 mV/K.

The *forward* voltage drop of a silicon diode has a negative temperature coefficient; thus it is often possible to combine one or more forward diodes with a positive temperature coefficient Zener to produce a compensated reference voltage. For example, three type 1N536 diodes in series with a 1N1604 Zener give a combined voltage drop of about 12.2 V with a temperature coefficient of only 0.5 mV/K.

Zener diodes are useful for the protection of various electronic components against excessive voltage. For example, a 10-V Zener in series with a small resistor, mounted across the terminals of a 10-V DC meter will prevent damage from higher voltage of the correct polarity or from any reverse voltage that might accidentally be applied. Zeners can similarly protect switch contacts from arcing, as when an inductive load is suddenly disconnected (Figure 3-27).

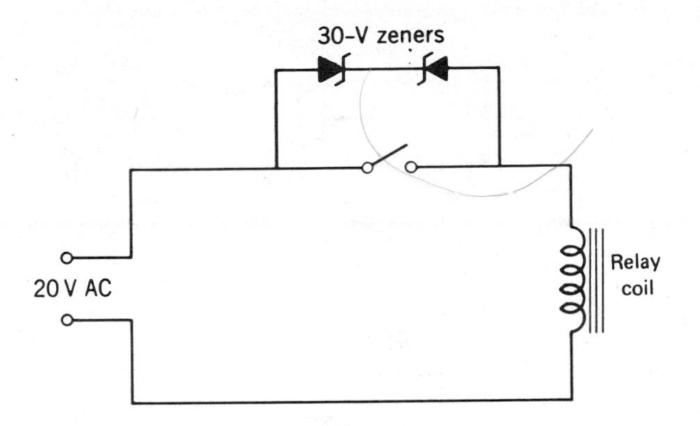

Figure 3-27. An AC relay circuit, with the switch protected by a pair of Zeners. In such a pair of back-to-back Zeners, for either polarity, one diode operates in the Zener region, and the other is forward biased.

ACTIVE DEVICES

The principal active devices for handling signals are transistors, of which several types are of importance to us. These are here described briefly as circuit components. Applications of active devices are discussed in further detail in Chapter VII.

The most extensive use of transistors is in integrated circuits, but discrete units are available for use in high-power or high-frequency circuits. Other active devices include thyristors and vacuum tubes. Vacuum tubes are remanents of a past historical period and are not treated here.

The Bipolar Transistor

This is the most common active device, often simply called a "transistor," without modifier. It consists of three layers or regions of silicon of different doping. Figure 3-28 shows the structure and symbols for the two complementary forms, called *npn* and

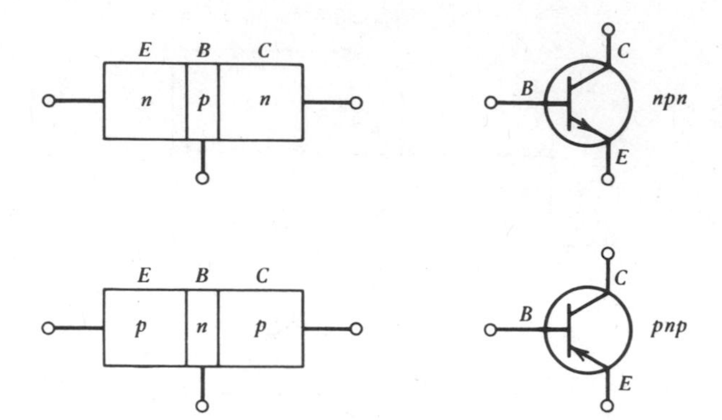

Figure 3-28. Structure and symbols for bipolar transistors.

pnp, respectively. The three areas in each form are the *emitter*, the *base*, and the *collector*. The base is a very thin region separating the emitter and collector.

Consider an *npn* transistor, with two batteries as power supplies, connected as shown in Figure 3-29. The currents are considered to be positive for the conventional (positive) current *entering* the transistor. Consequently, I_E is negative and I_C positive. (Remember that electrons move in the direction opposed to the conventional arrows.)

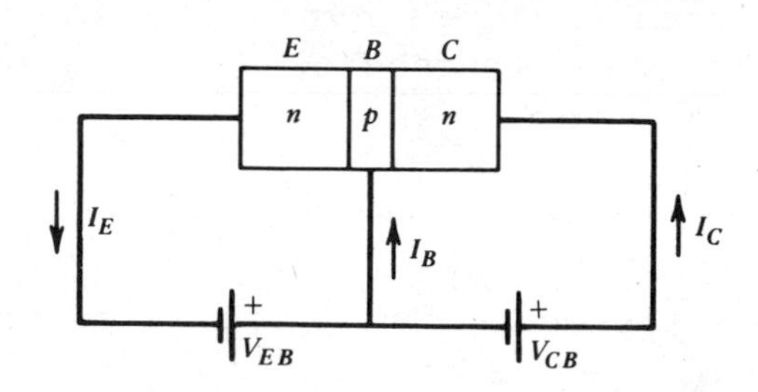

Figure 3-29. An *npn* transistor, showing voltage supplies and currents.

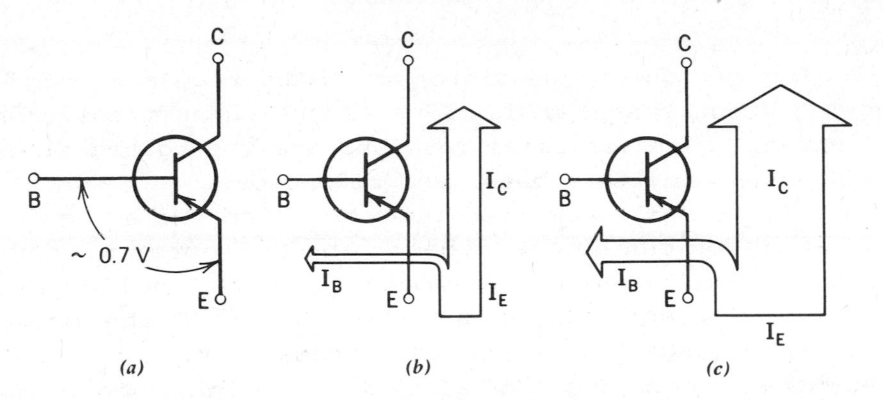

Figure 3-30. Representation of the voltage (*a*) and the current flow (*b, c*) in a transistor. Note that the ratio I_C/I_B is the same for large or small total currents.

The flow of electrons from emitter to collector within the transistor is subject to interaction with the base region, which acts as a controller. The base is so thin and sparsely populated with holes (lightly *p* doped) that most of the electrons pass through it without incident, but a small fraction are captured by holes. If base current is allowed to pass, an excess of holes will be maintained in the base region. The added holes increase the field sensed by electrons from the emitter, and thus flow more readily from emitter to collector. The relationships between the various currents in a transistor are illustrated in Figure 3-30.

The transistor is characterized by a factor β, the ratio of a change in collector current ΔI_C to the change in base current ΔI_B, causing it:

$$\beta = \frac{\Delta I_C}{\Delta I_B} \tag{3-17}$$

The β factor for small signal transistors is usually within the range 50 to 200.

The design of transistor circuits requires information about the behavior of the particular transistor selected. The variables involved are the voltages and currents at emitter, base, and collector.

Figure 3-31 shows two ways by which the action of a transistor can be described: the collector current I_C plotted either as a function of the collector-emitter voltage V_{CE} or as a function of the base-emitter voltage V_{BE}. For this transistor, the curves show that a change ΔI_B of 20 µA results in a change ΔI_C of about 4 mA, for a β of 200.

Consider the simple amplifier circuit shown in Figure 3-32, where the same transistor is shown with suitable connections. The currents are taken as positive flowing *into* the transistor. Because of the relationships between currents described previously, we can write

$$-I_E = I_B + I_C \tag{3-18}$$

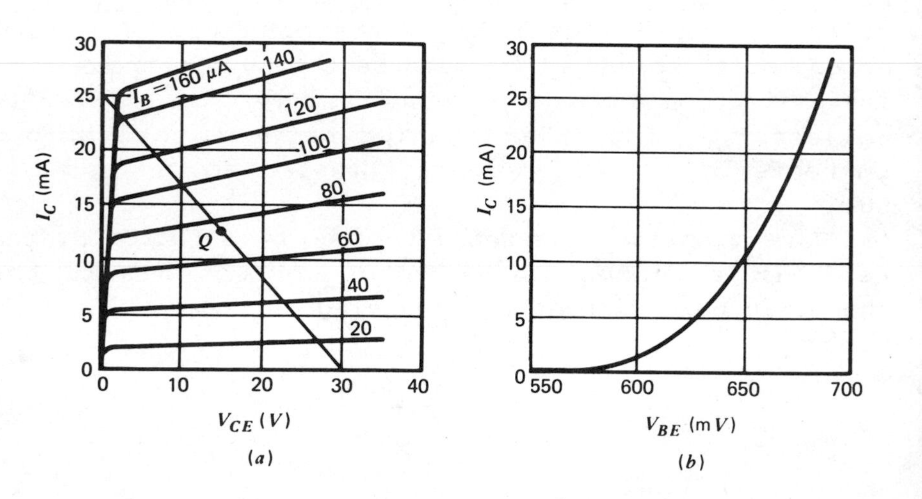

Figure 3-31. Current-voltage characteristics of the 2N4074 *npn* transistor.

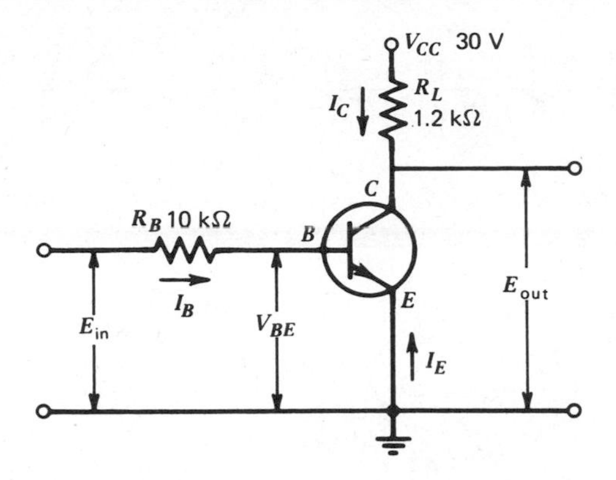

Figure 3-32. Basic circuit of a transistor amplifier, using the 2N4074.

where the minus sign indicates that the current I_E flows against the arrow. The current flowing into the base is given by the quotient

$$I_B = \frac{E_{in} - V_{BE}}{R_B} \tag{3-19}$$

Suppose that initially an input voltage E_{in} = 0.750 V produces a base current I_B = 0.015 mA. According to Eq. 3-19, the base-emitter voltage will be V_{BE} = 0.600 V. Figure 3-31*b* shows that this corresponds approximately to I_C = 1 mA. This current, passing through the load resistor R_L, causes a voltage drop of (1 mA)(1.2 kΩ) = 1.2 V, so that the output voltage is $V_{CC} - I_C R_L$ = 30 − 1.2 = 28.8 V.

Now let the input voltage be increased from 0.750 to 1.250 V. This increases the base current to 0.060 mA and gives V_{BE} = 0.650 V. Figure 3-31*b* shows that I_C is increased to 10 mA; hence the output is raised to 30 − (10)(1.2) = 12 V.

From Eq. 3-17 we can calculate that the β of the transistor is

$$\beta = \frac{(10 - 1)\ \text{mA}}{(0.060 - 0.015)\ \text{mA}} = \frac{9}{0.045} = 200 \qquad (3\text{-}20)$$

in agreement with our preliminary estimate. A voltage gain for the circuit can be defined as

$$-\frac{\Delta E_{out}}{\Delta E_{in}} = \frac{28.8 - 18.0}{1.250 - 0.960} = 37 \qquad (3\text{-}21)$$

The negative sign is required because increasing E_{in} causes a decreased E_{out}. It is sometimes convenient to regard the amplifier as converting a current signal to a corresponding voltage. In this case a useful parameter is the *transimpedance*, given by

$$Z_{tr} = \frac{\Delta E_{out}}{\Delta I_B} = \frac{10.8\ \text{V}}{0.045\ \text{mA}} = 240\ \text{k}\Omega \qquad (3\text{-}22)$$

The diagonal line superimposed on the curves in Figure 3-31*a* is the *load line*. Its terminals correspond to the open-circuit voltage (i.e., 30 V at zero current) and the short-circuit current (25 mA, with the output at ground potential). The ratio of these two quantities gives, by Ohm's law, the value 1.2 kΩ for the load resistor R_L. Such a line can be drawn for any proposed load and is useful in determining the result of any changes in it. At any moment, the current and voltage present in the circuit will define an operating point that must lie on the load line.

For optimum linear operation, the transistor must be kept within an appropriate portion of its characteristics. This requires a *bias circuit* to provide the proper current to the base. It is convenient to

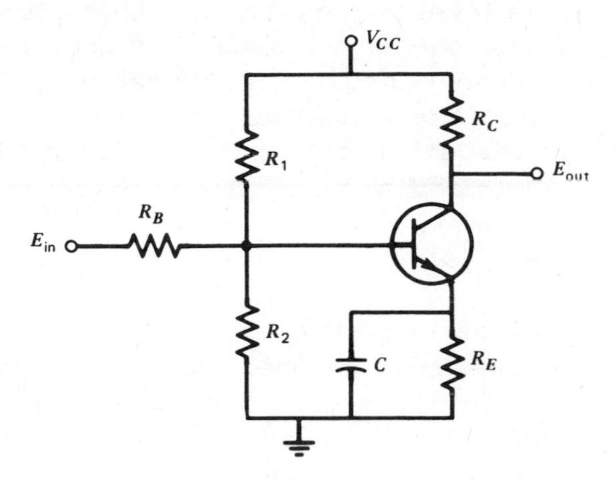

Figure 3-33. Transistor amplifier, with bias circuitry shown.

designate a particular operating point, such as Q in Figure 3-31*a*, as the ***quiescent point***, which will describe the currents through the transistor when no signal is present.

The required bias can be supplied by an RC network such as that shown in Figure 3-33. The voltage divi-

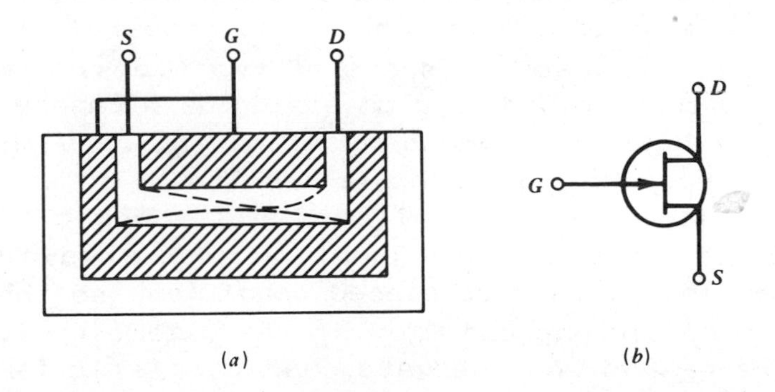

Figure 3-34. (*a*) Structure of an *n*-channel JFET and (b) its symbol.

der R_1 - R_2 serves to establish the Q point. The emitter resistor R_E (a few hundred ohms), increases the stability of the circuit. The capacitor across R_E reduces noise if the signal to be amplified is DC. It also establishes different operating conditions for AC and DC.

The Field-Effect Transistor (FET)

This type of transistor, also called a JFET (for junction FET), depends on the control of a current by a transverse electric field. The basic FET is diagrammed in Figure 3-34. It consists of a bar of *n*-type silicon through which a current is passed, analogous to the emitter-collector current in a bipolar transistor. Electrons are injected into the bar at one end (the *source*) and leave it at the other (the *drain*). The region between source and drain is called the *channel*.

On both sides of the channel are areas of heavily doped *p*-type silicon (p^+) forming a *gate*. A reverse potential applied to the gate-channel junction creates an electrostatic field that repels electrons, effectively narrowing the conductive channel and increasing its resistance. Thus the gate potential controls the flow of current through the channel.

Just as bipolar transistors appear in two forms (*pnp* and *npn*), so JFETs are of two types, the *n* channel, just described, and an analogous *p* channel, with a gate of heavily doped *n* material (n^-). In this type, the source-drain current is carried by holes injected at the source and annihilated at the drain. In both types the gate-channel junction must always be maintained in the reverse-biased condition, so that essentially no current can flow in the gate circuit. This is the reason why the gate characteristic is given in terms of potential, rather than current. The high gate impedance is the chief distinguishing feature of FETs, as contrasted with bipolar transistors.

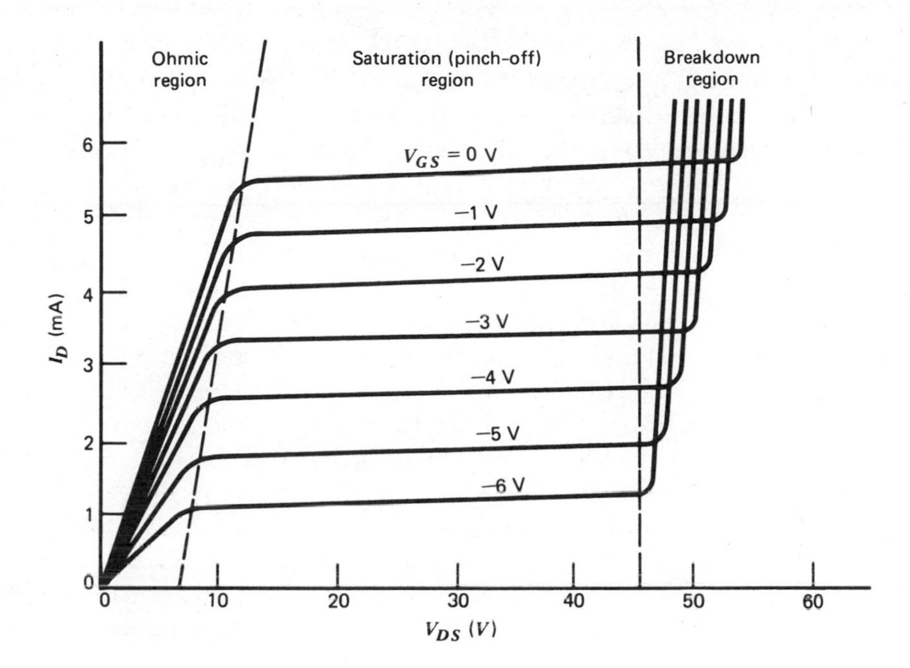

Figure 3-35. Current-voltage characteristics of an *n*-channel JFET.

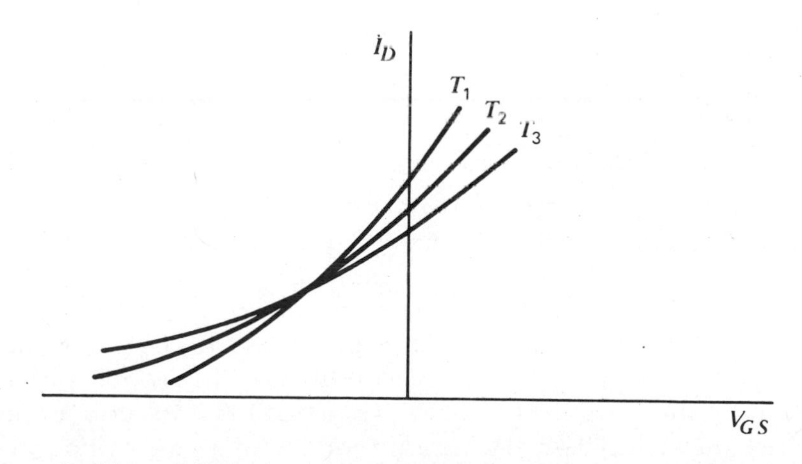

Figure 3-36. Effect of temperature on the characteristic curve of a depletion MOSFET ($T_1 < T_2 < T_3$). The point of zero temperature coefficient lies between 100 and 500 μA for small signal FETs.

Figure 3-35 shows a family of characteristic curves for an *n*-channel FET, in which the drain current I_D is plotted against the drain-source voltage V_{DS} for various values of gate-source potential V_{GS}. Consider first the curve for $V_{GS} = 0$. For potentials of less than about 5 V, the device acts as a simple resistor, obeying Ohm's law. However, when the current exceeds about 5 mA, the voltage drop through the resistance of the channel becomes sufficient to produce a depletion region, even without a potential applied to the gate. (This is the reason for the asymmetric depletion area shown in Figure 3-34.) This restricts the flow of current, which levels off into a region of saturation, called the *pinch-off* region. Eventually, if V_{DS} is increased further, breakdown will occur, similar to that of a Zener diode. For linear amplification, the FET is operated in the saturation region, where the drain current is nearly proportional to the gate-source voltage.

Another useful characteristic is the curve of drain current I_D as a function of the controlling gate voltage V_{GS}, shown in Figure 3-36. A unique property of FETs, evident in this graph, is the existence of a point where the channel resistance is independent of temperature. The curves for various temperatures intersect at a common point. This temperature independence suggests the usefulness of FETs in devices where thermal drift must be minimized.

An alternative construction, a metal-oxide-semiconductor FET (MOSFET), includes an insulating layer of silicon dioxide between the channel and the gate, as shown in Figure 3-37*a*. The gate now makes no electrical contact with the channel, but the field it produces is nevertheless effective in modulating the channel resistance. Clearly no current can flow through the gate, and thus it is permissible to give it either a positive or negative charge.

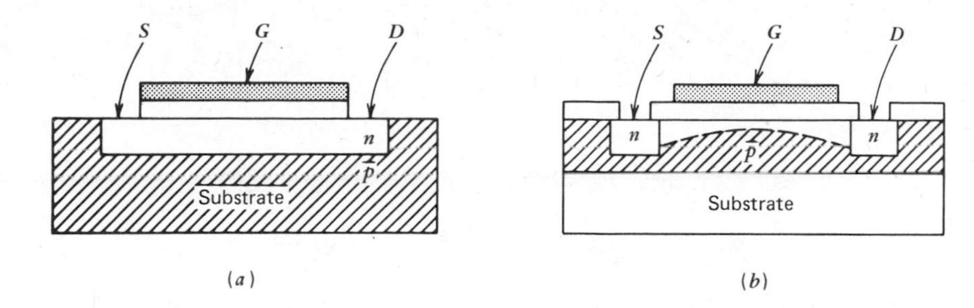

Figure 3-37. The structure of MOSFETs: (*a*) *n*-channel depletion, (*b*) *n*-channel enhancement.

The transistor shown in Figure 3-37*a* is called a *depletion-mode* MOSFET, as the field formed by the gate potential serves to deplete the channel of its majority carriers (i.e., electrons in *n*-channel, holes in *p*-channel types). Another variety, an *enhancement-mode* MOSFET, is diagrammed in Figure 3-37*b*. In this type, there is no channel at all until one is created by the action of the field. A block of *p* silicon has two *n*-doped areas that will become the source and drain. A positive charge given to the insulated gate attracts electrons into the space between source and drain, thus effectively producing a region of *n* doped material that constitutes a channel.

Each type of MOSFET can also exist in a complementary version in which *n* and *p* regions are interchanged. These are summarized for all types of FET in Figure 3-38.

FETs can be used in circuits similar to those of bipolar transistors. The chief difference is their large input impedance, which is important in the amplification of small currents flowing in high-impedance circuits. In addition, they can be used in the ohmic region as voltage-controlled resistors. An example of such an application is shown in Figure 3-39.

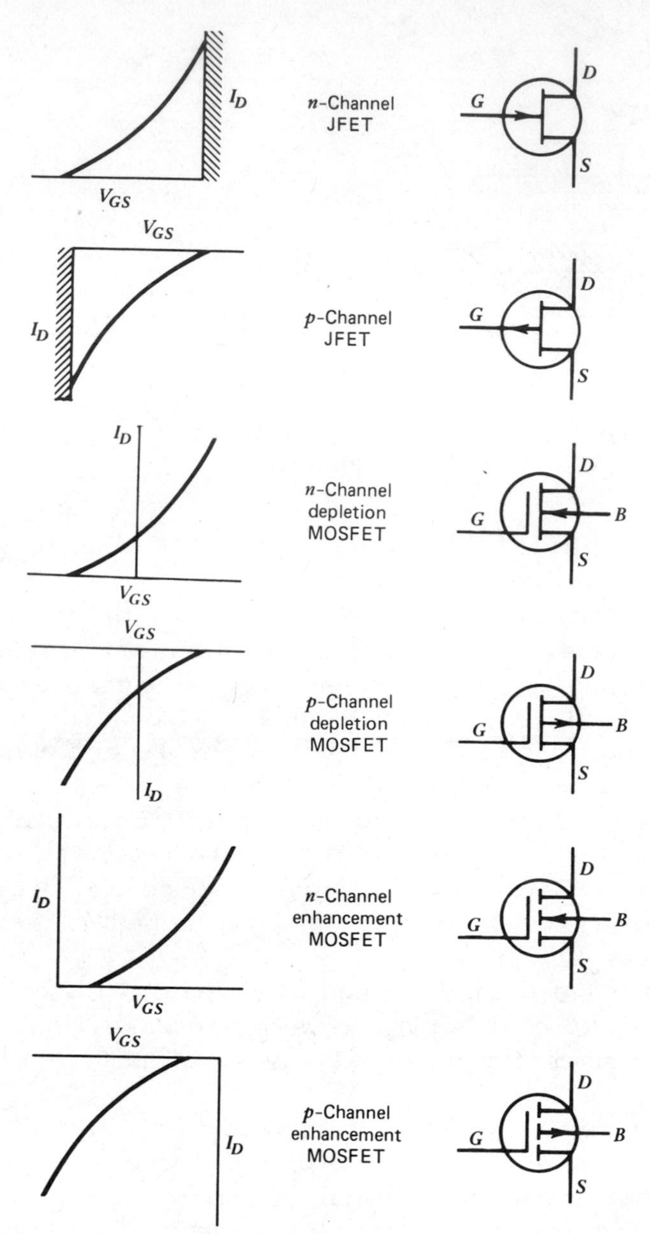

Figure 3-38. Summary of FET types. The shaded regions in the transfer plots for JFETs are undefined areas corresponding to forward gate bias (*S*, source; *G*, gate; *D*, drain; *B*, bulk substrate).

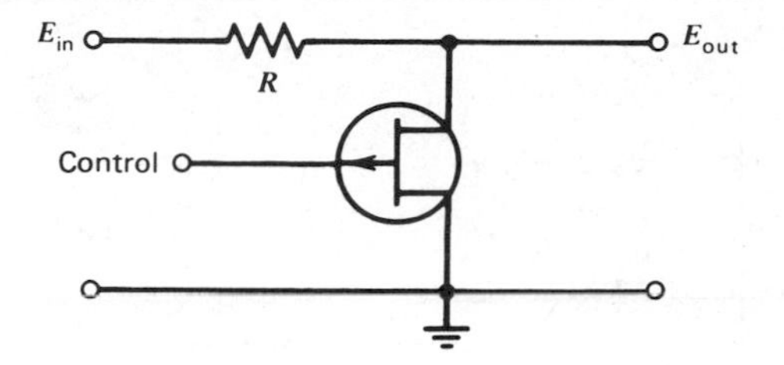

Figure 3-39. An automatic gain-control circuit using a JFET as a voltage-controlled resistor.

PROBLEMS

3-1. Determine the maximum voltage that should be applied across resistors marked as follows:

(a) 100 Ω at 2 W
(b) 10 kΩ at 0.5 W
(c) 22 MΩ at 250 mW

3-2. Ten identical resistors are connected in series. Each is marked 1.0 MΩ at 0.5 W. (a) What is the wattage rating of the string? (b) What is the maximum voltage that can be dropped across this string without exceeding the power dissipation limits?

3-3. Answer the same questions as in Problem 3-2 for a parallel connection of the same 10 resistors.

3-4. Calculate the overall resistance of the circuit shown in Figure 3-40.

3-5. In the circuit shown in Figure 3-5*b*, the input voltage is maintained constant at 15 V. Compute the following:

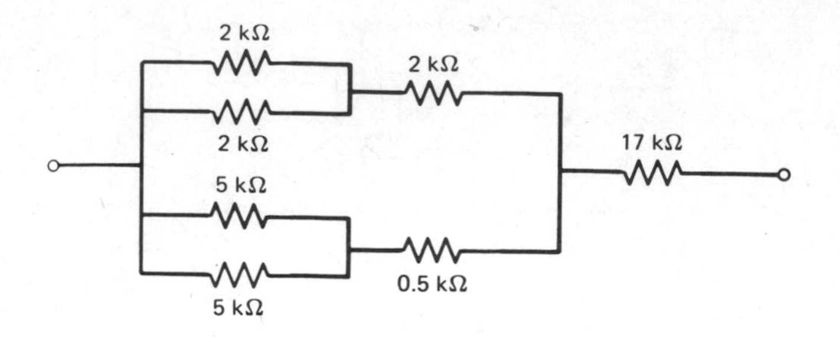

Figure 3-40. See Problem 3-4.

(a) E_{out}, I_{load}, and I_{in}
(b) Input impedance
(c) Voltage gain (in dB)
(d) Current gain (in dB)
(e) Power gain (in dB)

3-6. Determine the total impedance for each of the following combinations:

(a) Resistors of 100 and 151 Ω in parallel.
(b) A 33-pF capacitor in parallel with a 10 kΩ resistor at a frequency of 50 kHz.
(c) A 100-Ω resistor in series with a 100-mH inductor of negligible resistance, measured with DC.
(d) Same as (*c*), but measured at 0.1 MHz.

3-7. A capacitor and an inductor are connected in parallel and powered from an AC source, as shown in Figure 3-41. (a) At what frequency will the impedances of the two components be equal? (b) What is the net impedance at that frequency? You may assume that the DC resistance of the inductor is negligible.

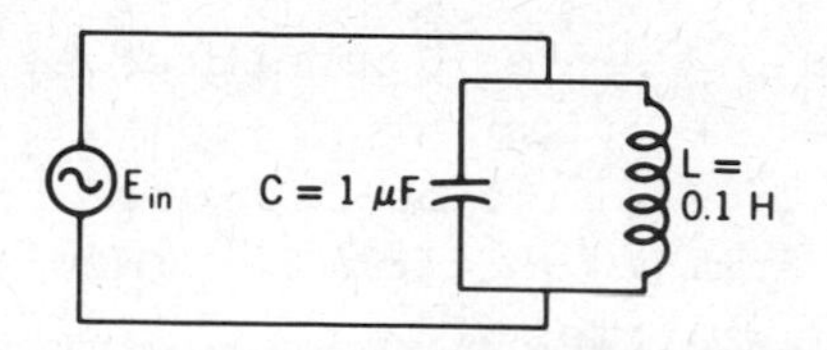

Figure 3-41. See Problem 3-7.

3-8. A meter movement is available with a full-scale sensitivity of 1 μA and internal resistance of 10 kΩ. Design a voltmeter using this meter, to have ranges of 0.1, 0.5, 1, 5, and 10 V.

3-9. A voltage divider is required that will attenuate a 10-V signal to 10 mV. The current from the source must not exceed 1 mA, and R_{out} must be less than 20 Ω. Design a circuit to fulfill these conditions.

3-10. What is the impedance of a 10 kΩ wire-wound resistor with an inherent inductance of 2 mH, when used at 1 kHz?

3-11. A particular choke coil has an inductance of 20 H, series resistance of 0.1 Ω, and stray capacitance of 30 pF. At what frequency will it resonate?

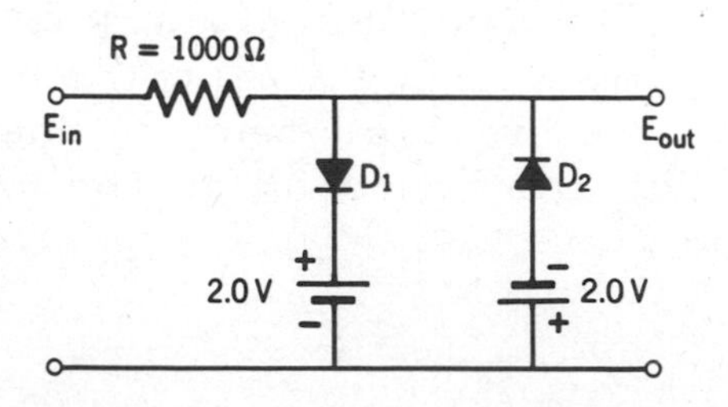

Figure 3-42. See Problem 3-12.

3-12. Diodes and batteries connected as in Figure 3-42 form a useful clipping circuit. What would be the value of E_{out} to be expected from each of the following input functions? (Neglect the forward drop of the diodes.)

(a) $E_{in} = +1.0$ V
(b) $E_{in} = -10.0$ V
(c) $E_{in} = 10 \sin \omega t$, where $\omega = 100$ rad/s
(d) $E_{in} = (10 - 2t)$ V

Sketch the outputs for (*c*) and (*d*) as functions of time.

3-13. A ±1 V zero-center meter with internal resistance of 10 kΩ, for use as a null indicator, can be shunted with a pair of crossed diodes. This serves to protect the meter against overloads and also extends the useful range while retaining its sensitivity near the null point. Explain each of these functions.

3-14. Design a low-pass filter with 1 kΩ input impedance and $f_0 = 500$ Hz.

3-15. A low-pass filter has an attenuation slope of 40 dB/decade and $f_0 = 3000$ Hz. The input consists of a composite signal containing a fundamental frequency of 2000 Hz (10 V RMS), with 2 V of second harmonic and 1 V of third harmonic. Compute the harmonic content of the output. You may neglect the rounding at the intersection on the Bode plot.

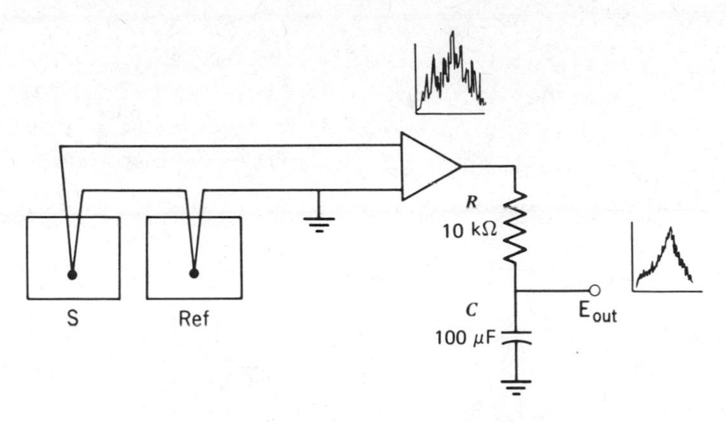

Figure 3-43. See Problem 3-16.

3-16. Consider the circuit shown in Figure 3-43. The signal is 100 mV (DC), together with 10 mV of 60-Hz noise. If R = 10 kΩ and C = 100 μF, what is the improvement in the S/N ratio?

3-17. Design a twin-T filter (Figure 3-15) with f_0 = 60 Hz. Assume R_{in} = 0.1 MΩ.

3-18. In the circuit of Figure 3-33, R_E = 120 Ω in parallel with a 0.10 μF capacitor. Calculate the impedance of the combination at (a) 0.1 Hz and (b) 100 kHz.

3-19. Suppose that the transistor described in Figure 3-31 is to be used in the circuit shown in Figure 3-33, with R_E = 120 Ω shunted by a capacitor of 10 μF, and R_C = 1.2 kΩ. V_{CC} = 30 V.

(a) What should be the values of V_B and V_E to force operation at the quiescent point Q? What will be the output voltage, E_{out}?

(b) Calculate approximate values for R_1 and R_2 consistent with the selected quiescent point.

(c) Calculate the effect on E_{out} of a low frequency AC input voltage that changes E_B by 0.1 V (peak-to-peak). Assume V_{BE} constant, and the effect of the emitter capacitor to be negligible.

IV

OPERATIONAL AMPLIFIERS

The modern development of integrated circuits (ICs) has made the operational amplifier (op amp) the single most important component in analog instrumentation. It is an amplifier characterized by very high gain, high input impedance, and low output impedance. Usually the gain is lowered by external circuitry including negative feedback. In addition to bringing the gain to a precisely desired value, negative feedback improves the overall behavior of the amplifier.

The basic configuration of an op amp with its necessary external components is shown in Figure 4-1a. The boxes marked "input" and "feedback" circuits are usually made up of resistors and capacitors, but other devices may be incorporated on occasion.

The op amp itself is symbolized by a triangle provided with two inputs, marked with plus (+) and minus (-), and a single output. A typical unit is limited in output to perhaps 10-mA current and ±10 V.* It is zero-crossing, meaning that its output can swing in both positive and negative senses, with zero input giving zero output. The output voltage is propor-

* Many commercial op amps are guaranteed to give at least a 10-V output, but actually will go to 12 V or more before saturating.

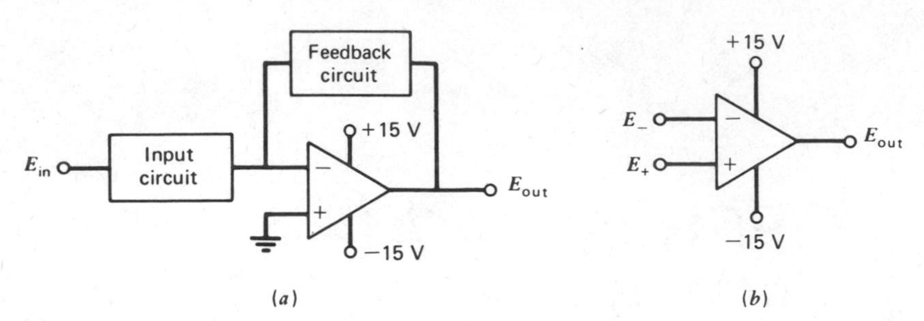

Figure 4-1. (*a*) Basic operational amplifier with ancillary circuits shown; (*b*) the op amp without feedback. The connections to the power supply (±15 V) seldom are shown.

tional to the *difference* between the potentials at the two inputs, according to the relationship

$$E_{out} = A(E_+ - E_-) \tag{4-1}$$

where E_+ and E_- are the potentials at the correspondingly marked inputs and A is the inherent gain.

The constant A in Eq. 4-1 is a large number, perhaps 10^5 or 10^6. If there were no external components, as in Figure 4-1*b*, the output would quickly reach its saturation limits for anything but the smallest input signals. For example, if $A = 10^6$, an input signal $(E_+ - E_-)$ any greater than about 12 μV would saturate the amplifier. Figure 4-2 shows a transfer plot for such an application, in which it is evident that for the vast majority of cases the output voltage would be either +12 or −12 V, depending on the sign of the input. In an application such as this, the op amp is called a *comparator* since it compares the relative values of E_+ and E_- and responds accordingly with positive or negative saturation. The comparator can be considered to be a digital device since it produces only two distinguishable output states. The levels thus established are not directly compatible with the usual levels of digital electronic systems, which are 0

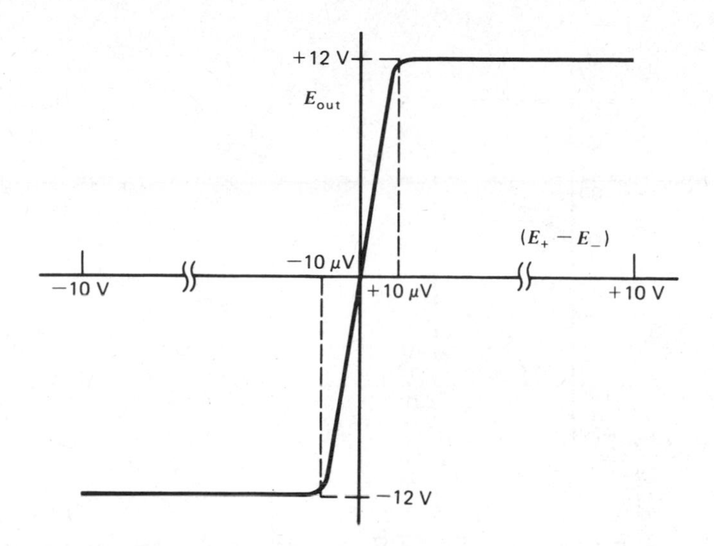

Figure 4-2. Transfer plot (output vs. input) of an op amp devoid of associated circuitry. The slope of the midportion is the inherent gain *A*. Note the change in scale on the horizontal axis.

and +5 V. The needed correction can be implemented with the use of a pair of diodes connected to a 5-V power source and to ground, as shown in Figure 4-3.

The gain of a comparator is a function of frequency, as illustrated in the typical Bode plot in Figure

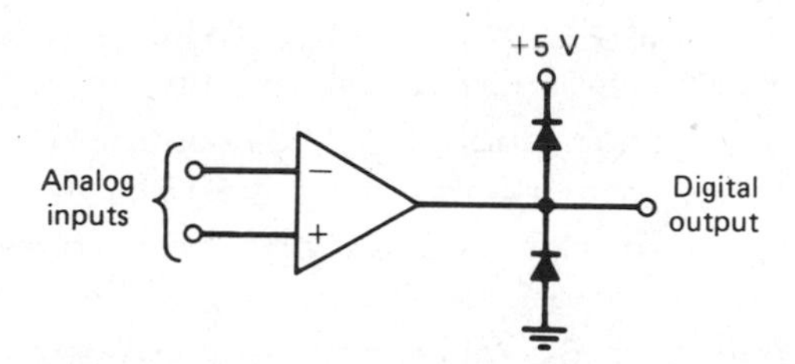

Figure 4-3. An op amp connected as a comparator to give outputs at +5 or 0 V. This circuit can be considered to be a simple analog-to-digital converter. For best results, a specially optimized comparator is to be preferred over a general-purpose op amp.

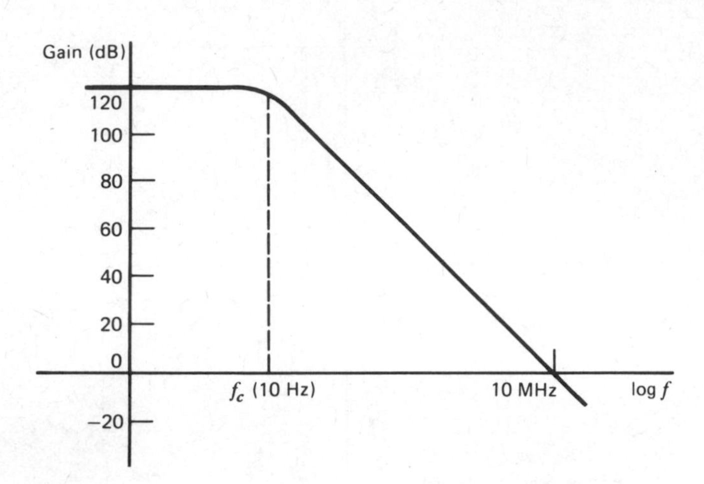

Figure 4-4. A Bode plot for an op amp. (A gain of 10^6 corresponds to 120 dB.) Note that DC (zero frequency) cannot be represented on this graph, as it would lie at negative infinity.

4-4. The response can be approximated by a horizontal line from DC to some rather low frequency, shown as 10 Hz in the figure, intersecting with another straight line that has a negative slope. Throughout the range covered by the sloping line (10 to 10^7 Hz), the product $A \cdot f$ is a constant, called the *gain-bandwidth product*. This is a useful figure of merit for an op amp.

The comparator is an example of a nonlinear application of an op amp. Many more circuits are possible if the amplifier is made to operate in a linear mode with an analog rather than digital output. This can be accomplished by incorporating *negative feedback*, a signal path between the output and the inverting input, best explained through an example. Consider the amplifier in Figure 4-5, with the noninverting input at zero potential (grounded) and resistors connected from the inverting input both to the signal (R_{in}) and the output (R_f, the *feedback resistor*). We know from Ohm's law that the current flowing through a conductor is given by the difference of potential between its

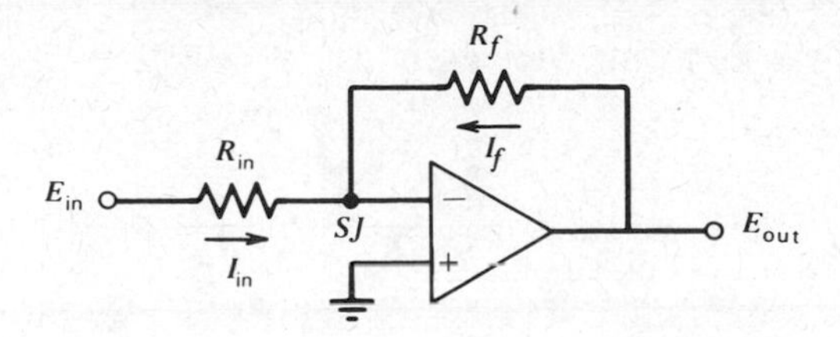

Figure 4-5. An amplifier with negative feedback. The conection to the inverting input is called the *summing junction* (SJ) (or *summing point*), as the two currents I_{in} and I_f are said to be "summed" at this point.

terminals divided by its resistance. Therefore, the input current I_{in} must equal $(E_{in} - E_{SJ})/R_{in}$, and the current I_f, the *feedback current*, is $(E_{SJ} - E_{out})/R_f$. Since the input impedance of the amplifier is very high, it follows that no current is lost into the amplifier; thus I_{in} and I_f must be equal, so we can write

$$\frac{E_{in} - E_{SJ}}{R_{in}} = \frac{E_{SJ} - E_{out}}{R_f} \tag{4-2}$$

We also know that $E_{out} = A(E_+ - E_-) = A(0 - E_{SJ})$, or $E_{SJ} = -E_{out}/A$. By substitution into both sides of Eq. 4-2 we obtain

$$\frac{E_{in} + (1/A)E_{out}}{R_{in}} = \frac{-(1/A)E_{out} - E_{out}}{R_f} \tag{4-3}$$

which simplifies to

$$\frac{E_{out}}{E_{in}} = -\frac{A \cdot R_f}{(A + 1)R_{in} + R_f} \tag{4-4}$$

This can be written as

$$\frac{E_{in}}{E_{out}} = -\left(\frac{A+1}{A} \cdot \frac{R_{in}}{R_f} + \frac{1}{A}\right) \tag{4-5}$$

or, after rearranging terms,

$$\frac{E_{in}}{E_{out}} = -\frac{R_{in}}{R_f}\left(1 + \frac{1}{\beta A}\right) \tag{4-6}$$

where the constant β is $1/[(R_f/R_{in}) + 1]$.

At higher frequencies, where A becomes small, or when β is small, the complete equation must be used. Otherwise the quantity $1/\beta A$ is negligible, on the order of perhaps 0.01%, and we can write

$$E_{out} = -\frac{R_f}{R_{in}} \cdot E_{in} \tag{4-7}$$

Note the absence of A from the equation, which indeed makes no reference to the particular amplifier. This emphasizes a great advantage of op amps: they obey simple mathematical formulas with great precision, irrespective of the identity and parameters of the individual unit. The only deviations that must be taken into account are those brought about by higher frequencies and certain offset errors discussed later in this chapter.

The mathematical treatment just given can be extended to other op amp configurations, but the following simpler approach will suffice. Let us assume that the gain A is large and that the input and feedback impedances are not excessive. For this case two closely approximated rules can be stated:

1. The input and feedback currents are equal and opposite in sign:

$$I_{in} = -I_f \qquad (4\text{-}8)$$

2. The input voltages as measured at the (+) and (-) inputs of the amplifier itself are equal, a relationship derived directly from Eq. 4-1.

Let us see how these rules apply to the circuit shown in Figure 4-5. We write

$$E_+ = 0 \text{ (grounded)} \qquad (4\text{-}9)$$

$$E_- = E_+ = 0 \qquad (4\text{-}10)$$

$$I_{in} = \frac{E_{in} - E_-}{R_{in}} = \frac{E_{in}}{R_{in}} \qquad (4\text{-}11)$$

$$I_f = \frac{E_{out} \quad E_-}{R_f} = \frac{E_{out}}{R_f} \qquad (4\text{-}12)$$

from which we can deduce that

$$\frac{E_{in}}{R_{in}} = -\frac{E_{out}}{R_f} \qquad (4\text{-}13)$$

or

$$E_{out} = -\frac{R_f}{R_{in}} \cdot E_{in} \qquad (4\text{-}14)$$

which is identical to Eq. 4-7.

These calculations apply only if the amplifier is in control of the circuit (i.e., in its active region). If the amplifier is saturated, it loses control, and the output is functionally unrelated to the input.

SINGLE-INPUT CIRCUITS

If an amplifier has its noninverting input connected to ground, it is said to be operating in the *grounded-reference* or *single-input* mode. The configuration in Figure 4-5 is a case in point. (In contrast, in the *differential* mode, both inputs are active.) In single-input operation, the summing junction is extremely close to ground potential, a condition described as *virtual ground.* This is a consequence of the virtual equality of E_+ and E_-.

An example of a grounded-reference circuit is given in Figure 4-6. Assuming, as before, no loss of current into the amplifier, one can write

$$I_1 + I_2 + I_3 = -I_f \tag{4-15}$$

where the minus sign reflects the fact that the I_f arrow in the figure points in the opposite direction to the others. The currents can then be expressed in

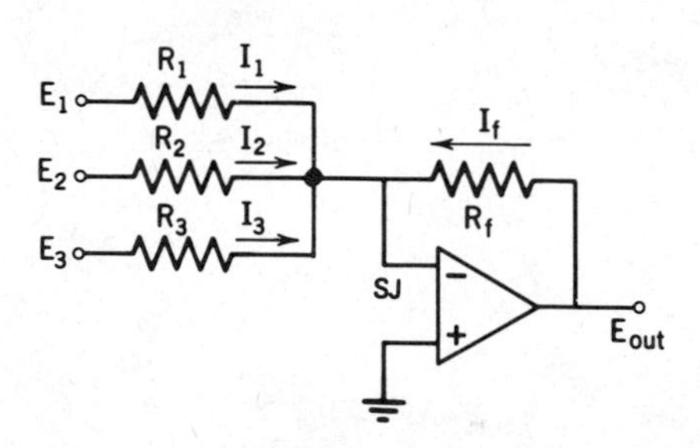

Figure 4-6. A voltage summer.

terms of the corresponding voltages, keeping in mind that the summing junction is at virtual ground:

$$\frac{E_1}{R_1} + \frac{E_2}{R_2} + \frac{E_3}{R_3} = -\frac{E_{out}}{R_f} \tag{4-16}$$

This can be rewritten in the form

$$E_{out} = -\left(E_1 \cdot \frac{R_f}{R_1} + E_2 \cdot \frac{R_f}{R_2} + E_3 \cdot \frac{R_f}{R_3}\right) \tag{4-17}$$

Thus this circuit sums all the input voltages while multiplying them by constant coefficients, with the usual sign inversion. This circuit is called a *summer*. If all the resistors are equal, the operation is the simple addition of voltages. If they are not equal, the circuit becomes a *weighted summer*. The voltages are in general time dependent, whereas the weighting coefficients are constants. Multiplication by variable quantities requires considerably more complicated instrumentation.

The equations derived in the preceding paragraphs have defined the gain of the amplifier in terms of resistance ratios, which can be implemented by a variety of resistors. There are a few restrictions of a practical nature, however. The maximum current in a typical op amp circuit is of the order of 10 mA; therefore, with ±10-V excursions, the minimum values of the resistors should be about 1 kΩ to avoid saturation. On the other hand, currents cannot be too small, since this would augment the effect of any stray currents that may be present. As a rough rule, currents should be larger than 10^{-6} A, corresponding to resistors no larger than $10^7 \Omega$. (This discussion applies to general-purpose op amps; models are available that extend the range of currents in either direction; for example, a FET-input op amp can handle much smaller currents.)

In practice, for a unity-gain inverter, it is well to use medium-value resistors for both R_{in} and R_f; 20 kΩ is a reasonable value. A large circuit gain, of course, requires one large and one small resistor; thus for a gain of 1000, it would be appropriate to choose R_{in} = 1 kΩ and R_f = 1 MΩ.

As Eqs. 4-11 and 4-12 indicate, the basic response of the amplifier and its network is to currents rather than voltages. The input resistors simply serve to establish the currents to be drawn from their respective signal voltages. If the desired information is already carried by a current, an input resistor is unnecessary, and a direct connection to the summing junction can be made, as in Figure 4-7*a*. This *current-to-voltage converter* is a useful configuration for an instrument in which small currents are to be meas-

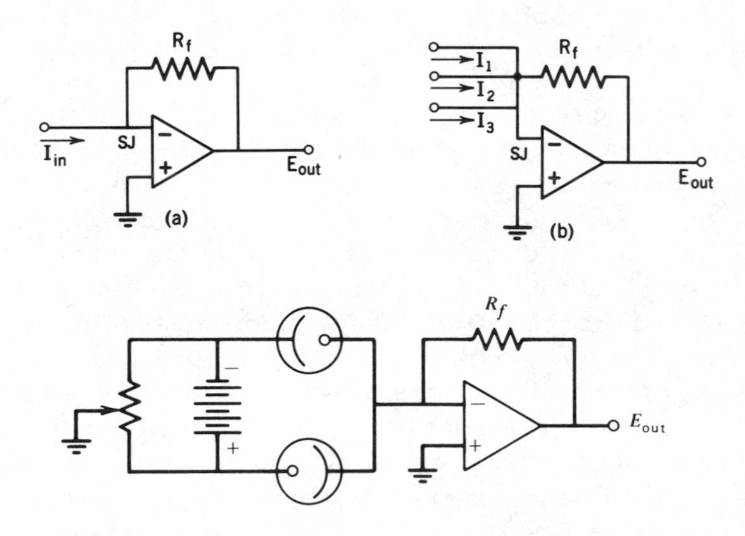

Figure 4-7. Current-to-voltage converters: (*a*) single-input; (*b*) multiple-input, serving as a current summer; (*c*) a current balance applied to a pair of phototubes.

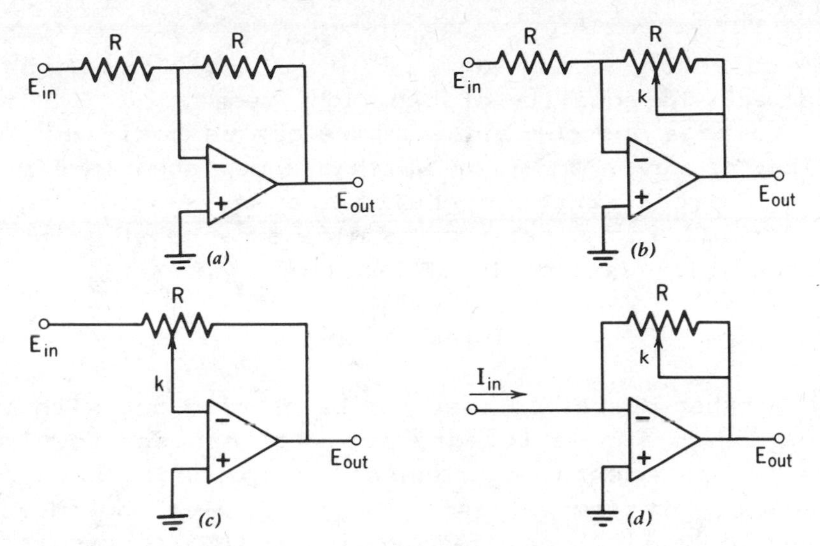

Figure 4-8. Examples of summer applications. The coefficient k is the fractional resistance exhibited by the section of each potentiometer between the variable contact and the output. The circuit in (*b*) is linear with respect to k and is usually preferred over (*c*).

ured. Thus for R_f = 1 MΩ, an input of 1 µA will produce an output of 1 V, a quantity much easier to measure than a very small current.

The circuit shown in Figure 4-7*b* is a *current summer*, in which the feedback current equals $I_1 + I_2 + I_3$. The output, by Ohm's law, is

$$E_{out} = -R_f(I_1 + I_2 + I_3) \tag{4-18}$$

The negative sign indicates that if the flow of positive charges at the input follows the arrows, the output will be negative. Since the summing junction, the point at which all currents meet, is at ground potential, there can be no interaction between the several current sources, an added advantage of this type of circuit.

An application of the current summer is shown in Figure 4-7*c*, where two photocells are connected so that their currents subtract. The amplifier acts like a null detector in a bridge. This circuit can be used to detect the equality of two light beams.

Voltage and current summers can be modified in a variety of ways, a few of which are depicted in Figure 4-8. Most of these circuits are self-explanatory; the circuit in (*c*) permits as wide a range of gains as the amplifier is capable of handling.

Integration

Another operation that can be carried out with the aid of an op amp is integration with respect to time. This is implemented by connecting a capacitor in the feedback of the amplifier. (In principle, an inductor in the input line would accomplish a similar result in AC circuit applications.)

An *integrator* circuit is shown in Figure 4-9. The current in the feedback loop charges the capacitor, and hence varies according to the relationship

$$I_f = C \cdot \frac{dE_{out}}{dt} \tag{4-19}$$

This must equal the negative of the input current:

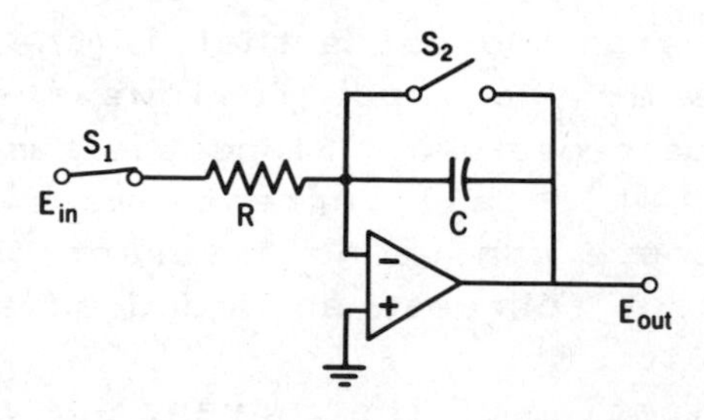

Figure 4-9. An analog integrator.

$$-\frac{E_{\text{in}}}{R} = C \cdot \frac{dE_{\text{out}}}{dt} \tag{4-20}$$

Rearrangement and integration gives

$$E_{\text{out}} = -\frac{1}{RC}\int E_{\text{in}}\, dt \tag{4-21}$$

where E_0, the integration constant, represents the initial output voltage. Thus the output becomes the time-integral of the input. Just as with filters, the product RC, having the dimensions of time, is called the *time constant*.

The integrator can be used to generate a DC *ramp* voltage, $E = kt$ (k being a constant), simply by holding E_{in} at some constant value. Then Eq. 4-21 becomes

$$E_{\text{out}} = -\frac{E_{\text{in}}}{RC}\int dt + E_0 = E_0 + kt \tag{4-22}$$

The output of the amplifier starts at E_0 and increases at a constant rate until it reaches its maximum voltage (saturation). Hence it is necessary to provide some means of discharging the capacitor to reset the integrator to zero, ready for the next integration. (This requirement is present for any DC integration, not only for a ramp generator.) Switch S_2 in Figure 4-9 serves to discharge the capacitor. On the other hand, switch S_1, when opened, stops the integration, and the capacitor retains whatever charge it has at that moment, thus holding E_{out} at a constant voltage. The three modes of operation are called, respectively, *integrate*, *reset*, and *hold*.

Also of interest is the integration of a sine wave, $E_{in} = B \sin \omega t$, to give a cosine wave:

$$E_{out} = -\frac{1}{RC}\int B \sin \omega t = \frac{1}{\omega RC} B \cos \omega t$$
$$= \frac{1}{\omega RC} B \sin\left(\omega t + \frac{\pi}{2}\right) \qquad (4\text{-}23)$$

Note that the output amplitude is a decreasing func-

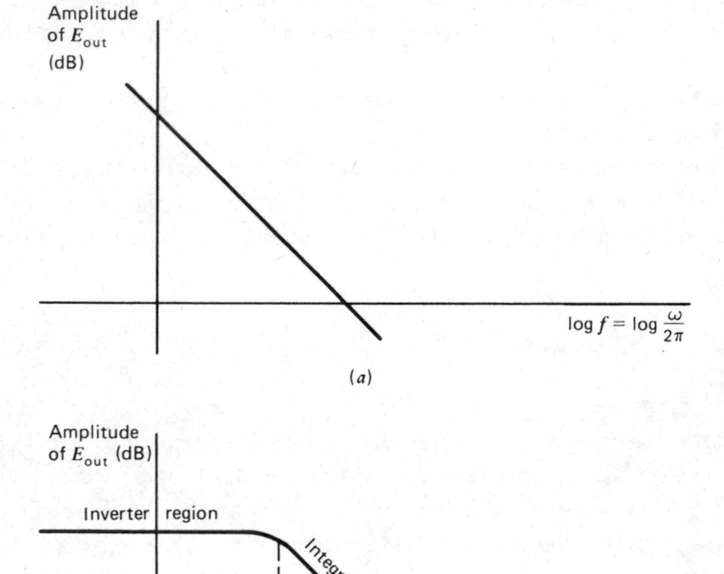

Figure 4-10. (*a*) The AC behavior of an integrator and (*b*) a modified integrator with limited gain at low frequency.

tion of frequency, and becomes equal to the input only for $\omega = 1/RC$; otherwise it has a gain of $1/\omega RC$. Such AC applications can also be described by means of the phasor notation. Equation 4-7 is valid in terms of impedances as well as resistances:

$$E_{out} = -\frac{Z_f}{Z_{in}} \cdot E_{in} \tag{4-24}$$

Substituting $\mathbf{Z}_{in} = R$ and $\mathbf{Z}_f = -j/\omega C$, we obtain

$$E_{out} = \frac{j}{\omega RC} \cdot E_{in} \tag{4-25}$$

This expression for the gain can be shown to be the same as that in Eq. 4-23. Recall that multiplication by j is tantamount to a vector rotation of $\pi/2$.

The frequency response of the integrator is shown in Figure 4-10*a*. Infinite gain at zero frequency, evidenced in the figure, is not compatible with AC circuits. This difficulty is overcome in the modified integrator circuit shown in Figure 4-11. The feedback can be described in terms of the parallel combination of R_f and C by the expression

$$Z_f = \frac{R_f Z_C}{R_f + Z_C} = \frac{R_f}{1 + j\omega R_f C} \tag{4-26}$$

Substituting this into Eq. 4-24, we obtain

$$E_{out} = -E_{in} \cdot \frac{Z_f}{Z_{in}} = -E_{in} \cdot \frac{R_f}{R_{in}} \cdot \frac{1}{1 + j\omega R_f C} \tag{4-27}$$

Note that the last factor becomes unity at very low frequencies, when $R_fC << 1$, and approaches zero at very high frequencies. Complete analysis indicates the frequency dependence shown in Figure 4-10*b*, the behavior of a low-pass filter. In terms of these circuits, the low-pass filter acts as an inverter with a gain of $-R_f/R_{in}$ for frequencies up to the transition point f_0, beyond which it acts as an integrator. The characteristic frequency f_0 corresponds to the condition $\omega R_fC = 1$, or

$$f_0 = \frac{1}{2\pi R_f C} \qquad (4\text{-}28)$$

A small feedback capacitor should be used routinely across the feedback resistor for all summers and inverters, in order to keep the high-frequency response of the amplifier within the band that contains the desired information. This improves the S/N ratio and increases the stability of the amplifier with respect to sudden voltage steps. As a general rule for DC or low-frequency operation, a feedback capacitor giving the product R_fC = 1 ms should be satisfactory.

The integrator often plays an important role in signal-processing applications. For instance, a laboratory instrument may produce a signal in the form of a series of peaks superimposed on a continuous background, as in Figure 4-12. If the quantity of interest is measured by the area enclosed beneath the peak, an integrator is needed. Thus the area beneath the second peak in the figure is measured by the height H_2 on the integral curve, given by

$$H_2 = -k \int_{t_1}^{t_2} E\, dt \qquad (4\text{-}29)$$

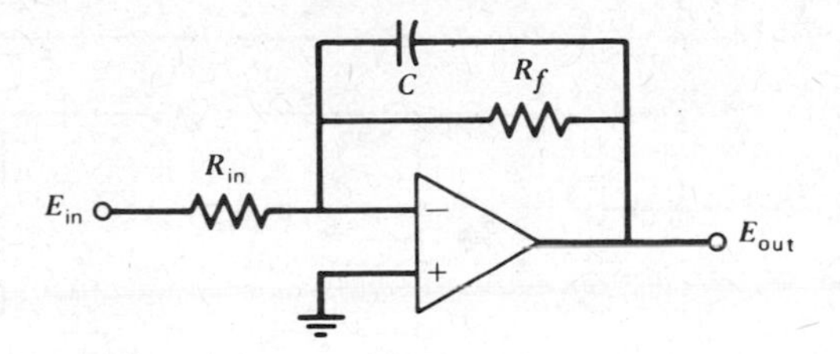

Figure 4-11. Modified integrator forming a low-pass filter. The circuit is not usable as an integrator at DC because the capacitor cannot hold a charge.

in which k is a scaling factor.

The design of an automatic integrator for signal processing requires two additional features: (1) A variable input gain is needed to accommodate differences in peak widths. Since the integral is a measure of the area under the peak, it changes with the width of the peak, even if the height remains constant. (2) The reset feature must be such that the output, upon reaching full scale, will automatically return to zero and resume integration.

These features are implemented in the circuit

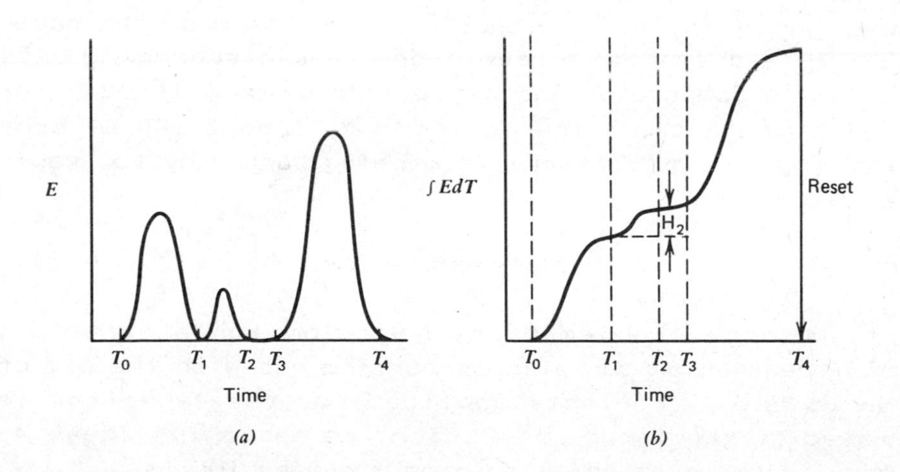

Figure 4-12. The integration of a series of peaks.

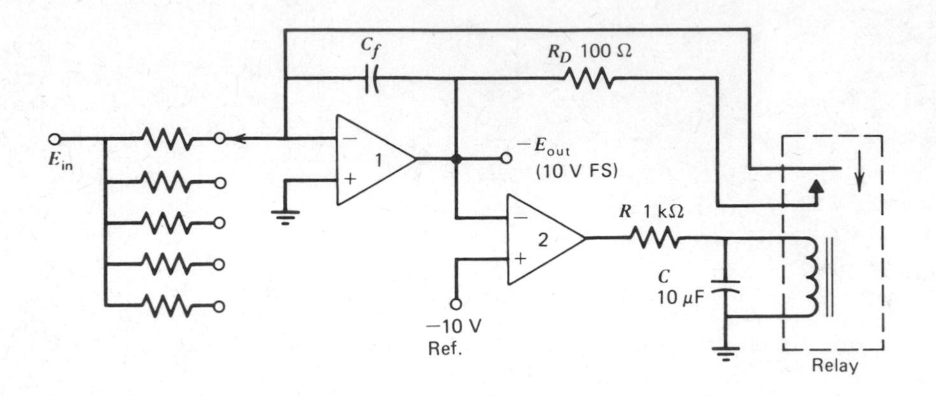

Figure 4-13. A practical integrator to perform the function required in Figure 4-12. Amplifier 1 should be a FET-input type.

shown in Figure 4-13. The values of resistors R_1 to R_5 depend on the duration of the peaks to be integrated. If specific advance information is lacking, a sequence of multiples of 10, from 1 kΩ to 10 MΩ may be used, with a feedback capacitor of 1 μF. The *RC* pair at the output of amplifier 2 serves to keep the relay closed long enough for the capacitor to discharge completely (a few milliseconds). Once charged, the capacitor is prevented by the diode from discharging back into the comparator and can discharge only through the relay contacts. The resistor R_D (about 100 Ω) protects the relay contacts from being burned by too rapid a discharge.

Differentiation

As the complementary function to integration, differentiation can also be carried out with the aid of an op amp. For this application, the capacitor is moved to the input connection, as shown in Figure 4-14*a*. The equation describing its behavior is

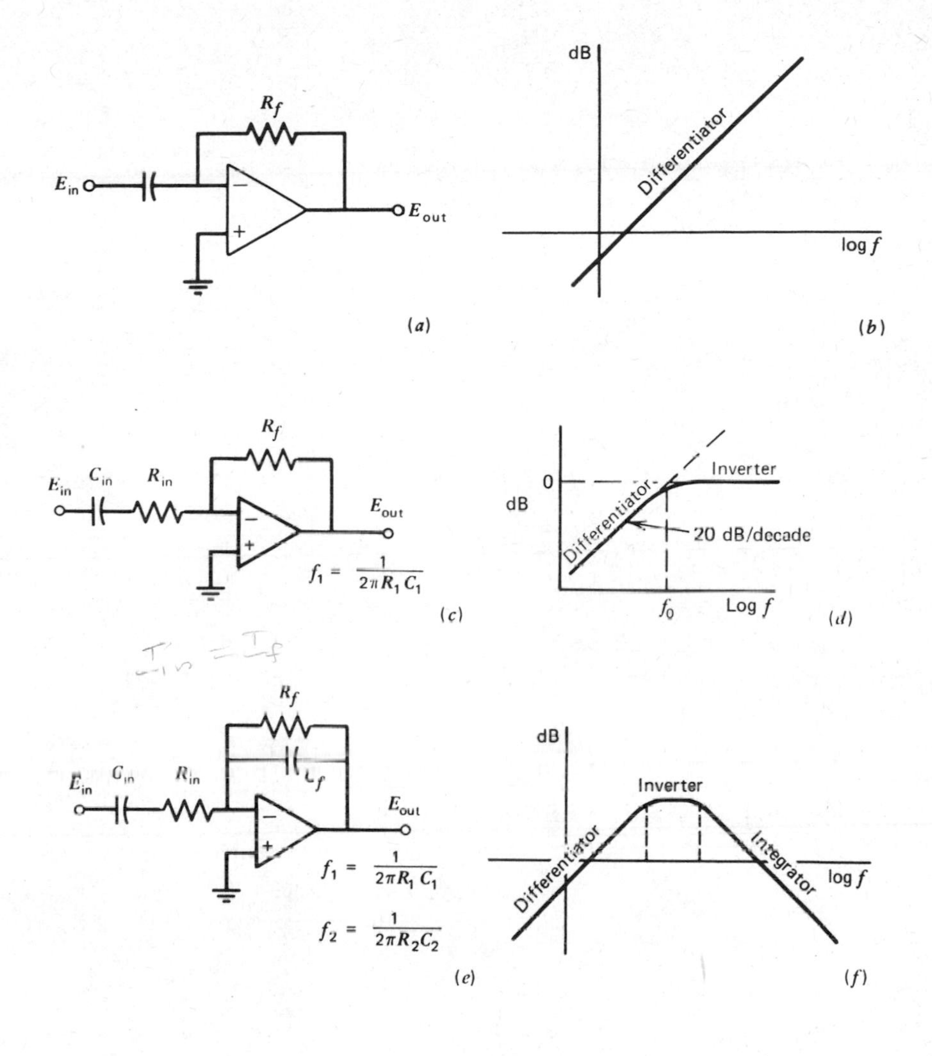

Figure 4-14. Various forms of differentiators, with their corresponding frequency response plots. In (*d*), $f_0 = 1/(2\pi R_{in}C_{in})$; in (*f*), $f_1 = 1/(2\pi R_{in}C_{in})$ and $f_2 = 1/(2\pi R_f C_f)$.

$$C \frac{dE_{in}}{dt} = - \frac{E_{out}}{R_f} \qquad (4\text{-}30)$$

which can be rewritten as

$$E_{out} = - R_f C \frac{dE_{in}}{dt} \qquad (4\text{-}31)$$

This differential equation finds a counterpart in phasor representation:

$$E_{out} = - \frac{Z_f}{Z_{in}} = - j\omega R_f C \qquad (4\text{-}32)$$

This indicates a phase angle of 270° and an AC gain linearly increasing with frequency (Figure 4-14*b*), a high-pass filter. This plot, as well as Eq. 4-32, predicts a very high gain at high frequencies. This is an undesirable feature, in that it accentuates the effect of noise. To avoid it, the circuit shown in Figure 4-14*c* should be used. Here

$$Z_{in} = R_{in} + \frac{1}{j\omega C} = \frac{1 + j\omega R_{in} C}{j\omega C} \qquad (4\text{-}33)$$

so that

$$E_{out} = - j \frac{\quad}{1 + j\omega R_{in} C} \cdot E_{in} \qquad (4\text{-}34)$$

For small values of ω, this reduces to Eq. 4-32, indicating operation as a differentiator. It can be seen that at higher frequencies, this expression reverts to

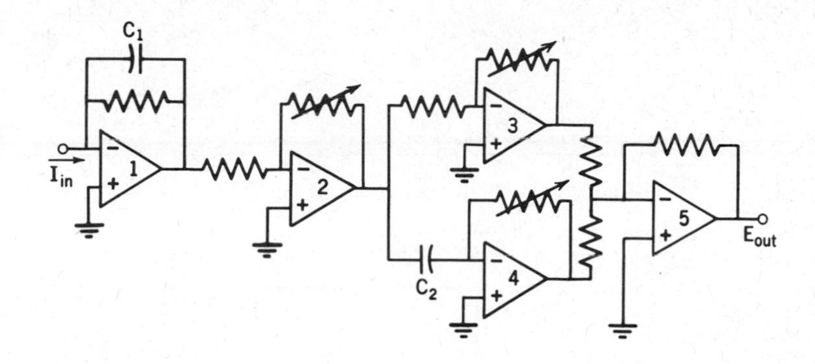

Figure 4-15. Circuitry for adding a derivative to the original signal. Amplifier 1 is a current-to-voltage converter (required for the particular application) in which capacitor C_1, a few picofarads, provides high-frequency attenuation. Amplifiers 2 and 3 permit variable gain, and amplifier 4 generates the derivative, which is added to the original signal at the summing junction of amplifier 5.

Eq. 4-7: $E_{out} = -(R_f/R_{in})E_{in}$. This general behavior is graphically described in Figure 4-14*d*, where the transition frequency f_0 corresponds to the condition $\omega R_f C = 1$, or $f_0 = 1/(2\pi R_f C)$.

Even more useful is the circuit in Figure 4-14*e*, in which high frequencies are eliminated by a feedback capacitor, leading to the behavior depicted in (*f*). This circuit is most useful if $R_f C_f < R_{in} C_{in}$. Note that the plots in (*d*) and (*f*) correspond to high-pass and band-pass filters, respectively.

Combinations of summers, differentiators, and integrators can be used for a variety of data processing tasks. An example of such a combination, which provides the sum of an original signal and its derivative, is shown in Figure 4-15; it has been used for improving the quality of pictures obtained from an electron probe microanalyzer.*

* K. F. Heinrich, C. Fiori, and H. Yakowitz, *Science*, 67, 1129 (1970).

Active Filters

Active filters are op amp circuits that serve to render the information content of signals more easily utilized by limiting the bandwidth.

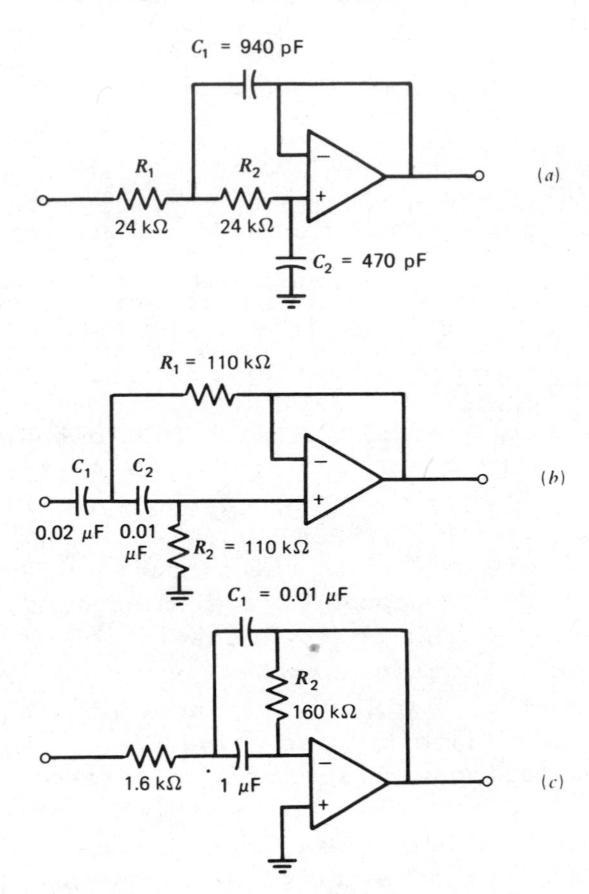

Figure 4-16. Second-order active filters: (*a*) low-pass, (*b*) high-pass, and (*c*) band-pass. The band-pass filter gives a sharp peak at the center frequency, rather than the wide band seen in Figure 4-14(*f*). (From E. R. Hnatek, *Applications of Linear Integrated Circuits*, John Wiley & Sons, Inc., by permission)

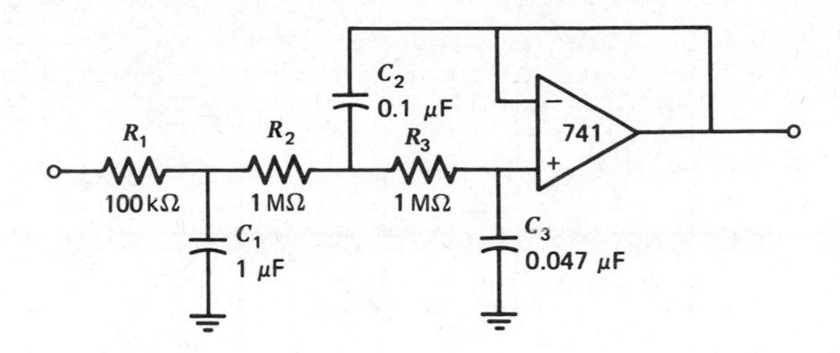

Figure 4-17. A second-order low-pass noise-rejection filter. Interchanging of resistors and capacitors will give a high-pass filter.

The three active filters shown in Figure 4-16 have an attenuation slope of 40 dB, twice those in Figures 4-10 and 4-14. The response of the band-pass unit (Figure 4-16*c*) can be tailored for various response curves. These filters are called *second order*.

There are many possible variants on second- and higher-order filters. An example of an easily constructed low-pass filter with f_0 = 1 Hz, which is very effective in removing noise from a DC signal, is given in Figure 4-17. This filter shows no appreciable attenuation from DC to about 0.5 Hz, but more than 90% attenuation at 2 Hz and complete extinction beyond about 10 Hz. Other values of f_0 can be obtained in this circuit merely by dividing all resistance values by the desired f_0.

The Twin-T notch filter described in Chapter III can be placed in the feedback loop of an op amp (Figure 4-18) to give the inverse of a notch, a sharply tuned band-pass filter, usually regarded as a *tuned amplifier*. The Twin-T network provides a low impedance for all frequencies other than f_0 and a high impedance at that frequency. Hence the closed-loop gain is high at f_0 and low elsewhere. The 22-MΩ resistor in parallel with the Twin-T limits the gain at resonance to avoid saturation.

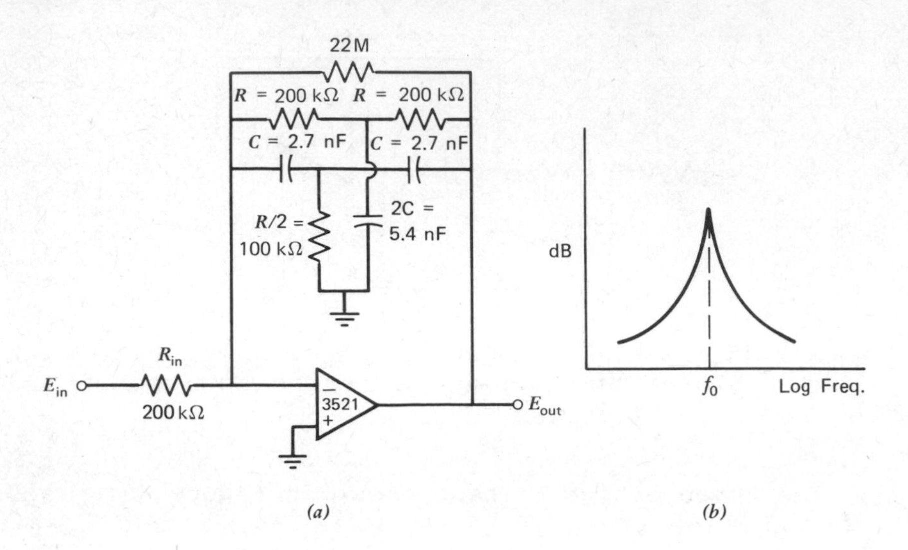

Figure 4-18. (*a*) A tuned amplifier using a Twin-T feedback filter. The values given correspond to a center frequency of f_0 = 300 Hz, determined from the formula $f_0 = 1/(2\pi RC)$. The resistors should be matched to ± 0.1% and capacitors to ± 1% or better. (*b*) The frequency response.

DIFFERENTIAL-MODE CIRCUITS

If both inputs of an op amp carry signals, the virtual ground condition is replaced by the virtual equality of the two inputs. Such differential operation is more versatile than the single-input mode. An example is the subtractor, shown in Figure 4-19. The output can be calculated in the following way. According to the voltage divider equation, the voltage E' at the noninverting input is given by

$$E' = E_2 \frac{R_2}{R_2 + R_4} \qquad (4\text{-}35)$$

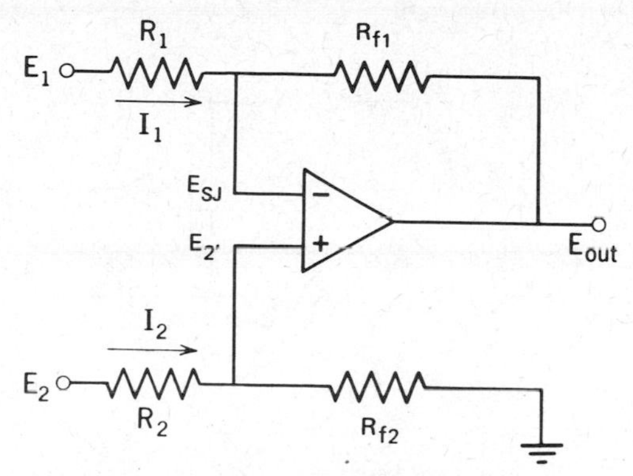

Figure 4-19. A circuit for the subtraction of one voltage from another.

The value of E_{SJ} is virtually equal to E', so that the current I will be

$$I_1 = \frac{E_1 - E_{SJ}}{R_1} = \frac{E_1 - E'}{R_1} \tag{4-36}$$

Thus the equality between the input and feedback currents can be written as

$$\frac{E_1 - E'}{R_1} = \frac{E_{out} - E'}{R_3} \tag{4-37}$$

By combining the last two equations and taking $R_3 = R_4 = R_f$, and $R_1 = R_2 = R_{in}$, we see that

$$E_{out} = \frac{R_f}{R_{in}}(E_2 - E_1) \tag{4-38}$$

This circuit takes the difference between two voltages and multiplies it by the factor R_f/R_{in}. The same result could be obtained by grounded-reference amplifiers, but two would be required rather than one.

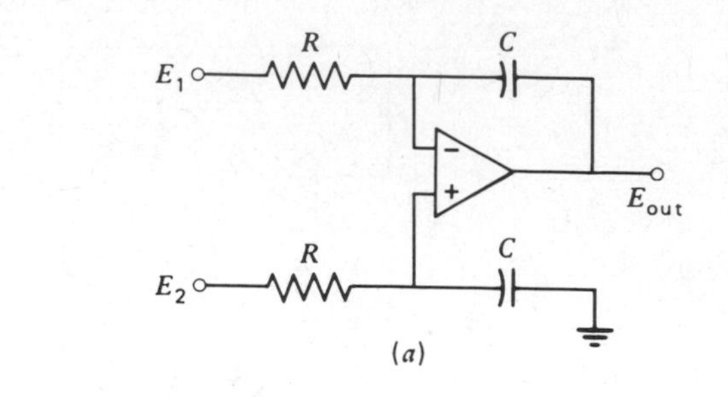

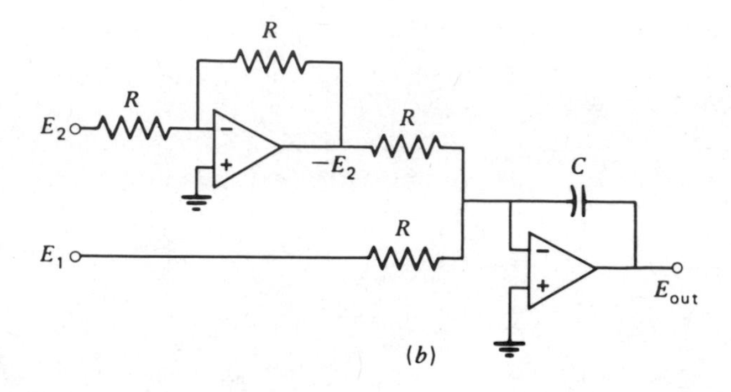

Figure 4-20. Difference integrators: (*a*) with a differential amplifier, and (*b*) a single-ended equivalent.

Along the same line, one can implement *difference integration*, as illustrated in Figure 4-20*a*. The output is given by

$$E_{out} = \frac{1}{RC} \int (E_2 - E_1)\, dt \tag{4-39}$$

Again, it is possible to use single-input amplifiers, as in Figure 4-20*b*. The latter alternative has the advantage of requiring only a single high-quality capacitor; the cost of a second, matched capacitor is certain to exceed the cost of a second amplifier. As a matter of fact, most difference operations can be implemented with pairs of single-input amplifiers just as well as with a differential amplifier, perhaps better.

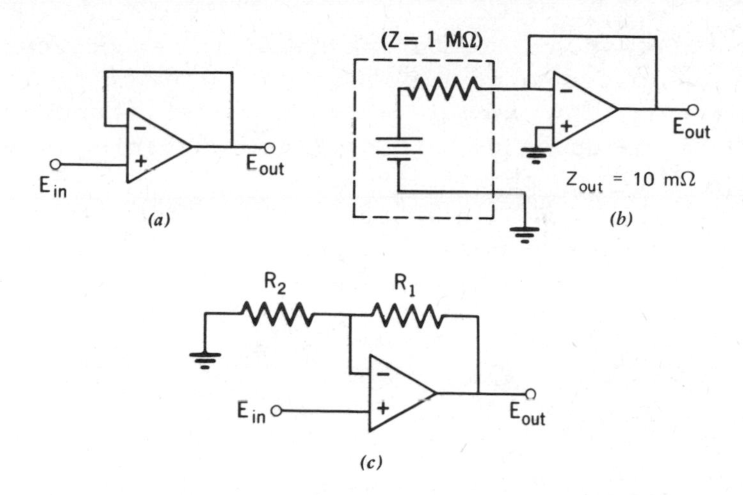

Figure 4-21. (*a*) A voltage follower. (*b*) The follower connected to a source represented by a battery and a resistor. (*c*) A follower with gain.

The Voltage Follower

A special case in which a differential amplifier is restricted to single-input operation is the *voltage follower* (Figure 4-21*a*). It uses the simplest form of feedback, a direct connection between the output and the summing junction. The voltage gain of this circuit is unity and noninverting. Its great merit is that it provides *impedance transformation*. Its input impedance is very large (typically 100 MΩ), and its effective output impedance is a small fraction of 1 Ω. In the example given in Figure 4-21*b*, insertion of the op amp corresponds to changing the impedance from megohms to milliohms.

Such an impedance transformation is very useful in a variety of applications. One example is shown in Figure 4-22, where the follower is interposed between a potentiometer and the load, to avoid the loading effects previously described (see Figure 3-5). This use of a follower (called *buffering*) also reduces the noise often produced by poor mechanical contact in a potentiometer; any variation in the resistance of the sliding contact will be totally negligible because no current will be passing through it.

A voltage follower can be provided with gain by means of the feedback circuit shown in Figure 4-21*c*. The governing relationship is

$$E_{out} = E_{in}\left(\frac{R_2}{R_1} + 1\right) \quad (4\text{-}40)$$

Note that gains less than unity cannot be implemented.

Another example of buffering is shown in Figure 4-23, a two-stage *RC* filter. In such cases interaction between successive stages of filtering is eliminated by the insertion of a follower.

A related circuit using a grounded-reference amplifier is shown in Figure 4-24*a*. Here a Weston standard cell (an electrochemical voltage source of 1.0183 V at 25°C) is placed in the feedback loop. The output of the amplifier is the precise voltage of the

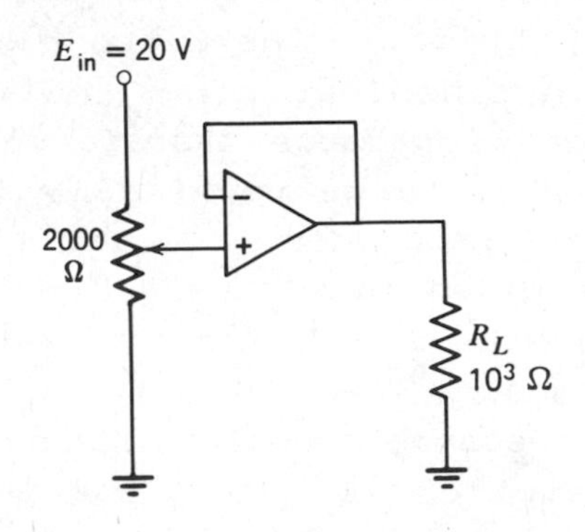

Figure 4-22. Unloading (buffering) of a potentiometer by means of a follower.

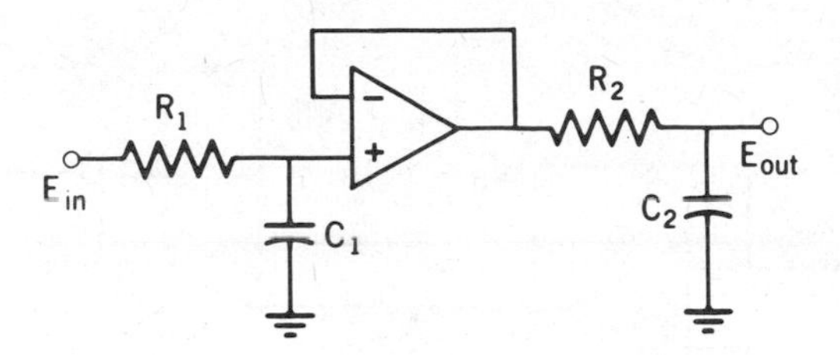

Figure 4-23. Buffering in a two-stage filter.

cell, and current can be drawn at this voltage without deleterious loading of the cell itself. Gain can also be built in, as depicted in Figure 4-24*b*.

A highly precise voltage source can be constructed as illustrated in Figure 4-25. If the resistors are of high quality, a precision of ± 0.1% percent is not difficult to attain. The Zener diode, which should be of the temperature-compensated type, plays the same role taken by the Weston cell in the previous example.

The *sample-and-hold* circuit shown in Figure 4-26*a* is an interesting application of the voltage follower. When the switch is closed, the voltage on the capacitor

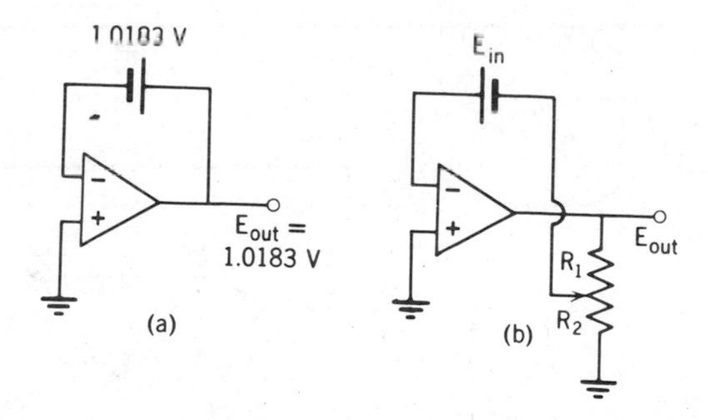

Figure 4-24. (*a*) A single-ended follower circuit used in conjunction with a Weston standard cell. (*b*) A single-ended follower with gain, in which the output is given by $E_{out} = [(R_1 + R_2)/R_2] \cdot E_{in}$.

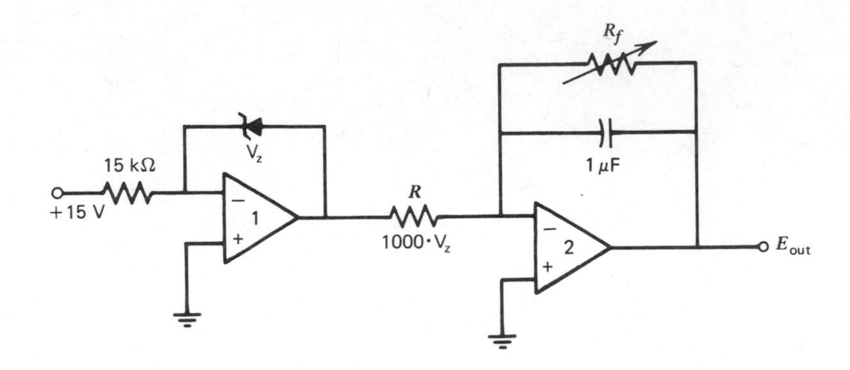

Figure 4-25. A high-precision voltage source (R_F is a 10-kΩ resistance decade box). The output in volts is numerically equal to the resistance in kilohms. The capacitor acts as a noise-rejection filter.

tracks E_{in}. When the switch is opened, the low-leakage capacitor simply retains (holds) its charge for a considerable period of time. The rate at which the capacitor is charged depends on the maximum current I_{max} that can be drawn from the source: $dE/dt = I_{max}/C$. To optimize this parameter, a two-amplifier circuit can be used, as in Figure 4-26*b*, where the capacitor is charged by the saturation voltage of the

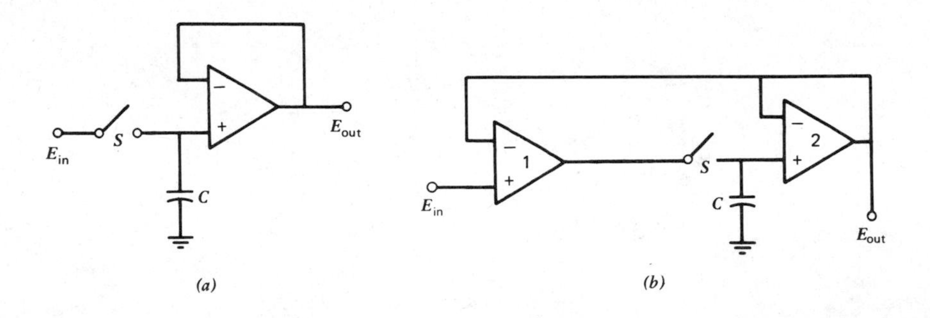

Figure 4-26. Sample-and-hold circuits: (*a*) a one-amplifier version; (*b*) an improved version. The switch *S* is ordinarily a FET gate.

first amplifier, rather than by E_{in} itself. On attaining equality with E_{in} the amplifier is removed from saturation. Complete sample-and-hold units are available as integrated circuits, lacking only the capacitor.

Another special-purpose IC is the multichannel programmable amplifier, such as the HA-2400 (Harris Semiconductor) shown schematically in Figure 4-27. This consists of an array of four input amplifiers, any one of which can be connected by an internal transistor switch to an output buffer amplifier. The choice of channel is controlled by a built-in selector (*decoder*, in digital parlance). Application of a 5-V control signal to one, both, or neither of the inputs D_1 and D_2 determines which input channel operates. Feedback must be provided from the common output to each input amplifier separately.

This module can be used in a programmable-gain mode by connecting different feedback networks around

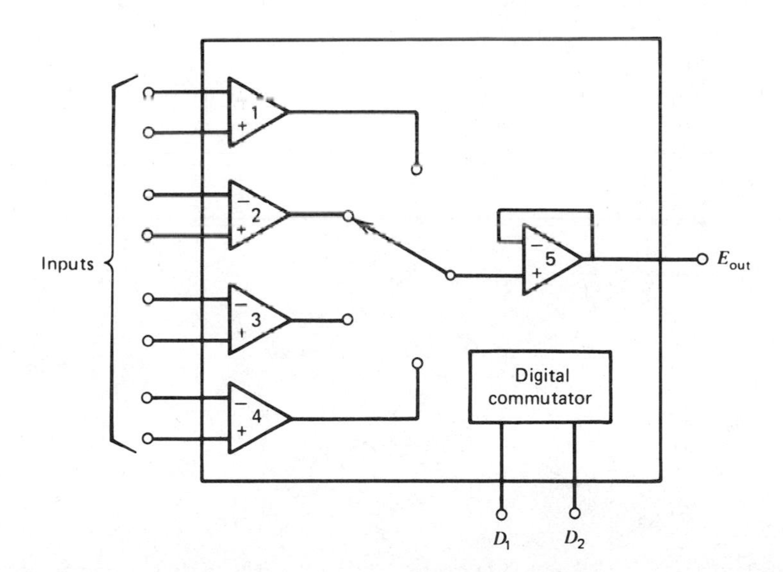

Figure 4-27. (*a*) The functional diagram of the HA-2400 programmable op amp. (*b*) The selection code; voltages of 0 or +5 applied to the *D* inputs (the address) will activate the designated channel.

the four amplifiers, selectable through the *D* inputs. It can also be used in a continuously sequencing mode of operation (*multiplexing*) such as the display of four functions at the same time on an oscilloscope. Because of the persistence of the oscilloscope screen, the four consecutive traces appear to be simultaneous.

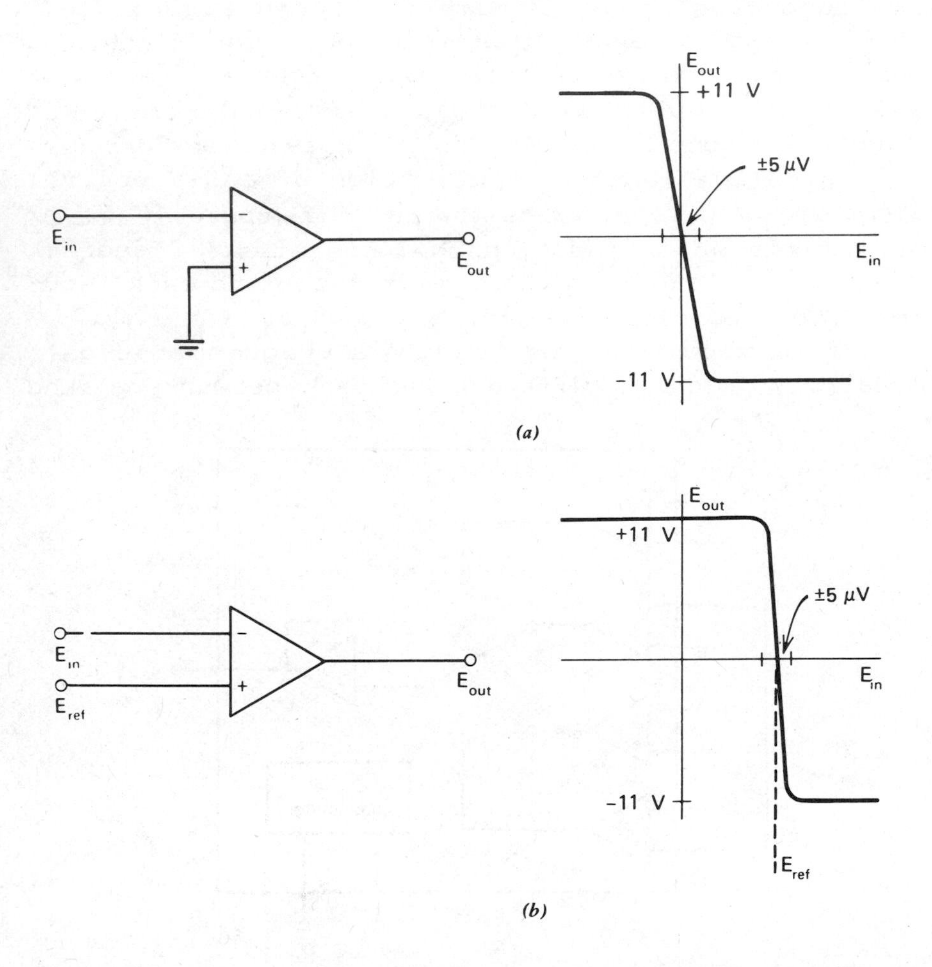

Figure 4-28. Comparators: (*a*) referred to ground and (*b*) referred to E_{ref}. Amplifiers optimized for comparator service have very rapid transition times, so that they pass through the linear region in less than 1 μs.

NONLINEAR CIRCUITS

The operations discussed so far--addition, subtraction, differentiation, and integration--belong to the category of *linear* operations. Linearity can be defined as the proportionality of cause and effect, conveniently stated in two forms:

1. Two input signals, E_1 and E_2, which, when applied individually, generate outputs E_3 and E_4, will produce, when applied together, an output $E_3 + E_4$.

2. If an input signal E_1 produces an output E_3, an input kE_1 will generate an output kE_3, where k is a constant.

Consequently, $E_{out} = kE_{in}^2$ is a nonlinear relationship since doubling E_{in} does not merely double E_{out}. In contrast, it can easily be proved that integration is a linear operation.

Resistors, capacitors, and inductors are linear devices, and the output of any op amp circuit using only these components is proportional to the input (assuming that saturation does not occur). On the other hand, nonlinear circuits are those that either cause the amplifier to saturate or contain nonlinear components such as diodes. The reader is already familiar with the comparator, a nonlinear circuit component that is presented again in Figure 4-28. The output assumes either the positive or negative saturation value, except for inputs in a very narrow region (a few microvolts). In this small region the behavior is linear. Figure 4-28*b* shows the corresponding use in the differential mode, permitting the comparison of two voltages.

An example of a nonlinear circuit using diodes is the *precision rectifier* (Figure 4-29*a*). When the

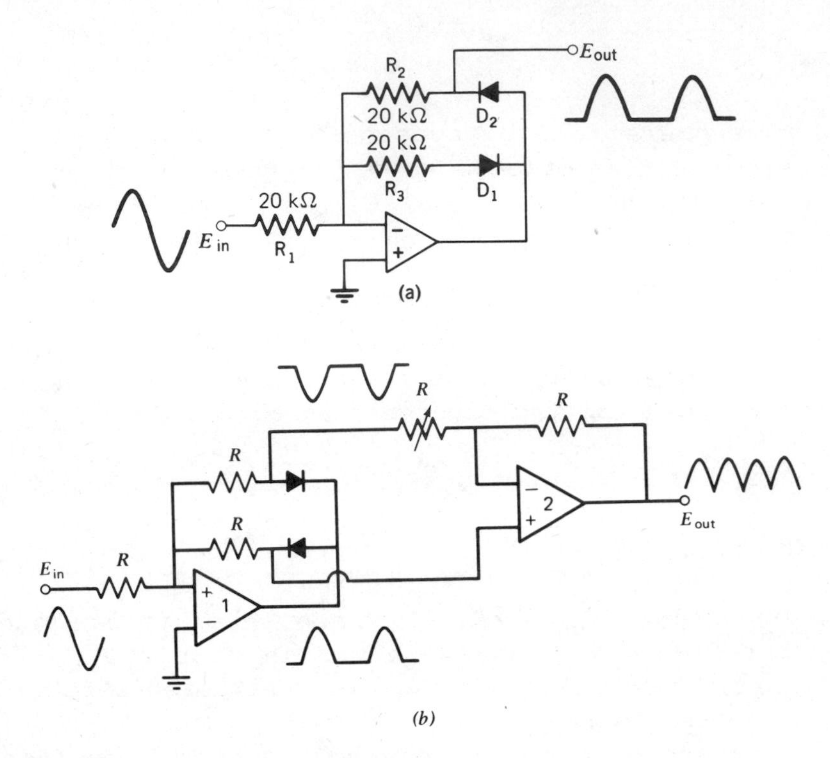

Figure 4-29. (*a*) A precision rectifier and (*b*) an absolute value circuit derived from it. The potentiometer *P* is used to balance the output.

input is *negative*, D_2 is forward biased and D_1 reverse biased. The equality of input and feedback currents requires that

$$\frac{E_{in}}{R_1} = - \frac{E_{out}}{R_2} \tag{4-41}$$

and since $R_1 = R_2$, it follows that $E_{out} = - E_{in}$. Note that the forward voltage drops are immaterial. The feedback path through D_1 becomes effective for positive inputs and takes over the feedback function exactly at zero volts, guaranteeing a sharp cutoff. Whereas a

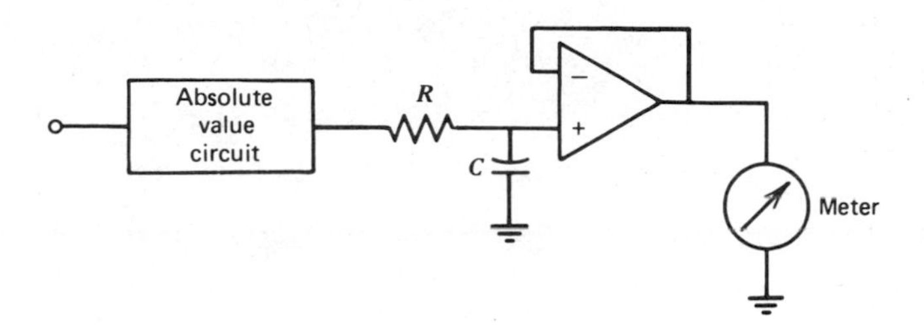

Figure 4-30. An electronic voltmeter using an absolute value circuit. It responds to DC as well as to AC, down to a few millivolts.

diode alone requires about 1 V for acceptable rectification, the precision rectifier operates with as little as 10 mV.

The rectifier of Figure 4-29*a* discards half of the available information contained in the signal (half-wave rectification). To utilize both halves (full-wave rectification), the circuit shown in Figure 4-29*b* can be used. This generates the *absolute value* of the input by adding together the positive half and the inverted negative half. If a low-pass filter is inserted at the output, such that $1/2\pi RC$ is less than the frequency of the signal, the result will be a DC voltage proportional to the *average* (not the RMS) value of the rectified AC. This circuit can be made to drive an indicating meter, as shown in Figure 4-30.

Logarithmic Function Generation

The current-voltage relationships for small signal diodes at room temperature is given by

$$E = B \log I + C \tag{4-42}$$

where B and C are constants depending on the particular type of diode. This equation is obeyed fairly well by all diodes and very closely by some specially selected types. This relationship suggests the use of diodes in the generation of both logarithmic and exponential functions.

An example is given in Figure 4-31*a*, for which one can write

$$I_f = -I_{\text{in}} = \frac{E_{\text{in}}}{R_{\text{in}}} \qquad (4\text{-}43)$$

This can be combined with Eq. 4-42 to give

$$E_{\text{out}} = B \log I_f + C = B \log \left(\frac{E_{\text{in}}}{R_{\text{in}}}\right) + C \qquad (4\text{-}44)$$

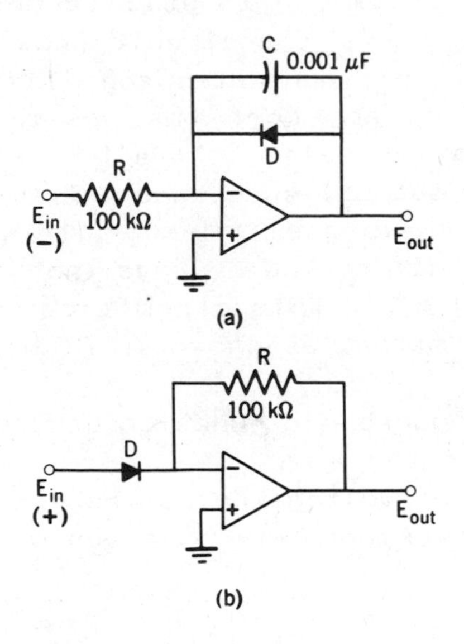

Figure 4-31. (*a*) A logarithmic converter and (*b*) an exponential (antilogarithmic) converter. The diode usually used is a silicon transistor with base and collector tied together (termed a *transdiode*). Some suitable types are 2N697, 2N1132, 2N2218, and 2N3900A.

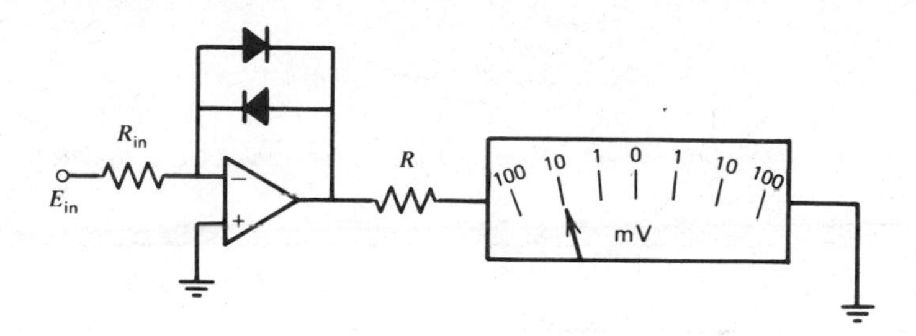

Figure 4-32. A logarithmic millivolt meter. For either polarity of input, one of the diodes is forward biased and produces the logarithmic function.

This equation is valid for a wide range of input voltages (typically 5 or 6 logarithmic decades). Observe that only one polarity of signal is accepted. If the diode is inserted in the input lead, as in Figure 4-31*b*, the inverse function, the exponential or antilogarithm, is obtained.

The availability of log conversion modules permits the implementation of a logarithmic meter for monitoring signals of large dynamic range, as depicted in Figure 4-32.

CHARACTERISTICS OF OPERATIONAL AMPLIFIERS

Operational amplifiers can be described by a number of parameters, some of which have been mentioned previously. The most important of these are listed in Table 4-1, with the desirable behavior of an "ideal" amplifier. Also given are the ranges of values normally found in commercial units. These parameters are defined as follows:

1. The *open-loop gain* (A or A_{VO}), also called the *large-signal voltage gain*, is the ratio of E_{out} to the voltage difference between the two inputs. As discussed previously, loss of gain at higher frequencies

TABLE 4-1

Operational Amplifier Parameters

Parameter	Ideal[a]	Typical	Units
Open-loop gain (A or A_{VO})	∞	20 - 1000	V/mV
Unity-gain bandwidth (BW)	∞	1 - 1000	MHz
Input impedance (Z_{in})	∞	0.1 - 100	MΩ
For FET input amplifiers	∞	10^4 - 10^6	MΩ
Common-mode input impedance (Z_{in-cm})	∞	1 - 10^6	Ω
Output impedance (Z_{out})	0	10 - 100	Ω
Offset voltage (V_{os})	0	0.1 - 10	mV
Temperature coefficient of V_{os} (TCV_{os})	0	5 - 500	μV/K
Input bias current (I_B)	0	10^{-4} - 10	μA
For FET input amplifiers	0	10^{-5} - 10^{-4}	μA
Input offset current (I_{os})	0	10^{-4} - 10	μA
Rise time (RT)	0	0.01 - 10	μs
Slew rate (SR)	∞	0.1 - 100	V/μs
Settling time, to 0.1% (t_s)	0	0.05 - 50	μs
Common-mode rejection ratio ($CMRR$)	∞	60 - 120	dB
Power-supply rejection ratio ($PSRR$)	∞	50 - 100	dB
Noise	0	0.01 - 0.1	$\mu V/\sqrt{Hz}$
Price (for small quantities)	0	0.5 - 20	$US

[a] The "ideal" is the limit approached by each parameter as the amplifier is optimized. Some of these limits are unattainable, or actually undesirable.

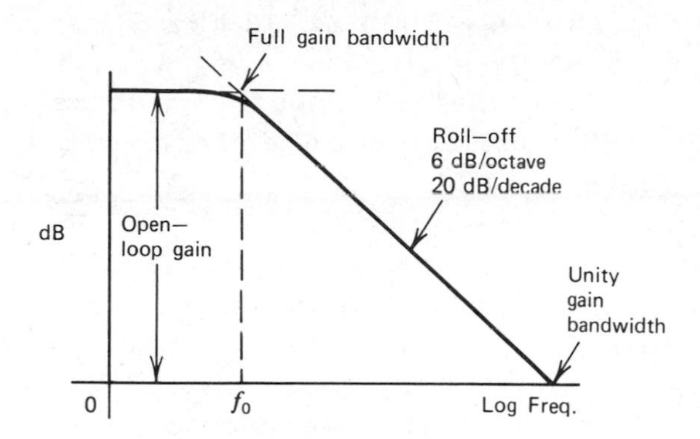

Figure 4-33. Bandwidth definitions in an op amp, illustrated on a Bode plot.

is a major limiting factor in the use of op amps. Since gain-generated errors are of the order of 1/*A* it follows that an amplifier requiring 0.1% accuracy at 20 kHz must maintain an open-loop gain of at least 1000 at that frequency. This requirement is valid for overall gains close to unity. As indicated in Eq. 4-6, a higher circuit gain requires a still larger open-loop gain for error-free operation.

2. The *unity-gain bandwidth* (BW) is the range of frequencies from DC to the point where the open-loop gain becomes unity (0 dB; see Figure 4-33). Since the descending portion of the frequency response has a slope of $-45°$, the BW is equal to the product of gain and frequency at any point on that slope. Thus the

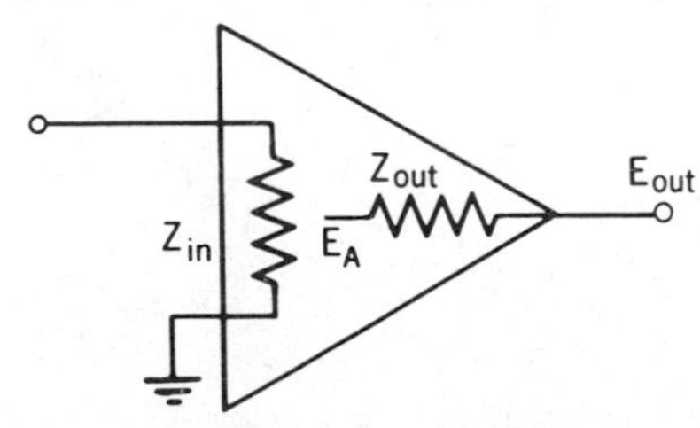

Figure 4-34. Input and output impedances of an amplifier.

requirement that $A = 1000$ at 20 kHz gives a product of 2×10^7, so that BW = 20 MHz.

3. The *differential input impedance* (Z_{in}) is the effective internal impedance from either input to ground (Figure 4-34).

4. The *common-mode input impedance* (Z_{in-cm}) is the internal impedance from either input to ground.

5. The *output impedance* (Z_{out}) is the impedance of the amplifier considered as a source of voltage driven by the input (Figure 4-34). This quantity becomes much smaller in the presence of negative feedback and is negligible in many applications.

6. The *offset voltage* (V_{os}) is an unwanted internal potential between the two inputs, appearing at the output as multiplied by the circuit gain. It can be minimized (nulled) by a trimming potentiometer used to balance the input circuit of the amplifier. The offset voltage may be temperature dependent; thus nulling at one temperature may not hold at another. The offset voltage is one of the major sources of error in op amp circuits. It is usually preferable to purchase a better amplifier than to attempt to correct an inferior one. Amplifiers with an offset of 0.1 mV can be aquired at moderate cost, often less than the cost of the additional components necessary to balance out the voltage offset.

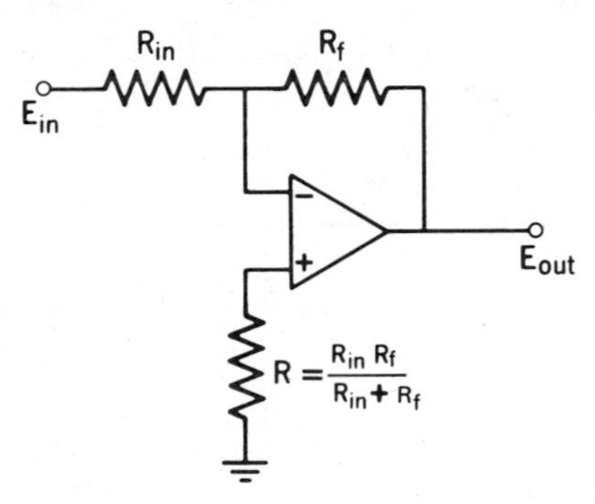

Figure 4-35. A method for minimizing bias effects.

7. The *temperature coefficient of* V_{os} is significant because ICs suffer from self-heating. It was formerly a dominant error factor. The effect was so important that thermostated op amps were actually produced commercially in the 1960s. Subsequently it was realized that correct placement of the transistors within the IC can lead to a greatly reduced temperature coefficient. Hence in modern op amps the effect is less important.

8. The *input bias current* (I_B) is the average of the residual currents flowing into the two inputs with no signal present. Figure 4-35 shows a method of correcting for the adverse effects of bias current in a summer. Resistor *R*, connected from the noninverting input to ground, is made equal to the parallel combination of all input and feedback resistors. In FET-input amplifiers the bias current is so small that this correction is not needed.

9. The *input offset current* (I_{os}) is the difference between the two individual input bias currents.

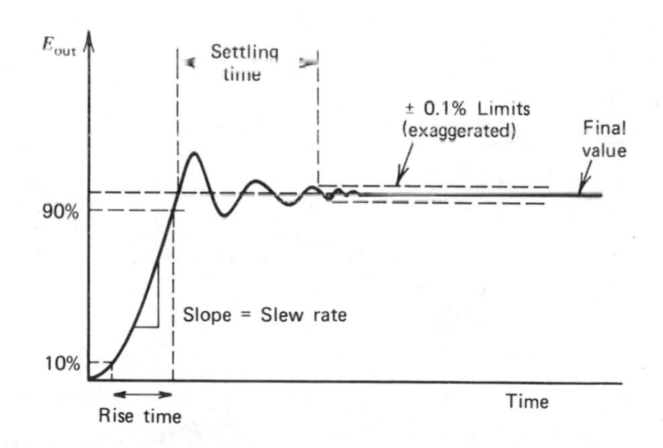

Figure 4-36. Illustration of the dynamic behavior of an op amp connected as a summer. The curve represents the output response to an input step. The oscillations around the final value are exaggerated for clarity.

10. The *rise time* (*RT*) is a different type of parameter, characterizing the transient response. It is illustrated, along with other time-dependent properties, in Figure 4-36. The rise time is the time needed for the signal to increase from 10 to 90% of a step. It is dependent on the height of the step. For small signals, the rise time is limited by the frequency response, becoming simply an expression of the inability of the op amp to pass the higher terms in the Fourier expansion of the step. In contrast, for signals of larger amplitude, it is limited by the maximum rate at which the output can change (the slew rate).

11. The *slew rate* (*SR*) is defined as the average slope of the response of the amplifier to a large step input. It also is calculated between the points at 10 and 90% of the total output excursion (Figure 4-36). During the rise, the amplifier is in a form of saturation, so that the customary relationships between input and output may not be obeyed. After reaching the correct output value, the amplifier first overshoots by a small amount, and then enters a process of recovery, settling to its final value.

12. The *settling time* (t_s) is the total time required for the recovery mentioned previously to take place and is measured from the application of the input step to the point where the output remains within specified limits of its final value, for instance, ±0.1% (Figure 4-36).

13. The *common-mode rejection ratio* (CMRR) can be defined as follows: An ideal op amp gives zero output when a voltage E_{in} is applied simultaneously to both inputs. In practice, however, the (+) and (−) inputs are not perfectly symmetrical, and a nonzero output results when such a common-mode voltage is applied. This output can be represented as E_{in}/CMRR (Figure 4-37).

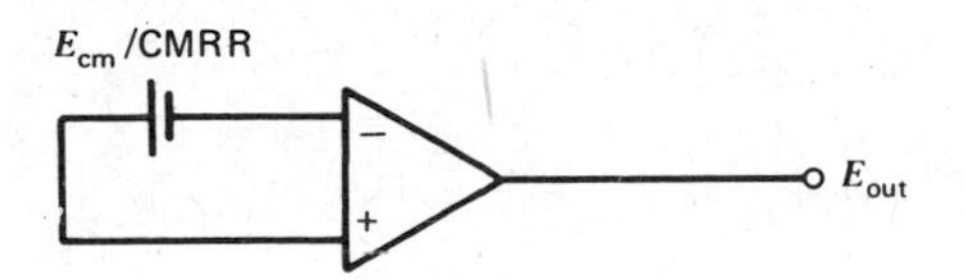

Figure 4-37. Model for common-mode voltage. The net effect is indistinguishable from an offset of E_{cm}/CMRR.

14. The *power-supply rejection ratio* (PSRR) is the ratio of a change in power-supply voltage to the resulting offset voltage of the amplifier. With regulated power supplies, this effect is usually negligible (a few microvolts). With batteries, it seldom contributes more than 1 mV to the voltage offset. The PSRR can be expressed either in terms of decibels or in μV per volt of change in the power supply.

15. *Noise* is usually specified only for low-noise amplifiers. As is the case with offsets, a distinction must be made between current noise (pA/$\sqrt{Hz}$) and voltage noise (μV/$\sqrt{Hz}$). Occasionally the noise is given over a frequency range, for example, 50 Hz to 10 kHz; in such a case, it is expressed in nA or μV, respectively.

Commercial op amps are made in many models and variations optimized for specific types of service. In most cases, general-purpose units should be selected, unless the requirements are particularly exacting. This might occur if unusually small currents or voltages are to be encountered or if extended frequency response is needed. Some units are specially optimized for use as voltage followers or as comparators.

In Table 4-2 the most important specifications are given for a selection of commercial amplifiers. Type 741 is one of the most widely used models at present. These are made by numerous manufacturers, some of which add prefixes to the basic type number. Thus μA741, MC1741, LM741, AD741, and CA741 are nearly the same

TABLE 4-2

Characteristics of Selected Operational Amplifiers

Parameter	741	LM301A[a]	LM308A[a]	OP-07[b]	CA3130B[c]	AD509[d]	Units
Open-loop gain	100	100	300	4000	320	10	V/mV
Input impedance	2	2	40	50	1.5×10^6	50	MΩ
Output impedance	75	70	500	60	--	--	Ω
Offset voltage	1	2	0.3	0.03	1	5	mV
Offset tempco	15	6	2	0.3	5	20	μV/K
Bias current	30	70	1.5	1.5	0.005	100	nA
Offset current	10	3	0.2	0.5	0.0005	25	nA
CMRR	90	70	110	120	100	80	dB
PSRR	90	70	110	100	32μV/V	--	dB
Bandwidth	1.5	1	1	0.6	15	20	MHz
Output current	25	25	6	15	22	5	mA
Slew rate	0.5	0.5	0.2	0.2	10	80	V/μs
Noise voltage	--	--	--	0.01	23μV[e]	0.03	$\mu V/\sqrt{Hz}$

a Models 101, 201, and 301 differ principally in the permitted temperature range. The same applies to Models 108, 208, and 308.
b Precision Monolithics, Inc.
c RCA Corporation; this is a FET-input type.
d Analog Devices, Inc.
e For *BW* = 0.2 MHz.

product. In addition, suffixes are used to indicate different performance classifications and packaging. Thus AD741KN designates a unit made by Analog Devices, with a maximum offset voltage of 2.0 mV (indicated by the letter "K") and in an eight-pin "minidip" package (indicated by the letter "N").

We recommend that a user standardize on three types of op amp. The 741 can be used for noncritical circuitry, a premium unit such as the OP-07 for applications calling for very low offset voltage, and a FET unit such as the CA3130 for high-impedance circuits. These three amplifiers are interchangeable in terms of pin assignments. A faster amplifier, such as the AD509, can be added to the list for special applications.

ERRORS

Although well-designed op amp circuits operate with very little error, it is important not to overlook the various residual errors that are possible. These may well become important when one is dealing with very weak signals or with high frequencies.

Noise

All op amps contribute a certain amount of noise power. This originates mostly in the input circuit and appears after amplification at the output. As a rule, noise is specified in terms of voltage and current rather than power. Thus at 1000 Hz, an OP-07 premium op amp exhibits about 10 $nV/\sqrt{Hz}$ and 0.1 $pA/\sqrt{Hz}$ of noise. The ratio of these two quantities is sometimes referred to as the "characteristic noise resistance" R_n; in the example, $R_n = 10^{-8} V/10^{-13} A = 10^5\ \Omega$. When the parallel combination of all resistors connected to the input, R_T, is larger than R_n, current noise predomi-

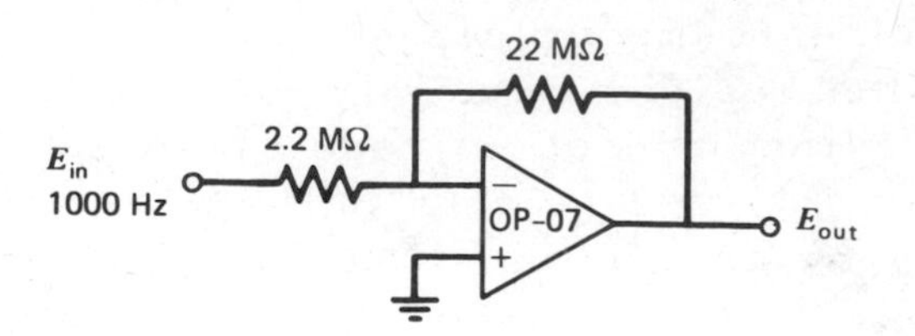

Figure 4-38. Circuit for illustrating the calculations of ultimate sensitivities. The current noise referred to the input is taken to be 0.1 pA/$\sqrt{\text{Hz}}$.

nates, and below that value the voltage noise is larger.

Let us determine the minimum amplitude of a 1000-Hz signal that can be handled by the circuit shown in Figure 4-38 in terms of the S/N ratio. Before starting, we need additional information about the bandwidth involved (assume it to be 100 Hz) and what S/N is considered acceptable (assume an S/N ratio of 10 to be the minimum). First let us calculate the combination of input and feedback resistors. This gives R_T as the parallel combination of 2.2 and 22 MΩ, namely, 2 MΩ. Since this is larger than R_n, as calculated previously, one can assume that the current noise dominates. We can calculate the voltage resulting when the specified current passes through this resistance, as

$$\left(0.1\ \frac{\text{pA}}{\sqrt{\text{Hz}}}\right)(100\ \text{Hz})^{1/2}(2 \times 10^{6}\ \Omega) = 2\ \mu\text{V}_{\text{RMS}} \tag{4-45}$$

Consequently, if the S/N ratio is to be greater than 10, it is necessary that the input signal be larger than 20 µV.

Note that the overall gain of the circuit does not affect the S/N ratio since it amplifies both signal and noise by the same amount. The preceding calculation does not take into account the noise contribution of the two resistors. If we consider them as the equivalent 2-MΩ resistor, the Johnson noise generated is, by Eq. 2-18,

$$e_{RMS} = (1.3 \times 10^{-10})(2 \times 10^{6})^{1/2}(100)^{1/2}$$

$$= 1.8\ \mu V \tag{4-46}$$

The amplifier noise and the resistor noise combine not directly, but as the square root of the sum of their squares. This is because they are random processes and follow the same rule as that used in the addition of standard deviations:

$$e_{Total} = (e_1^2 + e_2^2)^{1/2} \tag{4-47}$$

This gives, for our example, $(1.8^2 + 2^2)^{1/2} = 2.7\ \mu V$. It must be emphasized that this value is only a lower limit. Other types of noise, such as EMI, are usually present and further increase the total noise.

Offset and Bias Errors

The drift of the offset voltage and bias current also constitute a type of noise since the exact values and signs they will take cannot be predicted beforehand. In contrast, the *initial* offset and bias can be determined and thus should not be treated statistically. If we again use the symbol R_T for the combination of input and feedback resistors in an inverting circuit, we can write for the total error (referred to the input)

$$e_{total} = (V_{os} + I_B R_0) + \left[\sum(\text{drifts})^2 + \sum(\text{noise})^2\right]^{1/2} \tag{4-48}$$

where the drifts are determined by time fluctuations in V_{OS} and I_B and by random temperature variations. The low-frequency noise, say, up to 10 Hz, can contribute 1 to 10 μV to the total error.

Errors Caused by Finite Gain

The fundamental magic by which op amp circuits manage to keep errors negligible is ascribable to the negative feedback in connection with a basic amplifier of very high open-loop gain A, reduced to a small closed-loop value G. It was shown in Eq. 4-6 that the behavior of a summer is given by

$$\frac{E_{\text{in}}}{E_{\text{out}}} = -\frac{R_{\text{in}}}{R_f}\left(1 + \frac{1}{\beta A}\right) \tag{4-49}$$

so that the relative deviation from ideal behavior is of the order of $1/\beta A$. The constant β is given in terms of G by the relationship

$$\beta = \frac{1}{(R_f/R_{\text{in}}) + 1} = \frac{1}{G + 1} \tag{4-50}$$

The quantity βA is called the "gain margin," that part of the overall gain of the amplifier A that can be used to decrease deviations from ideality. This is appropriate because the quantity $1/\beta A$ represents indeed the residual fractional deviation from ideality in most circuits. It also represents the decrease in the effective output impedance ($Z_{\text{out}}/\beta A$). The same applies to AC distortion, which is also diminished by this factor. Consider an amplifier with an open-loop gain of 10^5 used as a gain-of-1000 amplifier. If the distortion inherent in the electronics is 5%, the distortion after feedback is $5\% \times [1/(1/1001 \times 10^5)] =$

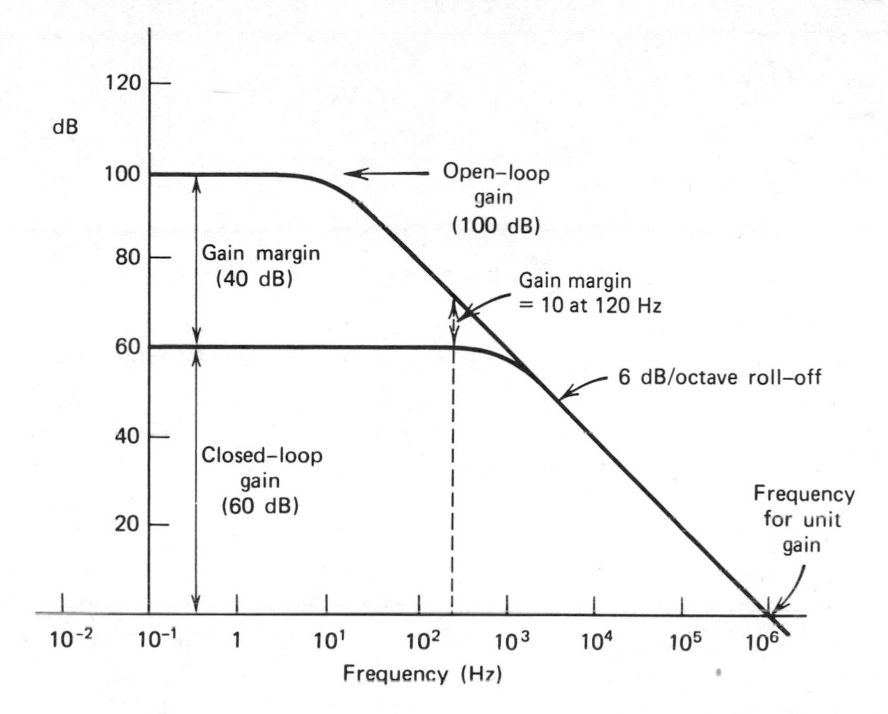

Figure 4-39. The relationships between various gain concepts.

0.05%, a negligible quantity.

Since the open-loop gain is a function of frequency, as illustrated in Figure 4-39, at higher frequency the gain margin may be lacking. Thus in the example shown, the value drops to only 10 at 120 Hz, which means that deviations from linearity up to 10% can be present. The distortion in the preceding example is decreased to only 0.5% at 120 Hz.

This points out an important caveat. One cannot have simultaneously high closed-loop gain *and* wide frequency range, without incurring considerable error. For audio circuits, where frequencies extend to 15 or 20 kHz, an amplifier with 1-MHz gain-bandwidth product should be used only with closed-loop gains less than about 10.

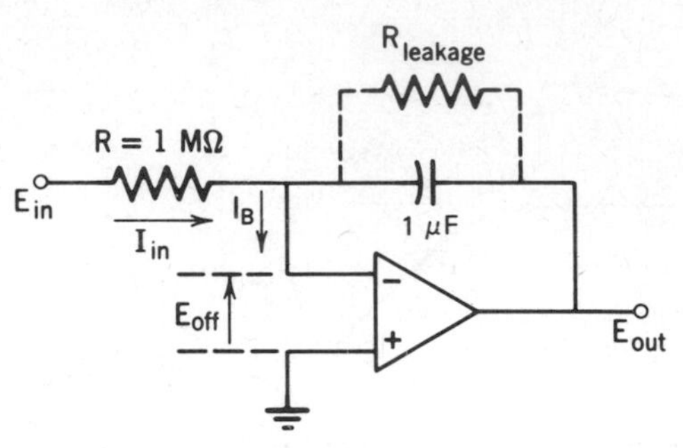

Figure 4-40. An integrator circuit showing the sources of error.

Integrator Errors

It is essential for the effective use of integrators to compute the extent of different offset and bias errors (Figure 4-40). For high-quality capacitors, such as should be used in integrators, the leakage resistance R_{leak} is about 10^{11} Ω/μF. Let us assume that a good-quality amplifier is selected, with E_{os} = 0.05 mV, and bias current I_B = 0.05 nA. The operation of the circuit requires that

$$I_{in} = -C \frac{dE_{out}}{dt} \tag{4-51}$$

In addition, there exists an error current that also charges the capacitor:

$$I_{error} = I_B + \frac{E_{os}}{R_{in}} + \frac{E_{out}}{R_{leak}} \tag{4-52}$$

For the data given in the preceding example, with E_{out} = 10 V, the error current* becomes

* This assumes the worst case, when the errors add rather than partially canceling each other.

$$I_{error} = (0.05 \times 10^{-9}) + \frac{0.05 \times 10^{-3}}{1 \times 10^{6}} + \frac{10}{10^{11}}$$

$$= 2 \times 10^{-10} \text{ A} = 0.2 \text{ nA} \tag{4-53}$$

This current will charge the capacitor at the rate of $(2 \times 10^{-10})/(1 \times 10^{-6}) = 0.2$ mV/s, independently of the signal integration. In practice, this means that, for a maximum error of 1%, the slowest permissible rate of integration is 20 mV/s, or about 1 V/min. This indicates that integrations over long periods of time (hours) are seldom feasible with op amps.

PROBLEMS

4-1. Design circuits to perform each of the following data-processing tasks: (The symbol t designates time.)

(a) $E_{out} = 5E_1 + 7E_2 - 14E_3$
(b) $E_{out} = (10 + 5t)E_{in}$
(c) $E_{out} = 20 \log (E_1/E_2)$

4-2. Write the operating equations for each circuit shown in Figures 4-41 through 4-46.

4-3. Demonstrate mathematically that both differentiation and integration are linear operations.

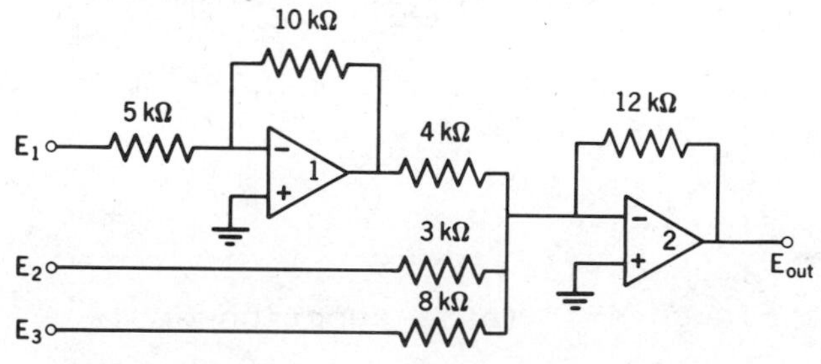

Figure 4-41. See Problem 4-2.

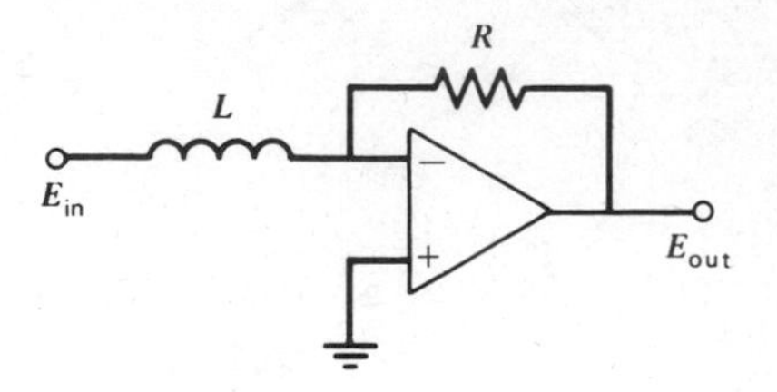

Figure 4-42. See Problem 4-2.

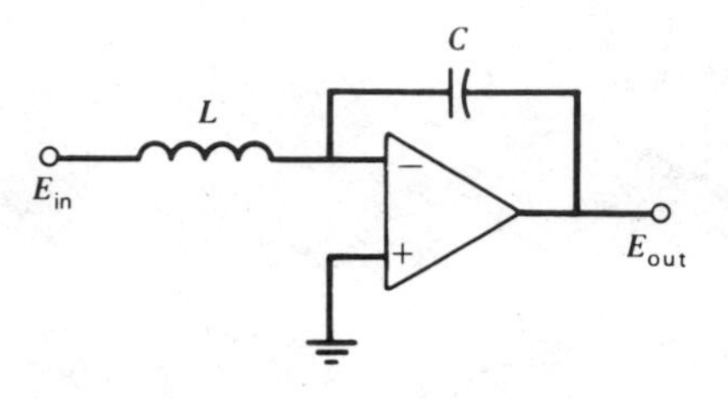

Figure 4-43. See Problem 4-2.

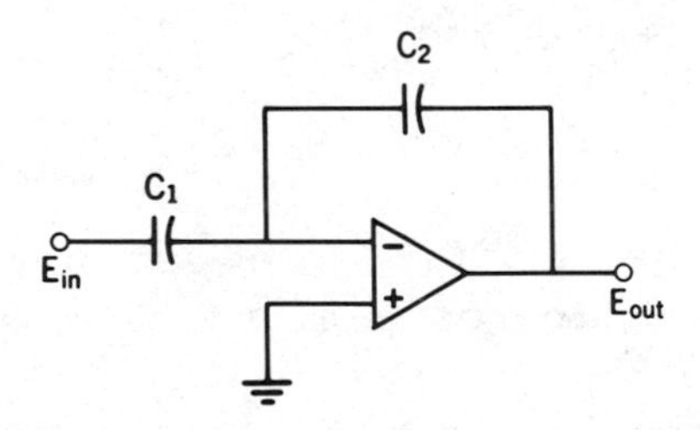

Figure 4-44. See Problem 4-2.

4-4. Calculate the error resulting from finite gain in the circuits shown in:

(a) Figure 4-6
(b) Figure 4-7*a*
(c) Figure 4-8*c* (as a function of *k*)

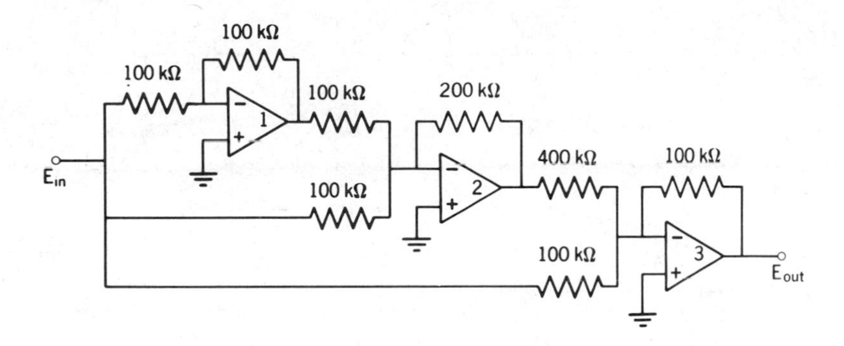

Figure 4-45. See Problem 4-2.

4-5. Figure 4-47 shows an alternative circuit for a perfect rectifer. Explain how it works.

4-6. Write the equations for the output as a function of input for the circuits shown in Figure 4-8, in terms of the parameter k.

4-7. What voltage will be present at the output of an integrator 5 min after the start of integration if the input resistor is 600 kΩ, the feedback capacitor is 0.5 μF, and the input voltage is 1 mV? Assume error-free operation.

4-8. Evaluate E_{out} in the circuit shown in Figure 4-46 for the component values marked as a function of time.

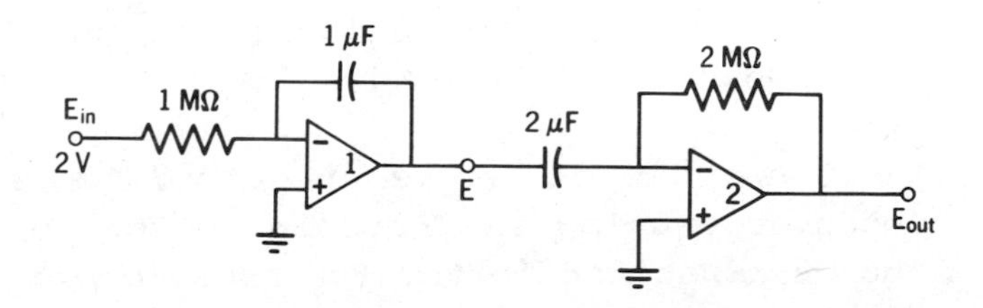

Figure 4-46. See Problems 4-2 and 4-8.

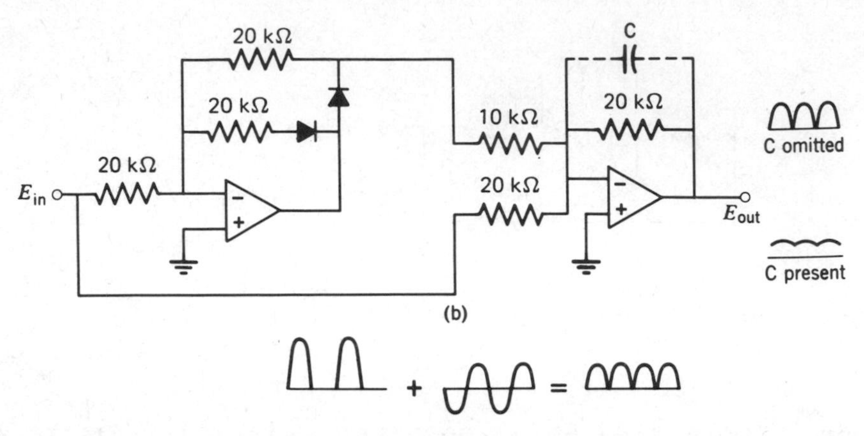

Figure 4-47. See Problem 4-5.

4-9. Design the input and feedback circuits for the HA-2400 amplifier shown in Figure 4-27 to give programmable gains of 1, 3, 10, and 30.

4-10. Describe and sketch the input-output relationships for the circuits shown in Figures 4-48 and 4-49. Assume that saturation occurs at ±10 V; neglect the forward voltage drops of the diodes.

4-11. Design an op amp circuit to produce two equal outputs exactly 180° out of phase at a frequency of 1000 Hz.

4-12. By the use of phasors, determine and plot the output phase as a function of R_{in}, for the circuit shown in Figure 4-14*e*.

4-13. What would be some disadvantages of an integrator using an inductor in the input rather than a capacitor in the feedback loop?

4-14. Design circuits by means of which an amplifier can measure its own CMRR and PSRR.

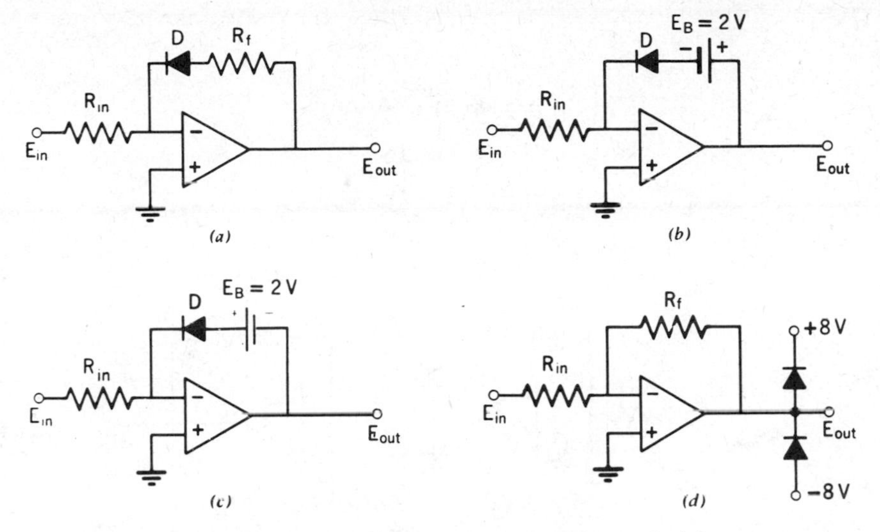

Figure 4-48. See Problem 4-10.

4-15. Consider an inverter with R_{in} = 10 kΩ and R_f = 1 MΩ, amplifying a 10-mV signal. What are the maximum permissible offset voltage and bias currents if the deviation from ideality is to be kept below 0.1%?

4-16. Prove that an attenuation of 6 dB per octave is equivalent to 20 dB per decade.

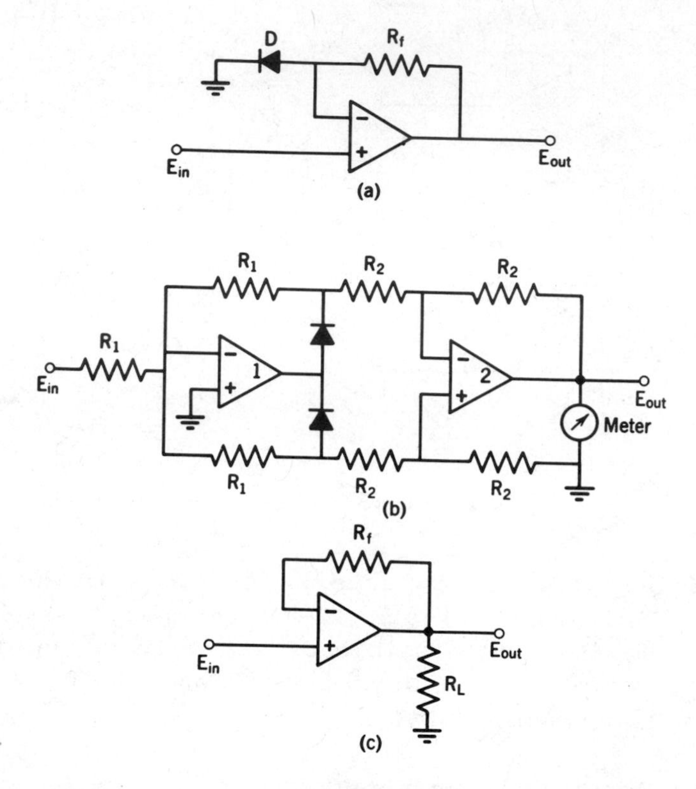

Figure 4-49. See Problem 4-10.

V

ANALOG INSTRUMENTATION

We now explore the use of operational amplifiers (op amps) and other analog modules in laboratory instrumentation. There are fundamentally three such applications to be considered: (1) *function generation*, the production of signals with which to excite or energize a physical or biological system; (2) *signal processing*, including, among other functions, measurement, scaling, multiplication, and integration; and (3) *simulation*, the electronic solving of the differential equations that describe a system.

FUNCTION GENERATION

Many laboratory experiments require the application of a particular program of voltages or currents to a system in order to excite it into producing a measurable response. Such time-dependent excitation signals, called *functions*, can often be generated by analog techniques.

Programmable Voltage Sources

The simplest excitation signal is a direct voltage, easily produced by a single op amp with a con-

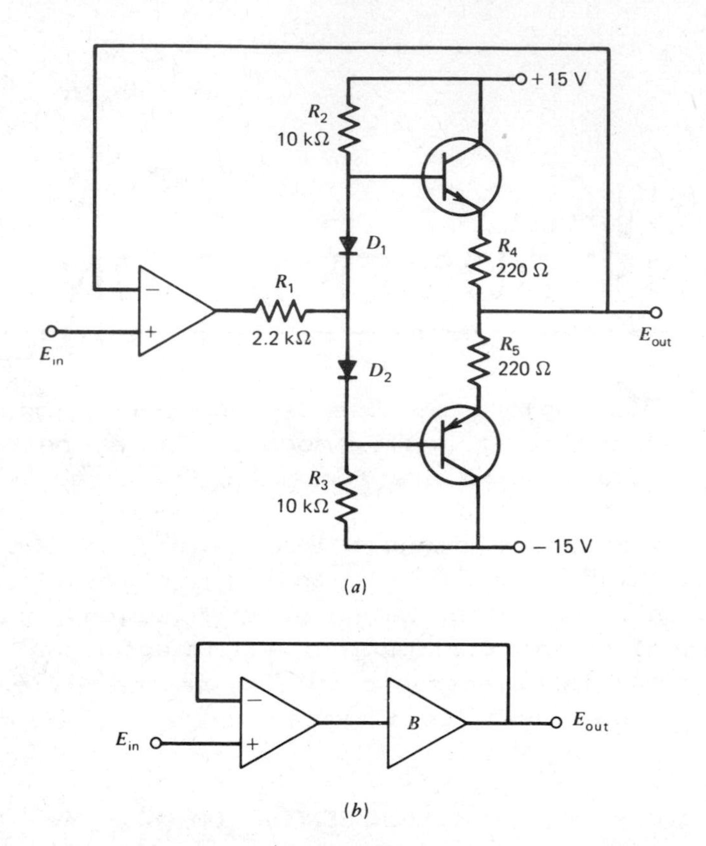

Figure 5-1. A voltage follower equipped with a booster to permit drawing increased current into the load. (*a*) Detailed circuit for currents up to 50 mA; for operation of one sign only, a single transistor is sufficient. (*b*) A conventional symbolic representation.

stant or variable reference potential as input. A reference voltage for this purpose can be fabricated with a Zener diode and resistor or a small mercury battery. A precision variable voltage source that can be used here is that shown in Chapter IV (Figure 4-25). It is advisable to use a voltage follower to prevent loading.

Boosters: For currents greater than the usual op amp can handle, it is necessary to include a *current boo-*

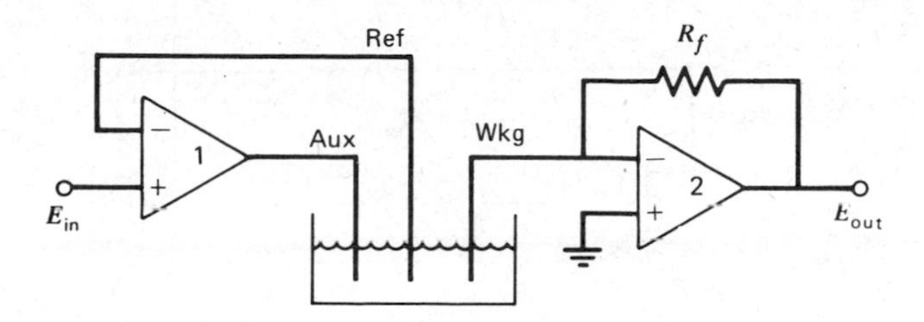

Figure 5-2. A potentiostat (amplifier 1) controlling the potential sensed by the reference electrode (Ref) in an electrochemical cell. Amplifier 2 is a current-to-voltage converter that measures the current carried by the working (Wkg) and auxiliary (Aux) electrodes.

ster. This is an added amplifier stage, included within the feedback loop, and often powered from a separate supply (Figure 5-1). The network formed by resistors R_2 and R_3 and the diodes serves to set the proper operating points for the transistors. Resistors R_4 and R_5 provide short-circuit protection and improve transient response.

Potentiostats: In electrochemical studies, where several metallic electrodes are inserted into an electrolytic solution, it is often necessary to control the potential between two of them while measuring the current flowing through the third. The potentiostat (Figure 5-2) is a device for accomplishing this function. It forces the potential between a reference electrode (Ref) and the working electrode (Wkg) to be exactly equal to the applied control voltage. The characteristics of the chemical interaction between the working electrode and active constituents of the solution can be explored by scanning the control potential while monitoring the current by means of the current-voltage converter. This circuit is at the heart of all polarographic instrumentation.

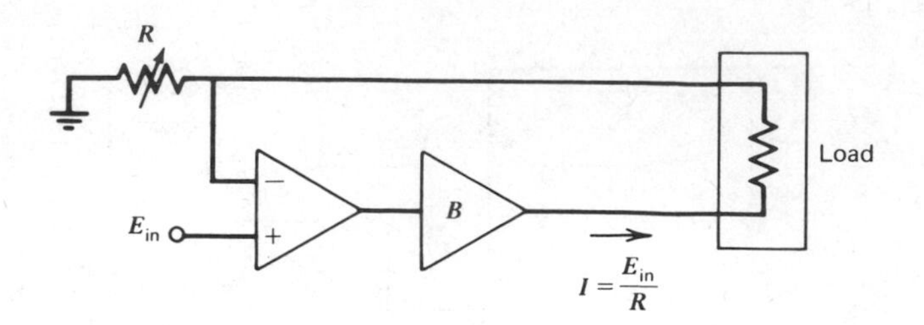

Figure 5-3. An adjustable current supply. Note that the load must be isolated from ground.

Programmable Current Sources

A current is easily driven through a load by placing it in the feedback of an op amp that is fed with a controlled voltage, as in Figure 5-3. The disadvantage to this circuit is that the load cannot be grounded. A circuit that avoids this difficulty, but at the expense of requiring several op amps is shown in Figure 5-4. The potential drop across resistor R_2, in series with the load, is sensed by the differential

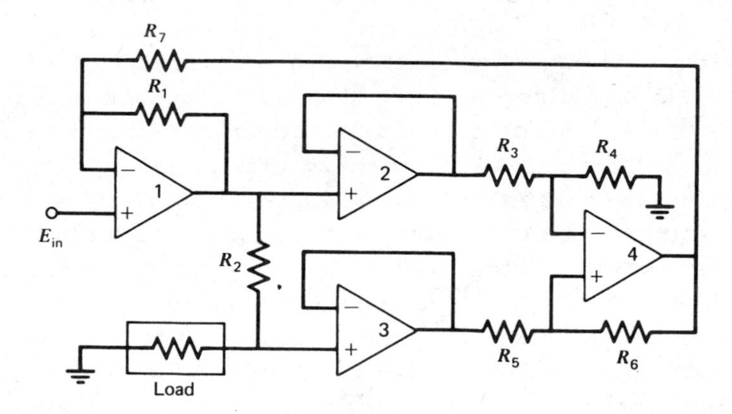

Figure 5-4. A feedback circuit for controlling the current through a grounded load. The circuit gain can be adjusted by varying the R_1/R_7 ratio.

amplifier (amplifier 4) and fed back to the summing junction of the controlling amplifier (amplifier 1) in such a way that any tendency for the current through the load to increase results in lessened output from the control amplifier, and vice versa. Amplifiers 2 and 3 are buffers to prevent loading the sensing resistor. This circuit allows the load current to vary by less than about 1% for a 100% change in the load resistance. The actual value of the current can be altered by changing the voltage at the noninverting input of amplifier 1 or by changing the sensing resistor. The sensitivity of the circuit can be increased by taking more gain at amplifier 1.

If a constant current, rather than a programmable one, is desired, it is much simpler to use an IC designed for this service, namely the current regulator described in Chapter VIII.

Alternating Current Sources

Sine-wave Oscillators: Many experiments call for an AC source. Sometimes the line frequency can be used, but

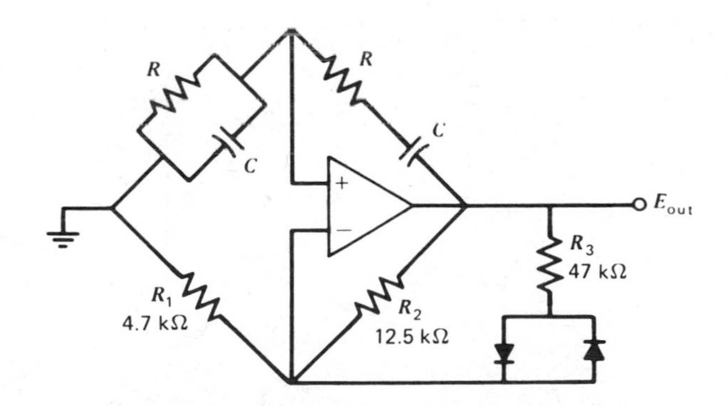

Figure 5-5. A sine-wave oscillator in which a Wien bridge is used for frequency determination. The frequency is given by $f = 1/(2\pi RC)$.

otherwise a sine-wave generator is required. Suitable units can be designed around op amps in several ways. The essential requirement is a network providing either positive feedback or lack of negative feedback selectively at a single frequency. Figures 5-5 and 5-6 show two such circuits.

In the circuit shown in Figure 5-5, called a *Wien-bridge oscillator*, the series and parallel *RC* networks connected to the summing junction of the amplifier determine the frequency, which is given by the formula $f = 1/(2\pi RC)$. The other two arms of the bridge, R_1 and R_2, provide the negative feedback. Some provision must be made for control of the amplitude. This is done with a pair of diodes, as shown in the diagram. If the output of the bridge is very small (less than a few hundred millivolts), neither diode will conduct. On the other hand, when the output is fairly large, the diodes will start to conduct on alternate half cycles, and R_3 will be placed in parallel with R_2, effectively increasing the negative feedback. An equilibrium is quickly reached that results in a sine wave of very

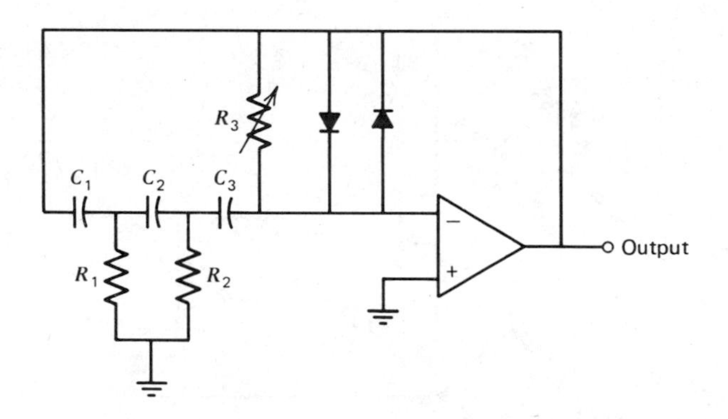

Figure 5-6. A phase-shift oscillator. The frequency is given by $f = 1/[2\pi\sqrt{6}(R_1C_1R_2C_2R_3C_3)^{1/3}]$.

stable amplitude.

Another way of producing a positive feedback for an oscillator is to incorporate a network that will return a portion of the output to the *inverting* input, but with its phase shifted by 180°. An oscillator so designed is shown in Figure 5-6. The phase-shifting network is composed of the several resistors and capacitors. The variable resistor permits frequency adjustment. If all resistors and capacitors are matched, the frequency is given by $f = 1/(2\sqrt{6}\pi RC)$. The crossed diodes, because of their nonlinear properties, serve to stabilize the amplitude of oscillations and decrease harmonic distortion. This oscillator is somewhat simpler than the preceding one but is less stable with respect to its frequency.

<u>Square-Wave Generators</u>: For some purposes a square wave is needed. This can be obtained most readily from a

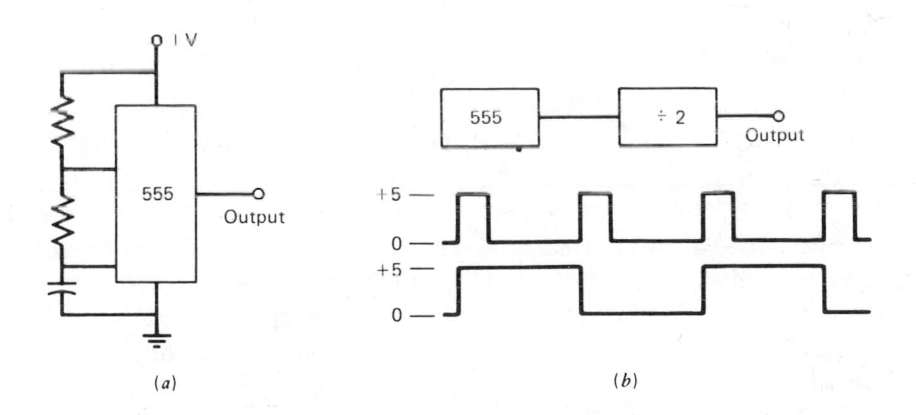

Figure 5-7. The 555 timer as a generator of square waves. (*a*) Standing alone; the frequency is given by the formula $f = 1.44/[(R_1 + 2R_2)C]$, but the symmetry is difficult to equalize. (*b*) Symmetry is perfected by the use of a component that divides the frequency by 2; this unit changes its output state each time the input *rises* from 0 to +5 V, as shown in the timing diagram.

special IC (often referred to as a *timer*) designated by its type number, 555. This operates on digital principles and is described in full in a later chapter. As a square-wave generator, its connections follow those shown in Figure 5-7*a*. The frequency of the output is given by $f = 1.44/(R_1 + 2R_2)C$. For best symmetry, the output of the 555 can be further conditioned by a "flip-flop" or "divide by 2" circuit (discussed in Chapter X). Figure 5-7*b* gives some details.

Another convenient IC is type 8038, which will simultaneously produce sine waves, triangular waves, and square waves, all at the same frequency. It is possible to control the frequency either by an applied voltage or a change of *R* and *C* components.

A network of diodes, resistors, and voltage sources can be designed to approximate any arbitrarily selected single-valued function by a sequence of straight-line segments. This is called a ***diode function generator***.

SIGNAL PROCESSING

The Analog Multiplier-Divider

In order to treat this subject, we will need to introduce a few components not previously discussed. One of these is the analog multiplier-divider, a unit (available as an IC) that will accept three variable inputs, *X*, *Y*, and *Z* (two at a time), generating the product of *X* and *Y*, to give

$$E_{out} = \frac{XY}{10} \qquad (5\text{-}1)$$

or the quotient of *X* divided by *Z*, giving

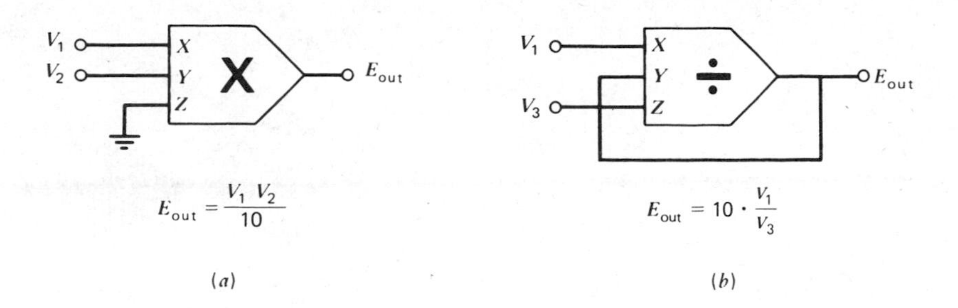

Figure 5-8. An analog multiplier-divider configured as (*a*) a multiplier and (*b*) a divider.

$$E_{out} = \frac{10X}{Z} \tag{5-2}$$

The factor of 10 is necessary to utilize fully the dynamic range of the associated op amps. If the two inputs in Eq. 5-1 are each 10 V (their maximum permissible values), therefore, the output will be 10 V, not 100 V, and similarly for Eq. 5-2. Figure 5-8 gives the appropriate connections for a multiplier or divider.

The multiplier has found many interesting applications, including the voltage-controlled amplifier (Figure 5-9). The gain can be adjusted by means of a voltage applied at *A*, which can itself be the output of

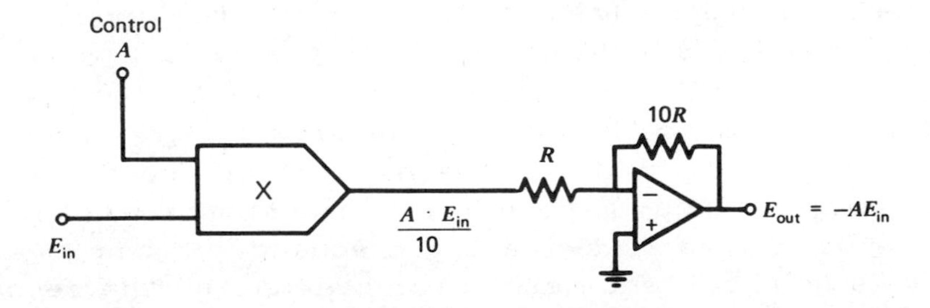

Figure 5-9. A multiplier used as a variable-gain amplifier. The op amp permits making the gain numerically equal to the value of *A* in volts.

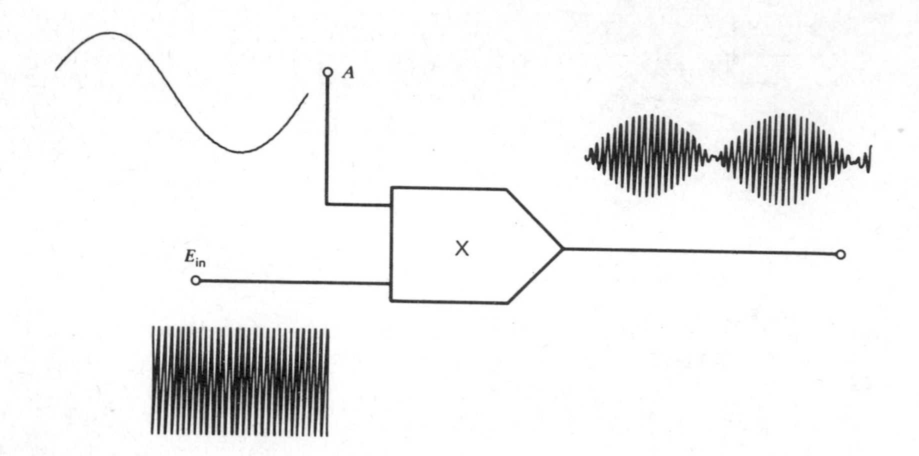

Figure 5-10. Modulation by multiplication. A so-called four-quadrant multiplier is needed to accommodate both positive and negative polarities.

some other device, either analog or digital. This could be useful, for example, in an autoranging voltmeter, to adjust the gain to match a signal to be measured.

If the control voltage at A is a sine wave, as in Figure 5-10, the effect is amplitude modulation. The resulting curve has an amplitude of E_{in} modulated by the control wave. Sine-wave modulation generates the sum and difference frequencies. If the two input frequencies are designated as f_1 and f_2, as in Figure 5-11*a*, the output will contain not only the original signals, but also the new frequencies $f_1 + f_2$ and $f_1 - f_2$. If the two inputs of the multiplier are fed with the same frequency Figure 5-11*b*, the output will contain a sine wave at twice the frequency of the input. This is an excellent method for generating the second harmonic of a sine wave.

A square wave is sometimes used in place of a sinusoid for modulation. If the control consists of a symmetrical square wave of alternating +1 and −1 V amplitude, the result will be a periodic sign inversion. If the square-wave frequency is made to be precisely the same as that of the signal input and coincident in phase with its zero crossing (Figure 5-12), the result is a type of full-wave rectification known as *synchronous detection* or *phase detection*. This is the heart of the "lock-in amplifier" described later in this chapter.

Logarithmic Amplifiers

Another analog component useful in signal processing is the logarithmic amplifier described in Chapter IV. Suppose we wish to design an instrument for the automatic generation of Bode diagrams. Figure 5-13 is a block diagram of an instrument that will do this. The AC input signal is divided into two parallel paths. One of these leads to a component that gives a DC output proportional to the frequency. The signal in the second path is converted to its absolute value by the precision rectifier. The signals in both

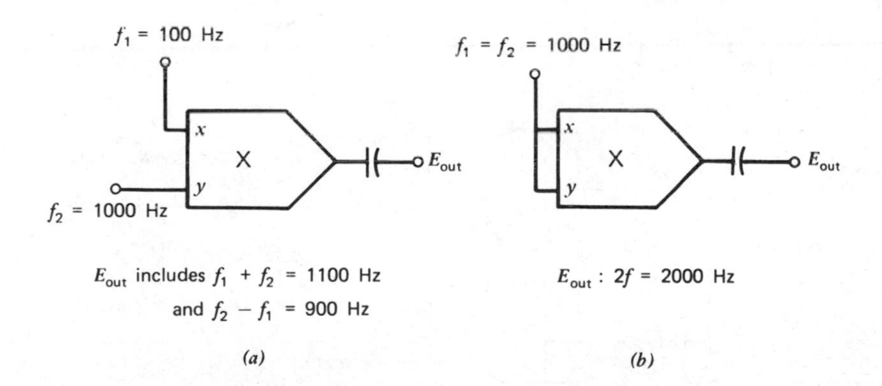

Figure 5-11. Multipliers used to synthesize new frequencies: (*a*) sum and difference; (*b*) second harmonic. The capacitor serves to remove a DC component generated by the process of multiplication.

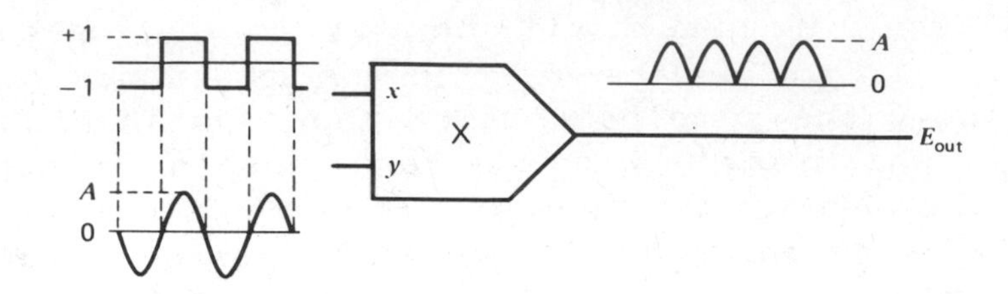

Figure 5-12. Synchronous detection.

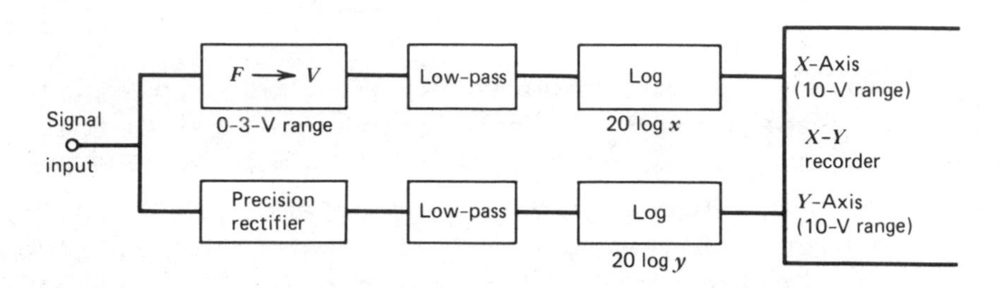

Figure 5-13. Block diagram of an automatic recorder of Bode plots.

paths are low-pass filtered and, after transformation into the corresponding logarithms, are sent to the x-y recorder.

Another application of logarithmic amplifiers can

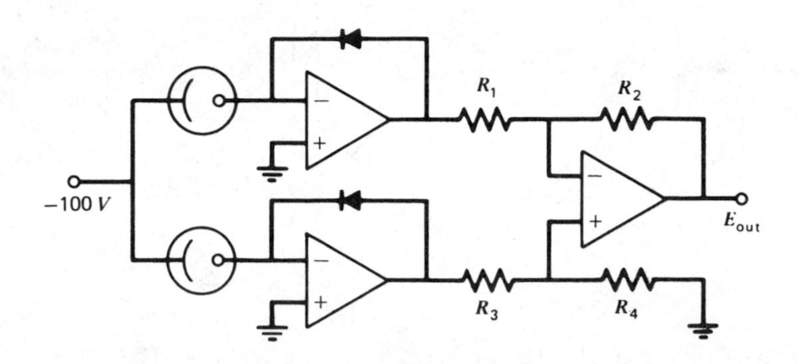

Figure 5-14. A dual-log circuit for taking the log ratio of the currents from two phototubes.

be encountered in spectrophotometry, where the power, P, of light in two beams is to be measured. Applicable theory (the Beer-Lambert law) calls for the quantity $E_{out} = B \log (P_1/P_2)$, where B is a constant. Figure 5-14 shows a circuit for making such a measurement.

SIMULATION

The use of operational amplifier techniques in the numerical solution of differential equations is known as *simulation* since the output voltage simulates physical systems that obey the given equation. Consider, for example, the pair of integrators followed by an

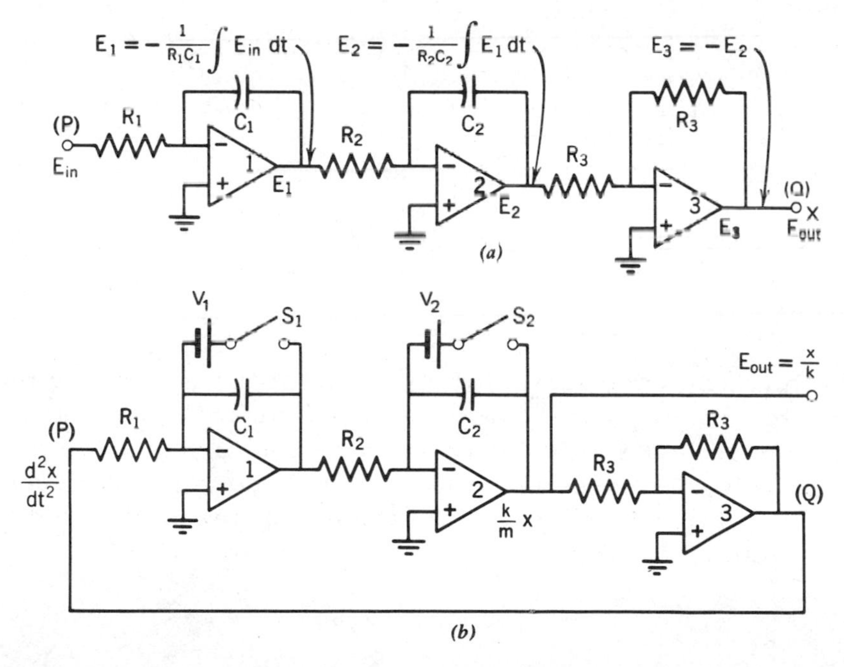

Figure 5-15. (*a*) A circuit for double integration with sign inversion. (*b*) An oscillator made by closing a loop from Q to P. The initial conditions should be $V_1 = 0$ (for integrator 1) and V_2 equal to the desired amplitude (for integrator 2).

inverter (Figure 5-15*a*). The output can be calculated by combining the individual expressions shown in the figure, which gives

$$x = -\frac{1}{(R_1C_1)(R_2C_2)}\iint E_{in}\,dt^2 + Mt + N \qquad (5\text{-}3)$$

in which *M* and *N* are constants of integration. This is equivalent to the relationship

$$E_{in} = -k^2\left(\frac{d^2x}{dt^2}\right) \qquad (5\text{-}4)$$

where $k = \sqrt{R_1C_1R_2C_2}$, a time constant. If a loop is now closed in the circuit between the points marked *P* and *Q*, a powerful restriction is introduced into Eq. (5-3). Of all values of x, only those are now possible for which $x = E_{in}$, or:

$$x = -k^2\left(\frac{d^2x}{dt^2}\right) \qquad (5\text{-}5)$$

We thus may conclude that by connecting *P* and *Q* we have forced the circuit to obey the differential equation, Eq. 5-5. The output of amplifier 2 describes the solution of the equation, multiplied by a constant, $1/k^2$. The solution of this particular differential equation is a sine wave:

$$x = A \sin(\omega t+\phi) \qquad (5\text{-}6)$$

Consequently the circuit simulates any physical system that oscillates in simple harmonic motion. The frequency can be shown to be $f = 1/2\pi k = 1/2\pi\sqrt{R_1C_1R_2C_2}$.

The circuit shown in Figure 5-15 can be used to generate a sine wave, with the frequency determined by the *RC* time constants.* The voltage sources marked V_1 and V_2 allow for impressing initial potentials on the integrators, causing the circuit to provide a numerical solution to the equation. These initial conditions relate to the integration constants *M* and *N* in Eq. 5-3 and to the constants *A* and ϕ in Eq. 5-6.

This approach can present a very convenient way to solve many differential equations. As an example, suppose that a physical system follows the relationship

$$\frac{d^3y}{dt^3} = -2\left(\frac{d^2y}{dt^2}\right) + 2y + 6 \tag{5-7}$$

Figure 5-16 shows that the equation can be solved by the use of two feedback loops leading into the same summer. The function $y(t)$ can be plotted at E_{out}.

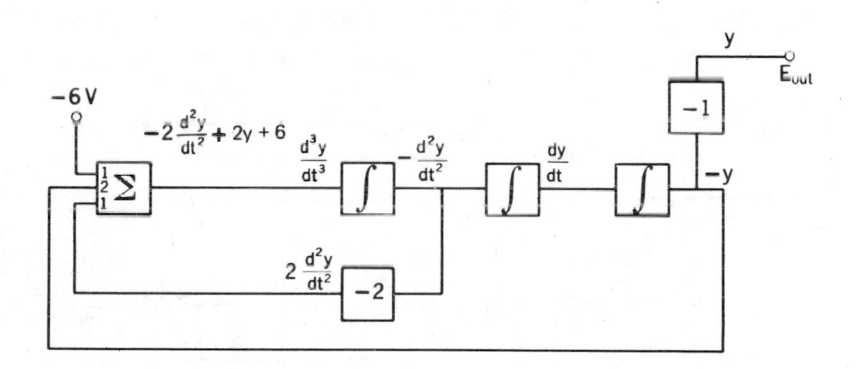

Figure 5-16. Solution of a third-order differential equation. The boxes represent integrators and inverters of the types previously discussed. The box marked "Σ" contains a summer with gains of 1, 2, and 1, respectively, at its three inputs.

* The circuit shown in Fig. 5-15*a* might be useful as it stands, for data processing purposes whenever the double integral of a signal is desired. It becomes a simulator only after the *P*-*Q* connection is made, thus solving a differential equation.

Linear differential equations of any order can be solved by this general method. An array of op amps and other components designed for this kind of application is known as an *analog computer*.

PHASE SHIFTERS

Many AC circuits require specific phase relationships between two potentials or currents, and hence it is essential to be able to vary the phase of one or both. This can be done most easily by means of an op amp in one of the differential circuits shown in Figure 5-17. Passage through an op amp, as we have seen, normally changes the phase of a signal by 180° (π radians). However, the *RC* pair connected to the noninverting input also shifts the phase by an amount depending on the frequency of the signal. The interaction of these two effects makes phase control possible. Either the resistor R_3 or the capacitor can be made the variable control. Such a circuit is useful, for example, with a lock-in amplifier, where signal and reference must be precisely in phase with each other.

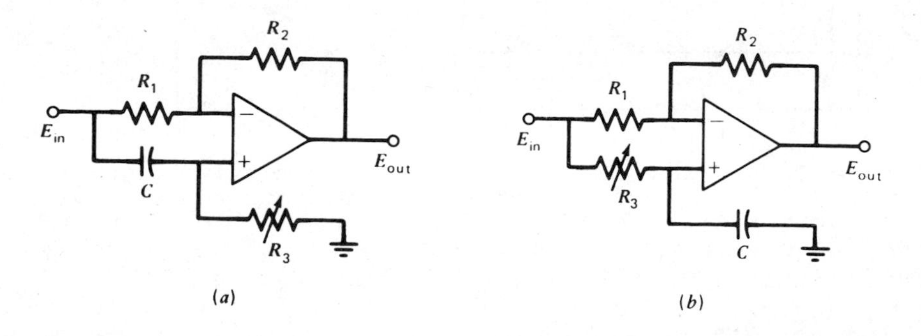

Figure 5-17. Phase-shifting circuits: (*a*) permitting a range from 0 to 180°; (*b*) 0 to -180°.

MODULATION

We have seen that in the region close to DC, noise increases substantially, following a $1/f$ variation with frequency. The S/N ratio can be improved drastically by a type of modulation that transfers the information to a higher frequency, thus bringing it into a relatively noise free region. (Note that whatever noise was introduced into the system prior to modulation is there to stay.) The two most widely employed forms of modulation are *amplitude* (AM) and *frequency* (FM).

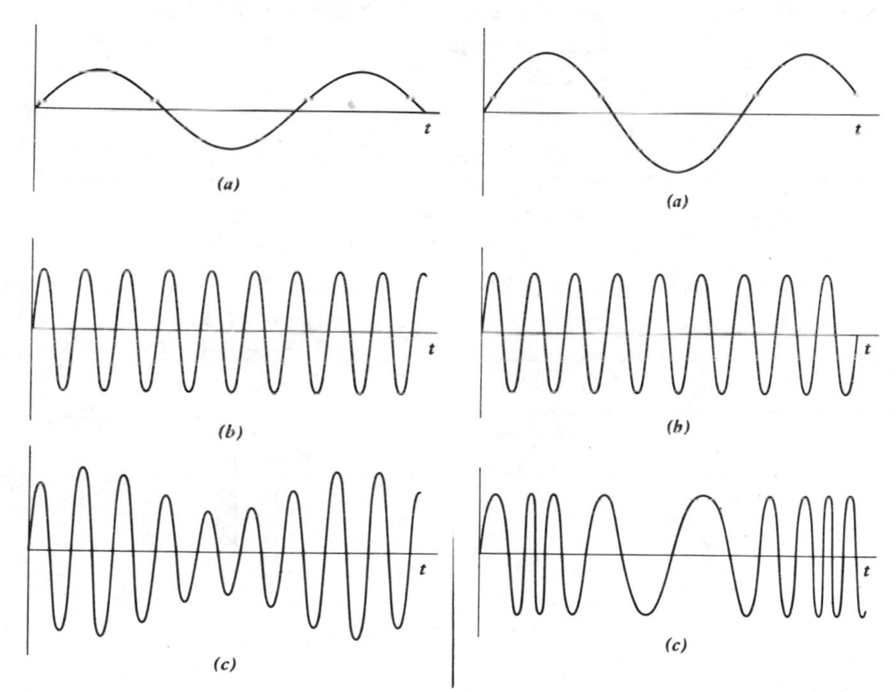

Figure 5-18. An example of amplitude modulation: (*a*) signal, (*b*) carrier, and (*c*) the modulated wave. The difference between the two frequencies is usually much larger than shown.

Figure 5-19. Frequency modulation: (*a*) signal, (*b*) carrier, and (*c*) the modulated wave.

Both of these are familiar in connection with radio broadcast techniques but have much wider fields of application. Waveforms corresponding to these two types of modulation are shown in Figures 5-18 and 5-19, respectively. Amplitude modulation is more common in instrumentation than frequency modulation, so the latter is not treated in detail here.

In communications, the frequency of the carrier wave is established by governmental regulation, but in laboratory instrumentation it can be freely selected to give the best possible S/N ratio in a given situation. A relatively low carrier frequency (1 to 10 kHz) may be appropriate. After suitable processing, the information can be returned to the original form by *demodulation*.

Numerous circuits have been devised for amplitude modulation. One of them has been seen in Figure 5-10. For higher frequencies than a multiplier can handle, the diode modulator shown in Figure 5-20 can be used. In both of these circuits, modulation amounts to the generation of product terms of the type $E_1E_2 \sin \omega_1 t \sin \omega_2 t$, which can be demonstrated as follows. The nonlinear current-voltage characteristic of a diode can

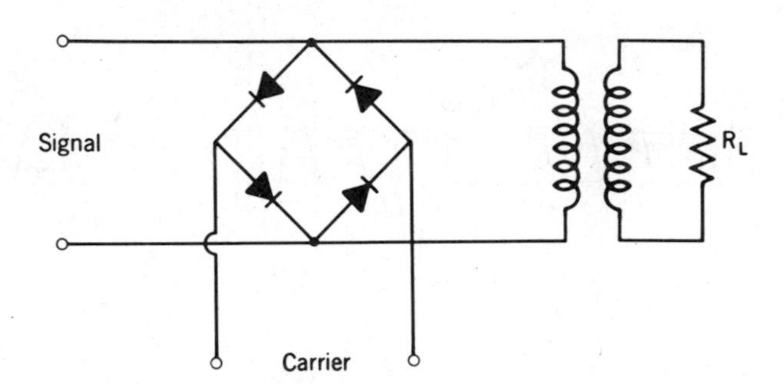

Figure 5-20. A diode modulator. A transformer is often used as shown for coupling the modulator to the load. Band-pass filters are usually added to reject the residual unmodulated signal and harmonics. The diode ring is available as an IC.

be expressed as a power series:

$$I = k_1 E + k_2 E^2 + \cdots \tag{5-8}$$

The first term describes the behavior of a linear device such as a resistor (where $k_1 = 1/R$). The second term is responsible for modulation. If two frequencies, ω_1 and ω_2, are present, the total current is given by

$$I = k_1(E_1 \sin \omega_1 t + E_2 \sin \omega_2 t)$$
$$+ k_2(E_1 \sin \omega_1 t + E_2 \sin \omega_2 t)^2 + \cdots \tag{5-9}$$

This can be rewritten in the form

$$I = k_1 E_1 \sin \omega_1 t + k_1 E_2 \sin \omega_2 t$$
$$+ k_2 E_1^2 \sin^2 \omega_1 t + k_2 E_2^2 \sin^2 \omega_2 t$$
$$+ 2k_2 E_1 E_2 \sin \omega_1 t \sin \omega_2 t + \cdots \tag{5-10}$$

Note the complete symmetry of these equations. Let us assign ω_1 to be the carrier and ω_2 the signal, with $\omega_1 \gg \omega_2$. The first two terms in Eq. 5-10 represent unchanged initial frequencies. The following two are found, by means of the trigonometric identity $\sin^2 x = 1/2(1 - \cos 2x)$, to consist of a DC current and the second harmonic (the term in 2ω) of the original frequencies. The presence of a DC component is to be expected in a rectifying system such as this. The term containing the product of sines represents the modulated wave, as shown in Figure 5-18. A tuned filter is necessary for elimination of the various unwanted harmonic frequencies. The unmodulated carrier frequency is usually left in, although some modulators, including the multiplier, suppress it. In the

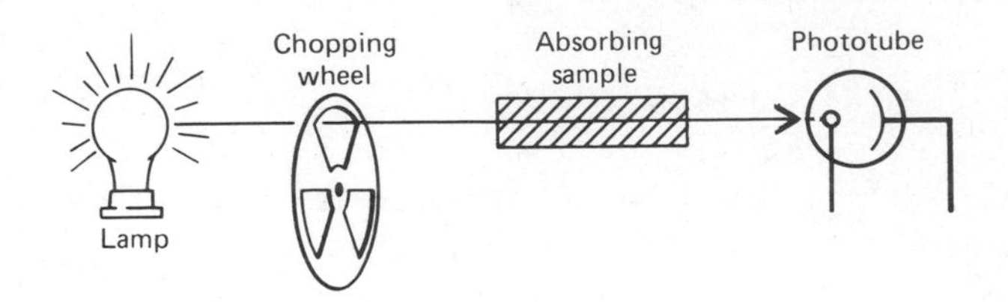

Figure 5-21. A single-beam absorption photometer. The light beam is converted to a string of pulses by a chopper wheel acting as a shutter; this helps to reduce noise but carries no information. The chopped beam is modulated by the selective absorption of certain wavelengths of light by components of the sample.

circuit shown in Figure 5-20, the DC component is eliminated by the transformer coupling.

The properties of the modulated signal now depend on the new frequency ω_1 rather than the original ω_2. This is advantageous in many ways. Most important is the considerably lower noise present at higher frequencies. Additionally, circuits at high frequencies require smaller capacitors, have easily designed amplifiers and more effective band-pass filters than those at low frequencies.

In measurements of high sensitivity, modulation must take place as early as possible, even before the signal reaches the transducer. An example taken from spectroscopy is shown in Figure 5-21, in which a chopped beam of radiation becomes modulated on passage through the sample. The emerging beam is equivalent to an amplitude- modulated AC carrier. The photocell transducer *P* converts it into the electrical domain for further processing.

The Diode Detector

A modulated signal must eventually be demodulated to extract the original information. An example of a simple demodulator is the *diode detector* shown in Figure 5-22. The diode, operating at higher voltage levels than in the modulator, rectifies the signal, producing a train of half-waves. The *RC* filter smooths this series of pulses to a faithful replica of the original signal. The fidelity improves with increased difference between the frequencies of the signal and the carrier. If, for example, the carrier is 10 kHz and the signal 1 Hz, there are 10,000 pulses in each cycle of the detected signal. Following the filter, the 1-Hz signal will be essentially free of residual 10-kHz ripple. For low to moderate frequencies, the precision rectifier makes an ideal demodulator and is the method of choice where applicable.

Another type of modulation sometimes used in laboratory instruments is *pulse modulation*. The repetition rate of a train of pulses can be considered analogous to the frequency of a sine wave. The pulse train can be modulated in either amplitude, duration, or position, as shown in Figure 5-23. Observe that the optical example presented previously involves pulse modulation.

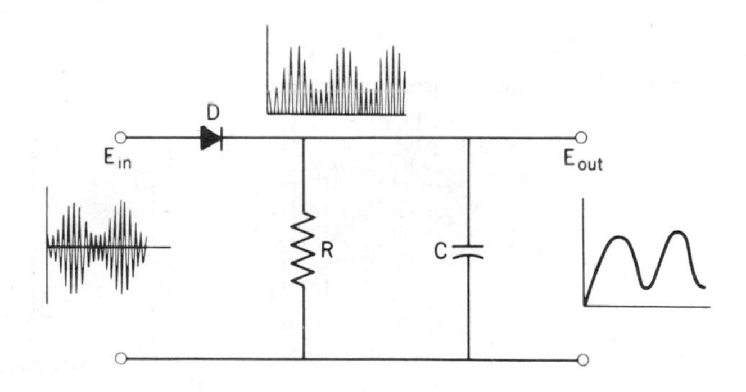

Figure 5-22. A diode detector. The diode rectifies the input signal, which is then smoothed out by the capacitor to give the envelope, which is the same shape as the original signal. The carrier must be of much higher frequency than the signal, so that the *RC* filter attenuates the former but not the latter.

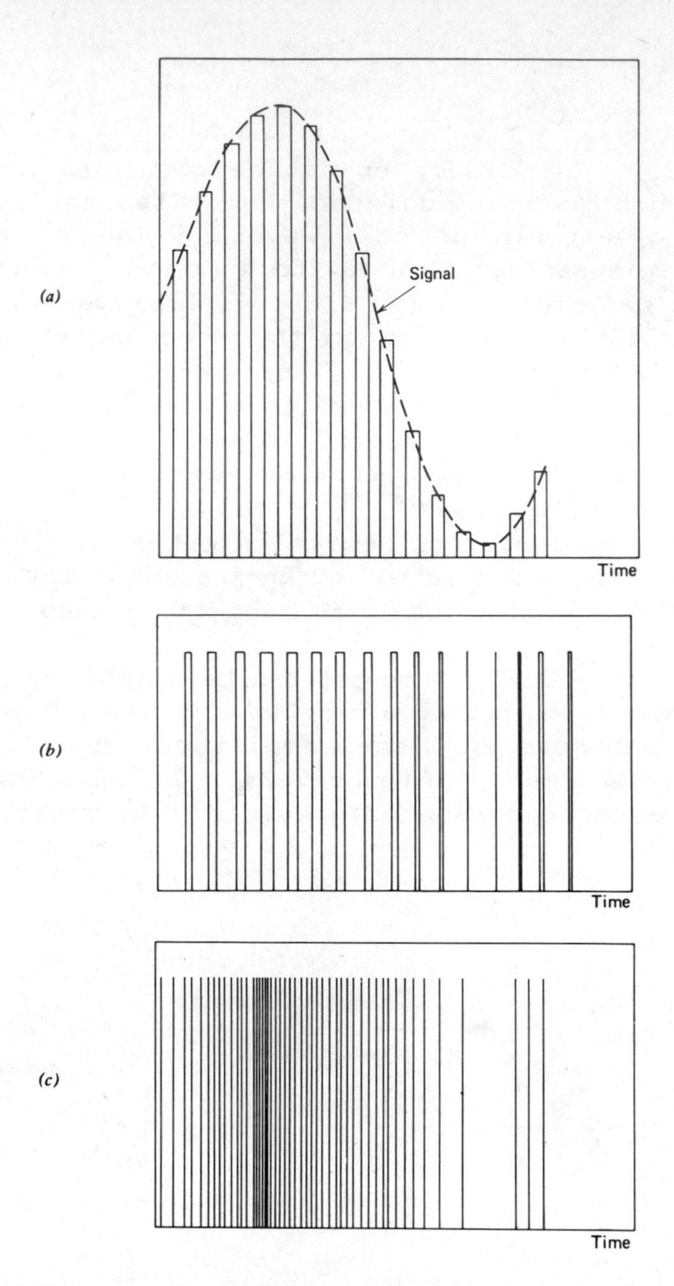

Figure 5-23. Various types of modulation: (*a*) amplitude modulation, in which all pulses are of the same duration; (*b*) pulse-width modulation, where the pulses vary in duration; (*c*) pulse position modulation, in which the pulses are uniform in duration but vary in spacing.

Pulse-amplitude modulation is used in the so-called *chopper-stabilized amplifier* (Figure 5-24). The complete system can be considered to be a single-input amplifier (the large triangle). It consists of two parallel channels: amplifier 1 is driven by a chopped version of the input and, after demodulation, feeds a very stable DC output into amplifier 2. A separate path, provided with a high-pass filter, brings the AC component of the signal directly into amplifier 2, where the two components are summed together to give the final output. The overall assemblage has a very high open-loop gain (>10^7) and an exceptionally low drift (<10 µV/year). On the other hand, this type of amplifier may take as long as several seconds to recover from saturation.

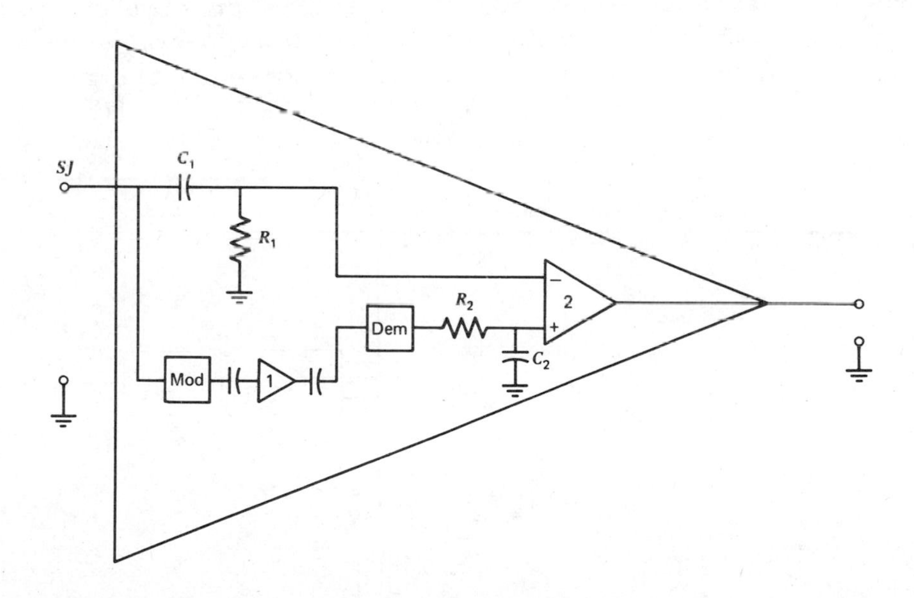

Figure 5-24. Chopper-stabilized amplifier. The sole input connection is the summing junction. External input and feedback circuitry is required.

LOCK-IN AMPLIFIERS

A special type of amplitude demodulation is used in the lock-in amplifier. This is an instrument with a very powerful ability to recover faint AC signals buried in noise. The lock-in amplifier requires, in addition to the signal itself, a reference that provides frequency and phase information for a synchronous detector such as that shown in Figure 5-12. The reference signal can be made quite strong, and hence immune from interference.

The diagram of a typical lock-in amplifier is shown in Figure 5-25. It consists of the following segments: (1) a low-noise preamplifier; (2) a band-pass filter tuned to the frequency of the carrier; (3) a phase shifter, which advances or retards the phase of the reference to allow for phase shifts or time delays in the system; (4) a sine-to-square wave-converter, which could be a simple comparator to sense the zero-

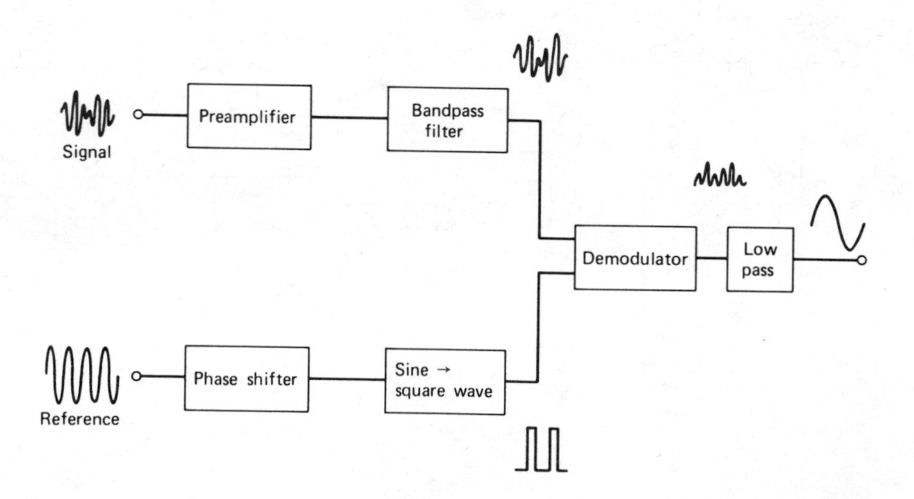

Figure 5-25. Block diagram of a typical lock-in amplifier.

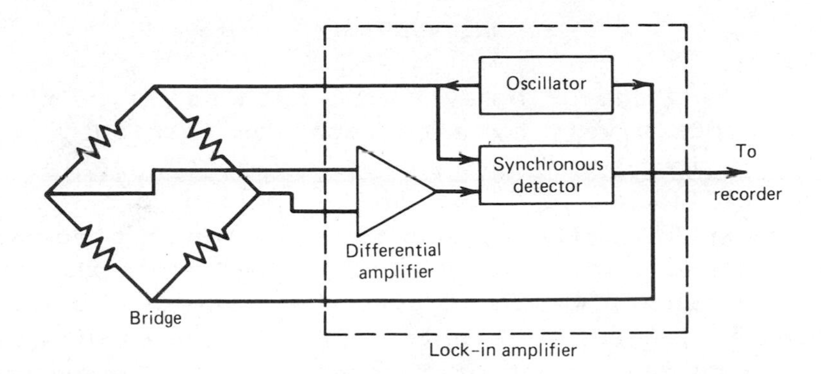

Figure 5-26. A Wheatstone bridge both energized and measured by means of a lock-in amplifier.

crossing of the wave; (5) the synchronous demodulator; and (6) a low-pass filter to eliminate the carrier frequency.

Many commercial lock-in amplifiers have a built-in oscillator that is used to supply excitation to an external system and to provide the needed reference signal. Figure 5-26 gives as an example a Wheatstone bridge that is both energized and measured by the same lock-in amplifier.

It is possible to use a diode detector or a precision rectifier in place of the lock-in amplifier unless the signals are too small. The diode will cut off when the signal is negative and cut in when it is positive, just as does the synchronous rectifier. If the noise should exceed the signal, however, the diode would cut off when the *noise* is negative, so that the noise rather than the signal is rectified and the signal actually attenuated. The merit of the lock-in amplifier is that, regardless of how large the interference may be, detection always takes place with the proper phase. This is possible because, unlike the signal, the reference is noise-free.

SIGNAL AVERAGING

In many measuring systems a process can be triggered repetitively by a suitable excitation. For example, in a luminescence measurement a series of identical flashes of light may be applied to a system, resulting in similar responses time after time. A suitable measuring instrument can superimpose the responses in such a way as to give their sum over a large number of sequential events. The result can then be divided by the number of flashes to give the average value of the responses. This reduces the effect of noise, since noise pulses are as likely to be negative as positive. It can be shown on statistical grounds to improve the S/N ratio by a factor equal to the square root of the number of repetitions. Digital computers can be used to average such results, and there are also analog instruments that can perform the same operation. Examples are the signal averager and the boxcar integrator.

The Multichannel Analyzer

The *multichannel signal averager* (Figure 5-27) consists of a series of channels--for example 500--each containing a capacitor and a switch. A timing circuit actuates the switches in sequence, always beginning at a fixed time after the trigger signal has been applied. Because of this synchronization, each channel will be fed information at the same phase of each successive signal. For example, channel 85 might be connected 85 ms after each triggering. As a result, the capacitors charge through R by a small amount each time and after n repetitions will contain n times the average response corresponding to a point on the curve. To retrieve the information stored on the bank of capacitors, the switch S_{in} is opened, and a slow sequencing is under-

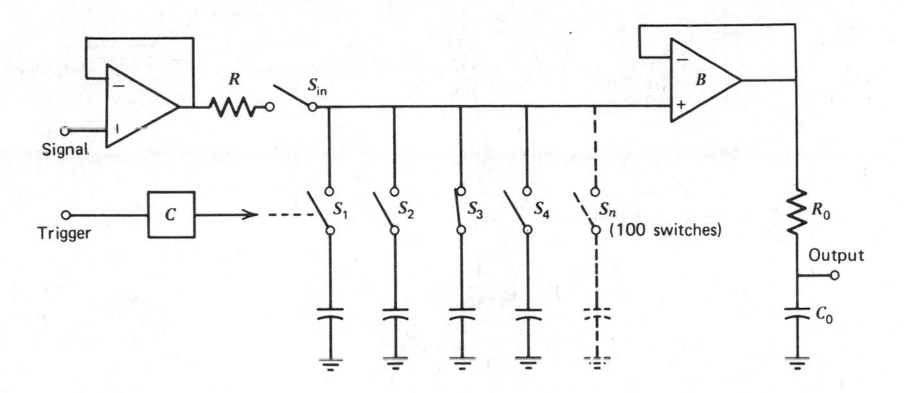

Figure 5-27. Schematic representation of a signal averager. Switch S_3 is shown as closed. Component C is a sequential device that closes one gate at a time in succession. The process starts again from the first gate each time a trigger pulse is applied.

taken. A recorder connected at the output will register the voltages of successive capacitors and produce the desired curve. The output *RC* network has the effect of smoothing the signal, which otherwise would be in the form of a series of steps from one channel to the next. Even though the result of a single run might seem to consist mostly of noise, the average of, say, 200 runs may be quite smooth in appearance.

The Boxcar Integrator

A somewhat simpler averaging device is the *boxcar integrator*, which, instead of having many channels in simultaneous operation, has only a single channel (Figure 5-28). The trigger pulse is used to start a time delay, at the end of which the switch is momentarily closed. The capacitor thus takes an average set of readings for one specific delay period. The process is repeated with increments of delay time until all the points are averaged. The operation requires more scans than the multichannel averager, but for signals

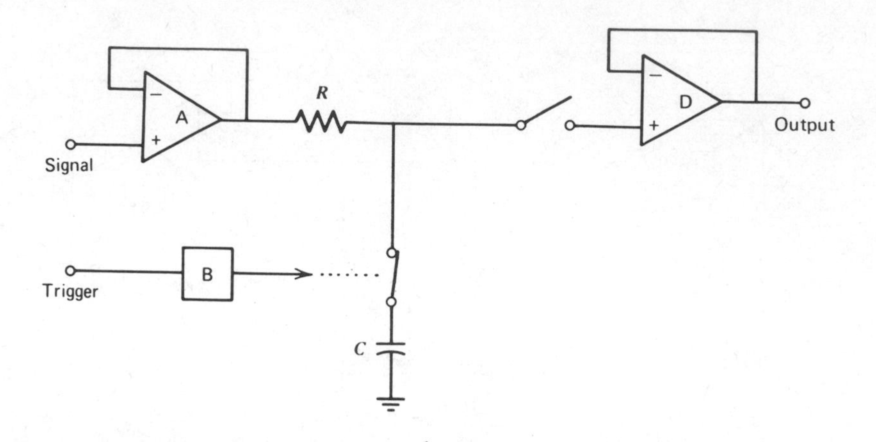

Figure 5-28. A boxcar integrator. Item *B* is an adjustable time-delay circuit.

of fast repetition rate, the increased time may not be objectionable.

PHASE-LOCKED LOOPS

The phase-locked loop (PLL) is a versatile unit for manipulating AC signals. It is made up of three functional components: a phase detector that measures the difference of phase between two AC voltages, a low-pass filter, and a voltage-controlled oscillator (VCO). These are connected so as to permit the VCO to form a feedback loop around the other two blocks, as shown in Figure 5-29.

The basic mode of operation of the PLL can be explained as follows. With no input signal, the VCO runs freely, generating a square wave at a characteristic frequency f_0, but with no output from the phase detector. When an AC signal is presented to the input at a frequency f_s within 10% or so of f_0, the phase detector will generate signal with a DC content that is proportional to the difference between f_s and f_0. The DC component, called the *error voltage*, is isolated by the filter and impressed on the VCO, changing its frequency so as to reduce the $f_s - f_0$ difference nearly to zero. This, in turn, brings the error voltage to zero, and the VCO frequency remains "locked" on to the signal frequency, differing only by a small phase angle.

The frequency span, centered around f_0, within which frequency locking can take place, is called the "capture range." Once locked, the signal can move further away from f_0 in either direction without losing lock, over an extent called the "lock range."

There are many applications for the PLL, of which we mention only those most useful in laboratory instrumentation. In signal conditioning, the PLL can be used as a lock-in amplifier, to extract a repetitive signal from a noisy background. The VCO will duplicate the incoming frequency but greatly attenuate the noise. Since the output of the VCO follows the signal amplitude, the PLL can be considered to function as a *tracking filter*.

The PLL makes a convenient demodulator for amplitude-modulated signals. For this purpose an analog multiplier is needed, with the signal connected to one of the inputs and the VCO to the other. Integrated circuit PLLs are available that have a multiplier on the same chip for this service.

Another application is in signal generation. A digital binary counter (described in a later chapter), connected between the VCO and the phase detector (Fig-

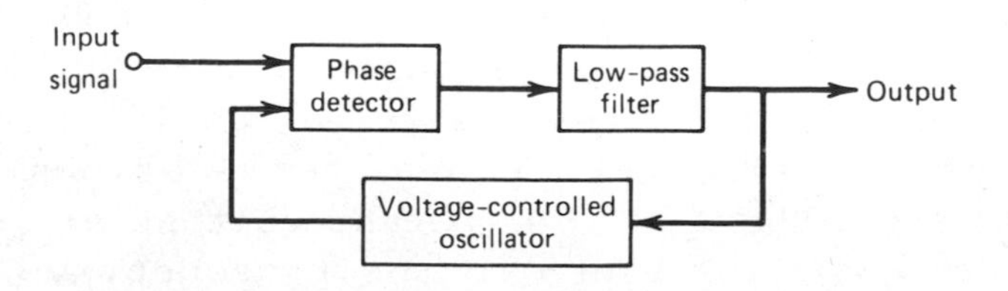

Figure 5-29. Basic circuit of a phase-locked loop.

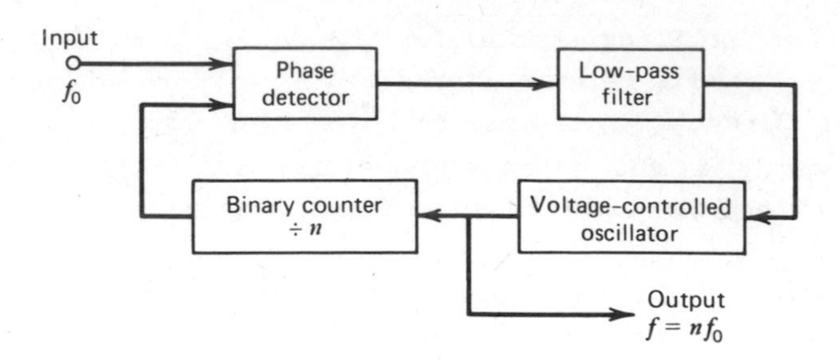

Figure 5-30. A phase-locked loop as a harmonic generator.

ure 5-30), will effectively force the VCO to run at a higher rate. If the counter produces one pulse for each four that it receives, for example, the oscillator will have to run four times as fast to keep up with the incoming signal. This is an effective mode of frequency multiplication. It is possible to utilize a low-frequency signal from an external source--say, at 100 Hz--and to arrange a series of counters so that the ratio of pulses can be varied by a switch control from 2 to, say, 20. This enables the circuit to produce frequencies from 200 to 2000 Hz in convenient 100-Hz steps.

GROUNDING AND SHIELDING

Most circuit schematics have numerous ground connections as indicated by the conventional symbol, and the matter is usually given very little thought. Yet inadequate grounding is a major source of improper instrument operation, especially in low-level systems.

By definition, all ground terminations must be connected together. This guarantees that all currents can find a way to return to their respective sources and ensures a common voltage reference, a zero level. These connections, however, cannot be made in a random way without the likelihood of impairing the precision of measurement.

An example of improper grounding is shown in Figure 5-31, where the signal ground has simply been connected to the output return line on its way back to the power supply. In such a case, the section of the wire used in common by the two circuits might have a resistance of 20 mΩ and carry a current of 50 mA. This would introduce an error of 1 mV, about 10% of the input signal.

To avoid grounding problems, two rules should be obeyed: (1) the ground wires should be as short as possible, of heavy gauge, and firmly connected; and (2) each ground line should be connected by a single wire to *one* master ground point. Whether this point is itself connected to the earth is less important. The whole ensemble should resemble a star, with no branching or looping (Figure 5-32). In low-current lines, branching is less objectionable.

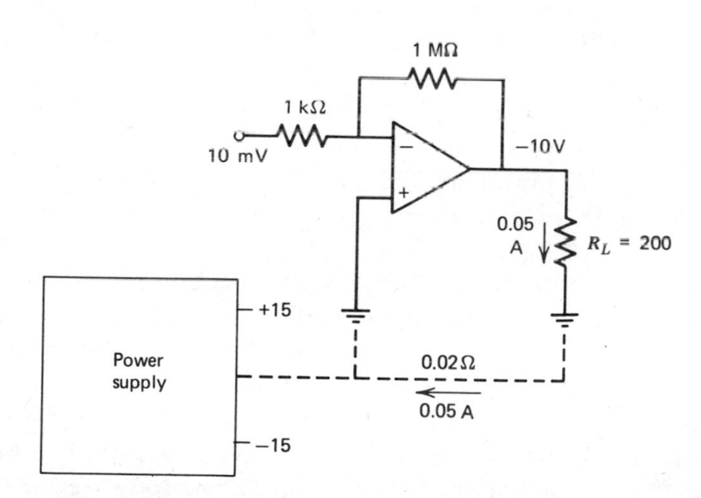

Figure 5-31. The effect of improper grounding. An input error of 1 mV (10%) is produced. Note that the resistance of a 10-cm length of No. 28 copper wire is about 20 mΩ.

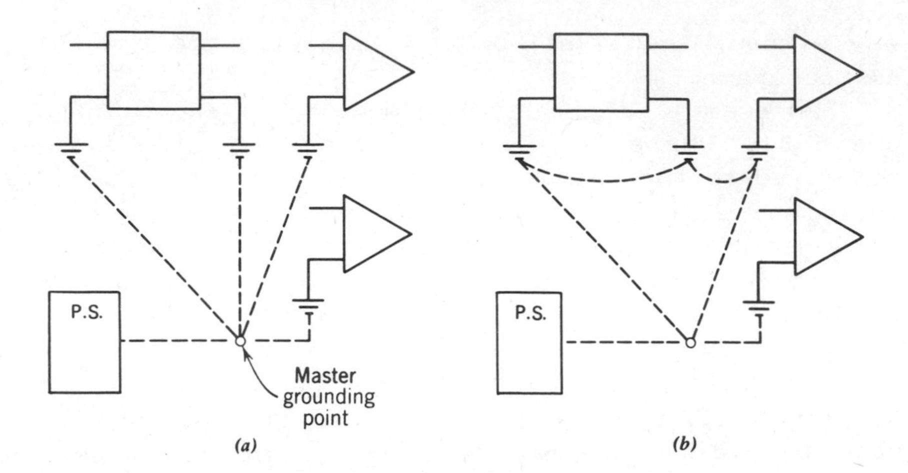

Figure 5-32. (*a*) A grounding system with all ground connections brought to a single point in a star configuration. (*b*) An improperly designed system, including undesirable ground loops.

If a single ground point is not practicable because of the large number of ground connections, a compromise can be made by using one master center point to which several heavy wires are connected, each leading to a satellite serving a segment of the instrument. For instance, one satellite might take all the ground connections for the reference inputs of op amps, another the ground terminals of any filters and voltage dividers, yet another the power supply common, and so on.

Even for properly grounded systems, interference from external fields may still be strong. If the impedances involved are very large, electrostatic fields can be especially troublesome since they consist of high voltages originating in very high impedance sources. Fields of hundreds of volts per millimeter may easily be present, either of atmospheric origin or simply from rubbing one's shoes on an insulating carpet. Electromagnetic fields, generated by transformers, fluorescent lamps, radio stations, and the like, induce currents in conductors, which in turn generate voltage drops. In general, one can assume that stray fields are apt to become important whenever impedances larger than about 1 MΩ are involved.

Interference from these sources can be minimized by observing the following rules, which are particularly significant in low-level measurements:

1. The AC power section of an instrument should be physically remote from the signal section.
2. Any line carrying high current, especially AC, should have the two insulated conductors twisted together so that electromagnetic fields largely cancel out.
3. High-input-impedance amplifiers should be connected to low-impedance sources, if possible.
4. Differential-input devices should be employed to eliminate common-mode noise.
5. Shielding should be incorporated where appropriate, as described in the following paragraphs.

The principle of differential operation is shown in Figure 5-33. The amplifier depicted responds to

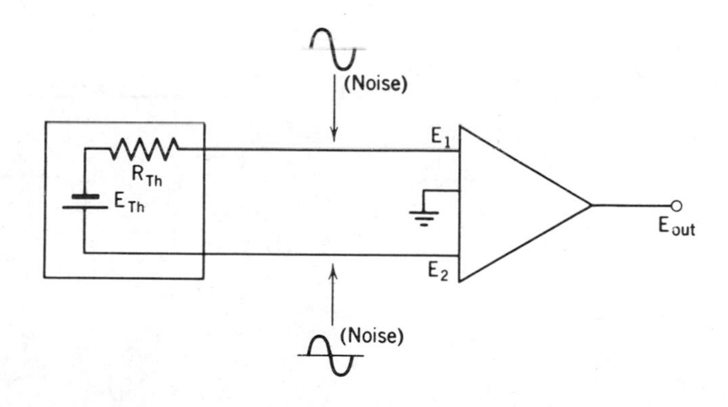

Figure 5-33. A differential system. Noise, assumed to be AC, enters into both inputs with the same phase (common mode) and is largely rejected.

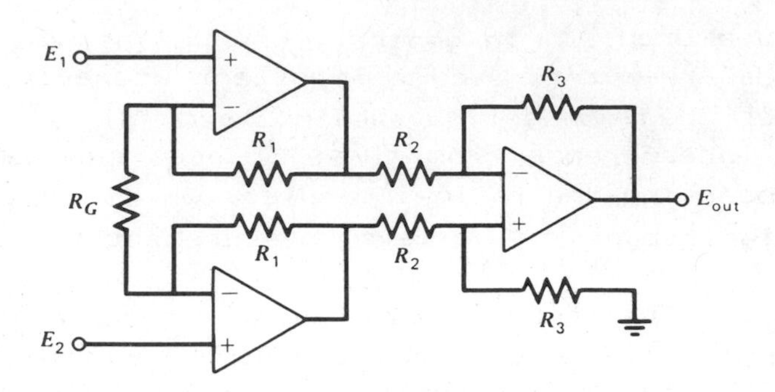

Figure 5-34. An instrumentation amplifier. Corresponding resistors must be closely matched.

the difference $E_2 - E_1$. If the noise is impressed equally on both inputs (common mode), it is largely subtracted out. The ability of an amplifier to eliminate such interferences is measured by its CMRR.

Specially designed amplifiers, such as the Analog Devices AD522 or the Burr-Brown 3626, have a well-balanced differential input with the very high CMRR of 110 dB, together with high input impedance. The feedback is internal, and the gain (1 to 1000) is controlled by a single external resistor. These units, called *instrumentation amplifiers*, are recommended for use

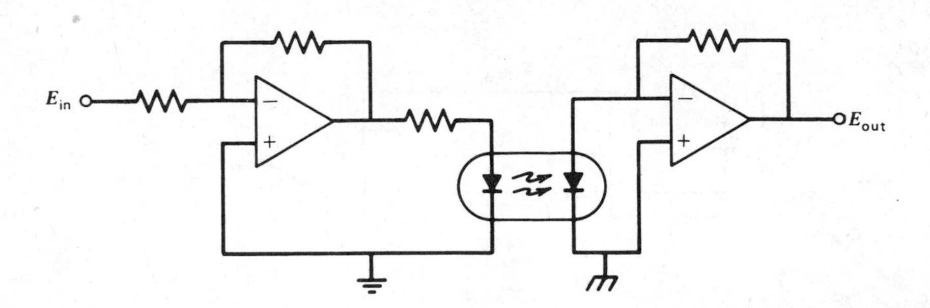

Figure 5-35. An isolation amplifier, in which the input and output parts are separated by an optical linkage. The two portions have distinct grounds and power sources.

in noisy environments and for floating measurements such as bridge readouts. The equivalent of an instrumentation amplifier can be assembled from three high-quality op amps (Figure 5-34). The necessary high-input impedance and high CMRR are ensured by having both inputs lead directly to the summing junctions of voltage followers. Circuit analysis shows that the gain is controlled by the single resistor R_g.

Another important device is the *isolation amplifier*, which is unique in that the input and output circuits are referenced to separate grounds (Figure 5-35). This type of amplifier is much used for reasons of safety in monitoring hospital patients.

Shielding consists of surrounding either the source of interference or the sensitive stages of an instrument by a metal enclosure, which can be made of copper sheet or screen. The enclosure provides a low-impedance path to ground for the various currents induced by stray fields. The power section of an instrument should be shielded separately from the other components (Figure 5-36). The shields must be connec-

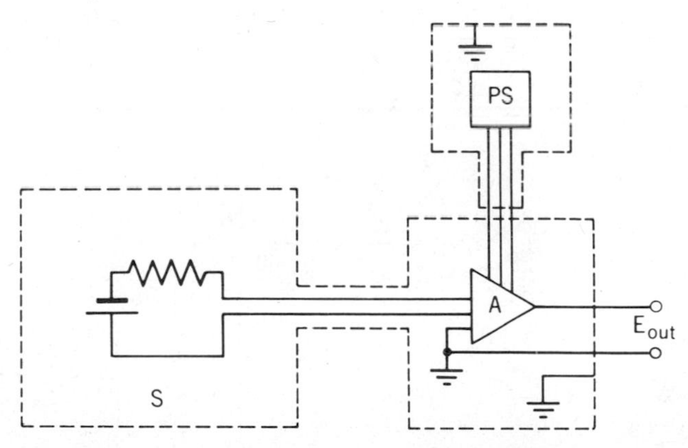

Figure 5-36. A shielding system. The source *S* and amplifier *A* have a common shield, connected at a single point into the ground system. The power supply *PS* has its own shield. The only ground connection between the two shielded segments is through the common line of the power-supply output.

ted to the master ground at only one point. Shielding should extend to any connecting wires or cables, in the form of braided metal covering ("shielded cables"). Shielding against magnetic fields requires special materials of high magnetic permeability, such as "mu-metal". This is inconvenient and expensive to fabricate, and is used only in special cases.

In conductors carrying very small currents, leakage to the shield may produce significant errors. In this case the *guard* method can be used effectively. The guard is a second shield connected to a voltage close to the signal level, rather than to ground. An example of a device for the measurment of very high resistances is given in Figure 5-37. The 300-V source

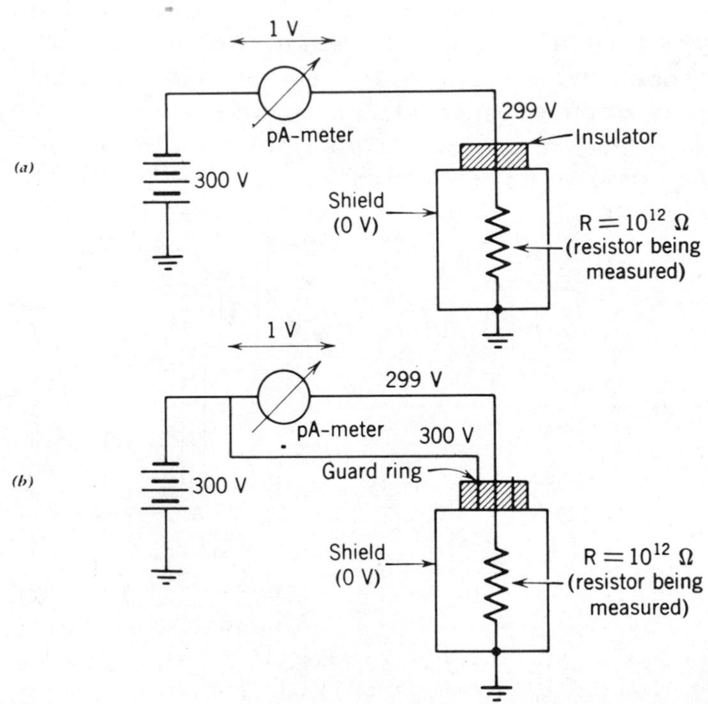

Figure 5-37. Devices for measuring very high resistances, using (*a*) a simple shield and (*b*) a guarded shield.

drives a current through R, to be measured by the meter. In (*a*) of Figure 5-37, the insulator between the shield and the measuring line is subjected to the entire 300 V, and surface leakage can cause significant errors. By contrast, in (*b*), the guard, placed between the conductor and shield, is at the potential of the conductor, so any leakage to ground will pass through the guard without affecting the measurement process. Many commercial measuring instruments are provided with separate connections so that a guard can be used when needed.

PROBLEMS

5-1. Redesign the circuit given in Figure 5-2, with amplifier 1 configured as a summer rather than a follower. Insert an additional op amp as a follower in the "Ref" line. Do you see any advantage to the new circuit with respect to frequency response?

5-2. Redesign Figure 5-3, using a summer as control amplifier.

5-3. Show by an impedance analysis of Figure 5-5 that the frequency of a Wien-bridge oscillator is indeed given by $1/(2\pi RC)$.

5-4. Redesign the circuit shown in Figure 5-7*b* to give an output frequency of 10 Hz. Include proper values for all capacitors and resistors, and pin numbers for interconnections. Reference to the manufacturer's literature will be necessary.

5-5. Explain the functioning of the phase shifter circuit shown in Figure 5-38.

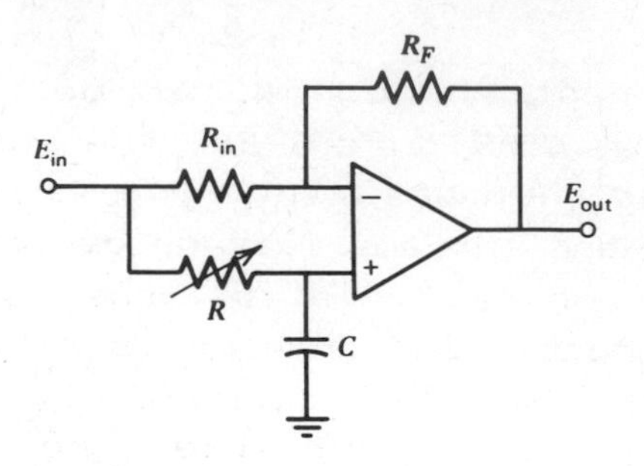

Figure 5-38. See Problems 5-5, 5-20, and 5-22.

5-6. Sketch the output waveforms for the circuit given in Figure 5-8 when the two inputs are a square wave and a sine wave of the same frequency. Consider the phase angles to be (a) 0°, (b) 90°, and (c) 180°. Use the sine wave for phase reference.

5-7. A description of a spring and weight system (Figure 5-39) must include the effect of friction, which is a force proportional to the velocity dx/dt andofthe acceleration d^2x/dt^2, as well as the spring force kx. The overall equation of motion becomes

$$m \frac{d^2x}{dt^2} = -kx - L \frac{dx}{dt}$$

Show how this equation can be simulated with op amps.

5-8. Design a circuit to solve the equation describing radioactive decay:

$$\frac{dx}{dt} = -kx$$

where x is the amount of active isotope remaining at time t seconds, and k is the decay constant. Let $k = 0.40 \times 10^{-6} s^{-1}$ and $x_0 = 2.0 \times 10^{18}$ atoms.

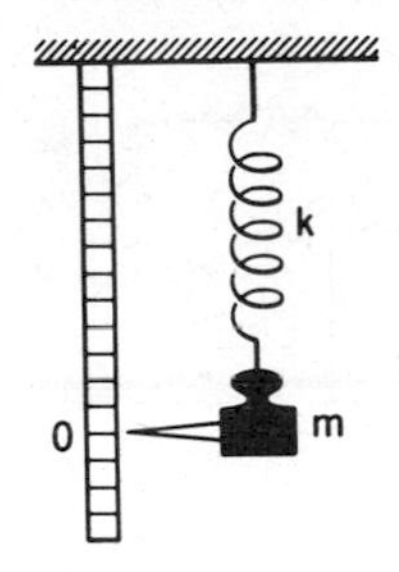

Figure 5-39. See Problem 5-7.

5-9. A function multiplier can be designed based on the algebraic identity $(x + y)^2 = x^2 + 2xy + y^2$, provided squaring circuits are available. Draw a block diagram for such a multiplier.

5-10. Show that a capacitor in the circuit shown in Figure 5-27 is charged to 63% of its steady-state value after an elapsed time of *RC* seconds.

5-11. Show that a divider can be constructed by placing a multiplier in the feedback of an op amp.

5-12. Two AC signals of frequencies 100 and 1000 Hz, both with amplitudes of 2 V, are multiplied together.

(a) Sketch the resulting waveform.
(b) Sketch the waveform that would appear if a bias of 2 V DC is added at the 100-Hz input.
(c) Do the same for 2-V bias applied to both inputs.

5-13. Explain the circuit shown in Figure 5-40, and give the overall equation. What use could you find for it?

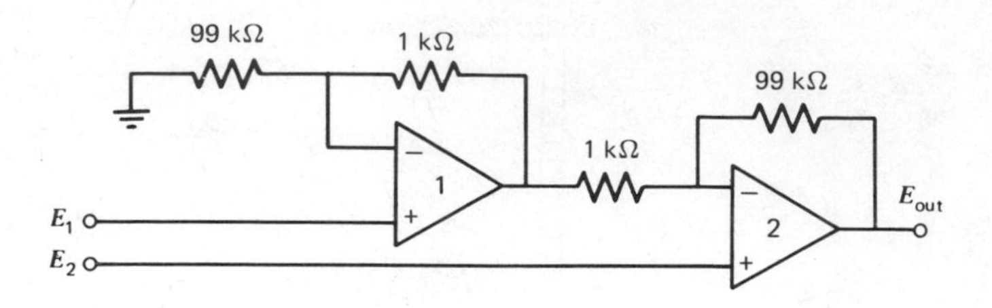

Figure 5-40. See Problem 5-13.

5-14. Two chemicals, *X* and *Y*, with overlapping optical absorption spectra can be determined simultaneously by means of the following equations, where *A* and *B* represent the absorbances of the mixed solution at two wavelengths and the *a* values are the absorptivities (specific constants of the system):

$$A = a_{11}x + a_{12}y$$

$$B = a_{21}x + a_{22}y$$

Show how the calculation for x and y can be carried out with op amp circuitry (as in an analog computer).

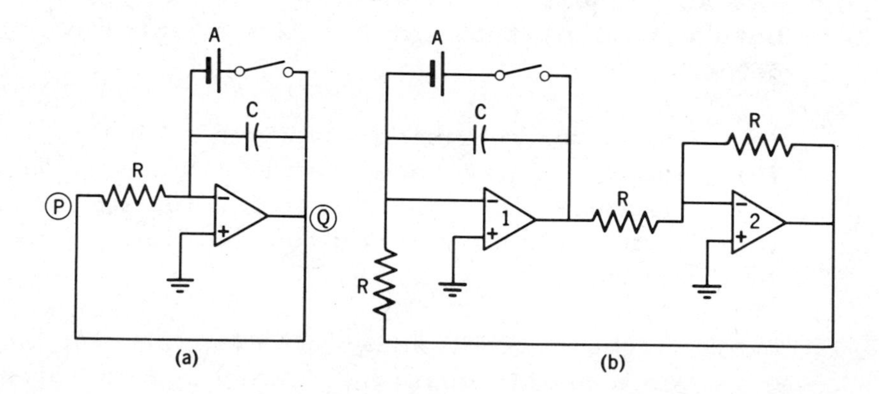

Figure 5-41. See Problem 5-15.

5-15. Find the differential equations solved by the circuits shown in Figure 5-41.

5-16. Set up analog circuits to solve the following differential equations:

(a) $(d^2x/dt^2) = -(6/7)(dx/dt) + (1/2)x$

(b) $(d^2x/dt^2) = -10.62x + 6$

5-17. Referring to the sine-wave generator shown in Figure 5-15*b*, indicate the output, in both phase and amplitude, for the following sets of initial conditions (in volts): (Assume that $R_1C_1 = R_2C_2 = 1$ s.)

(a) $V_1 = 0$, $V_2 = 2$
(b) $V_1 = 2$, $V_2 = 0$
(c) $V_1 = 2$, $V_2 = 2$
(d) $V_1 = 0$, $V_2 = 0$

5-18. Describe qualitatively the waveforms of the output voltage for the two extreme settings of the potentiometer R_Q in Figure 5-42.

5-19. Devise an op amp circuit for obtaining waveforms that are 180° out of phase.

5-20. Design in detail the circuit shown in Figure 5-29, using op amps (including the circuit given in Figure 5-38) together with a multiplier.

5-21. There is considerable resemblance between the phase-locked loop (PLL) and the lock-in amplifier. Explain the difference and the types of application to which each is best suited.

5-22. Design a phase-locked loop utilizing the synchronous (phase) detector shown in Figure 5-12, the phase shifter in Figure 5-38, and an 8038 IC used as a voltage-controlled square-wave generator.

5-23. Explain the meaning of the term "tracking filter."

5-24. The op amp connection shown in Figure 4-19 can be used to measure the difference in potential of two points in a circuit. In what way does this differ from an instrumentation amplifier?

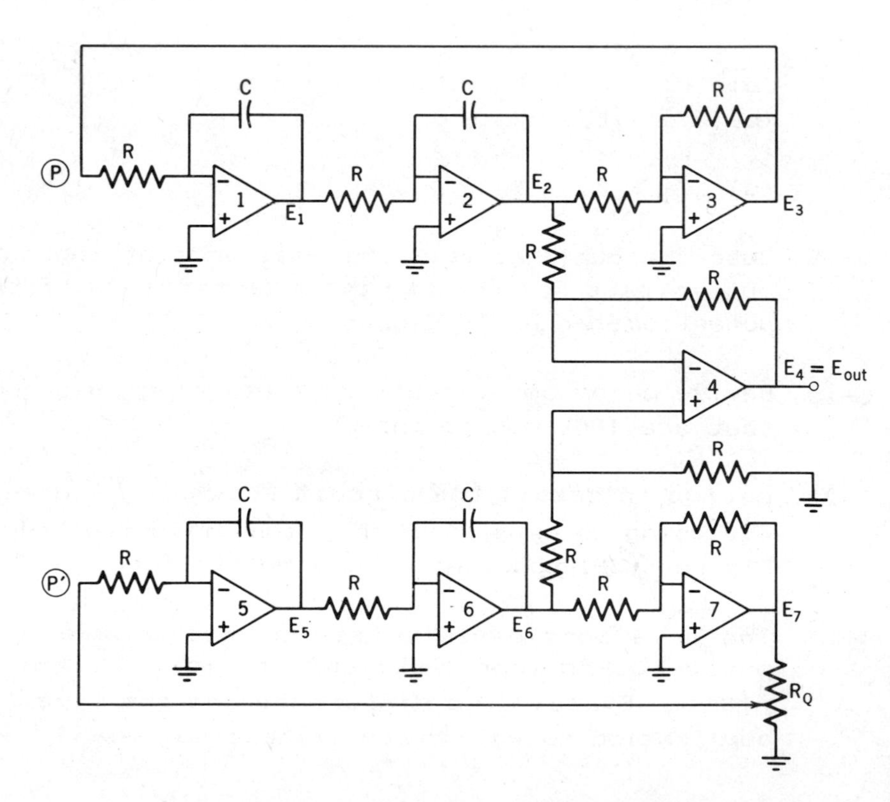

Figure 5-42. See Problem 5-18.

5-25. The optical link shown in Figure 5-35 is commonly called an *optical coupler*. Explain how it works and its chief areas of application.

5-26. Draw all the ground connections for the circuit described in Problem 5-9, including the power supply common and the line ground. Show your strategy for avoiding ground loops.

VI
CIRCUIT ANALYSIS

In this chapter we discuss once more, in some detail, the mathematical description of electronic circuits. We will start with the phasor method of representing steady-state AC currents. This duplicates previous material to some extent in order to give a unified presentation. We then continue with other aspects of circuit analysis.

COMPLEX VARIABLES AND PHASORS

A complex quantity is defined as a pair of numbers, one real, a, the other imaginary, b, which can be written as $\mathbf{W} = (a + jb)$, where $j = \sqrt{-1}$.* Observe that $j^2 = -1$, $j^3 = -j$, and $j^4 = 1$. When these properties are taken into account, most of the operations with real numbers are found to apply to complex numbers as well. In the final result of a computation with complex numbers, it is standard practice to collect together all the terms without j (the real part) and all terms containing j (the imaginary part) and to express them in the conventional format. For example,

* In electronics the operator j is used rather than i to avoid confusion. Boldface symbols denote vector quantities.

$$(5 + j3) + (5 - j) = (10 + j2) \tag{6-1}$$

and

$$(2 + j3) \cdot (2 - j3) = (13 + j0) = 13 \tag{6-2}$$

Of special interest in electronics is an alternative description of complex numbers--the *vectorial* or *polar* representation. Figure 6-1 shows that a vector of length A that makes an angle ϕ with the abscissa defines a point with the coordinates (a,jb)--in other words, a pair representing a complex number.

This equivalence of representation can be written as

$$\mathbf{W} = a + jb = A \angle \phi \tag{6-3}$$

The relationship between these notations is illustrated in Figure 6-2. There are advantages to both the polar and cartesian forms. In the cartesian form, numbers can

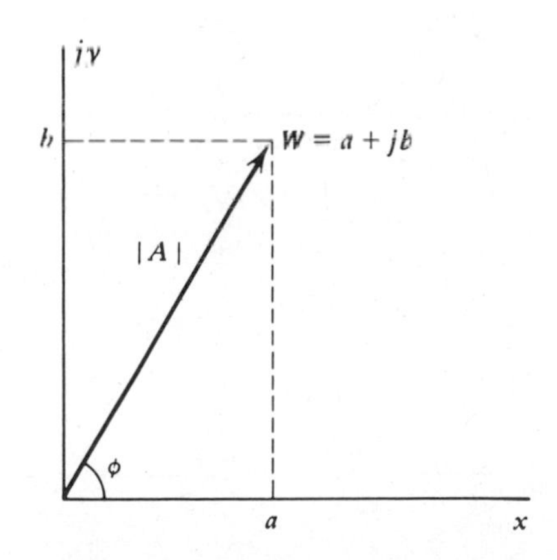

Figure 6-1. Polar notation for a complex number. The symbol **W** represents the end point of the vector of length A.

be added and subtracted with ease, whereas multiplication and division are particularly simple in the polar form. Cartesian addition and subtraction follow the pattern

$$(a + jb) + (c + jd) = (a + c) + j(b + d) \tag{6-4}$$

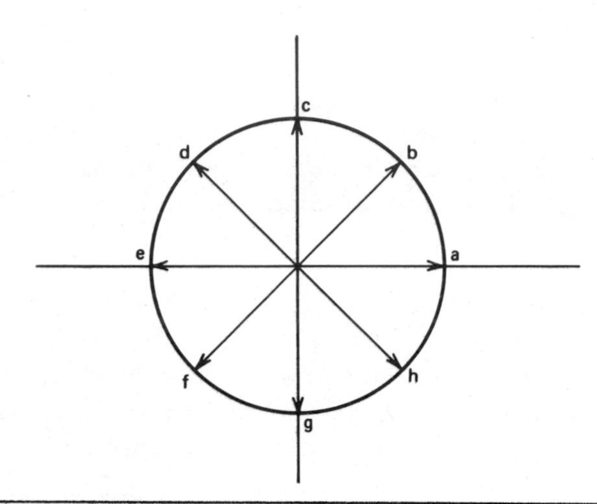

Vector	Cartesian	Polar (degrees)	Polar (radians)
a	$1 + j0$	$1 \angle 0^{\circ}$	$1 \angle 0$
b	$1/\sqrt{2} + j/\sqrt{2}$	$1 \angle 45^{\circ}$	$1 \angle \pi/4$
c	$0 + j$	$1 \angle 90^{\circ}$	$1 \angle \pi/2$
d	$-1/\sqrt{2} + j/\sqrt{2}$	$1 \angle 135^{\circ}$	$1 \angle 3\pi/4$
e	$-1 + j0$	$1 \angle 180^{\circ}$	$1 \angle \pi$
f	$-1/\sqrt{2} - j/\sqrt{2}$	$1 \angle 225^{\circ}$	$1 \angle 5\pi/4$
g	$0 - j$	$1 \angle 270^{\circ}$	$1 \angle 3\pi/2$
h	$1/\sqrt{2} - j/\sqrt{2}$	$1 \angle 315^{\circ}$	$1 \angle 7\pi/4$

Figure 6-2. Unit vectors in several alternative notations. The polar form can be written in terms of either degrees or radians.

with the expected sign changes in subtraction. For multiplication the formula is

$$(a + jb) \cdot (c + jd) = (ac - bd) + j(bc + ad) \quad (6\text{-}5)$$

which can easily be verified in the light of the known properties of j. Multiplication in polar form is written

$$(A_1 \angle \phi_1) \cdot (A_2 \angle \phi_2) = A_1 A_2 \angle (\phi_1 + \phi_2) \quad (6\text{-}6)$$

whereas division obeys the rule

$$\frac{A_1 \angle \phi_1}{A_2 \angle \phi_2} = \frac{A_1}{A_2} \angle (\phi_1 - \phi_2) \quad (6\text{-}7)$$

Observe that the angles add and subtract whereas the absolute values multiply or divide. For example, multiplication by ($2 \angle 60^\circ$) involves an increase in absolute value by a factor of 2, with a simultaneous rotation of 60° in the counterclockwise direction. Similarly, division by ($1 \angle 90^\circ$) means a clockwise rotation by 90° with no change in amplitude.

Division in cartesian notation is somewhat more complicated. The numerator and denominator are both multiplied by the *complex conjugate* of the denominator. (The complex conjugate is obtained by replacing j by $-j$ wherever it occurs.) Thus

$$\begin{aligned}
\frac{a + jb}{c + jd} &= \frac{(a + jb)(c - jd)}{(c + jd)(c - jd)} \\
&= \frac{(ac + bd) + j(bc - ad)}{c^2 + d^2} \\
&= \frac{ac + bd}{c^2 + d^2} + j\left(\frac{bc - ad}{c^2 + d^2}\right) \quad (6\text{-}8)
\end{aligned}$$

The interconversion between the two modes of representation can be performed by means of the following equalities:

$$x = A \cos \phi \tag{6-9}$$

$$y = A \sin \phi \tag{6-10}$$

$$x + jy = A(\cos\phi + j\sin\phi) \tag{6-11}$$

$$A = \sqrt{x^2 + y^2} \tag{6-12}$$

$$\phi = \arctan\ (y/x) \tag{6-13}$$

For example, the number $\mathbf{W} = (3 - j3)$ can be rewritten in polar form by obtaining the angle and the absolute value as follows:

$$\phi = \arctan\ (-3/3) = -45^\circ \tag{6-14}$$

$$A = \sqrt{3^2 + 3^2} = 3\sqrt{2} \tag{6-15}$$

This is a vector of length $3\sqrt{2}$ pointing 45° below the positive x-axis. The reverse transformation can be accomplished by calculating

$$x = 3\sqrt{2}\ \cos\ 45^\circ = 3\sqrt{2}(\sqrt{2}/2) = 3 \tag{6-16}$$

$$y = 3\sqrt{2}\ \cos\ (-45^\circ) = 3\sqrt{2}(-\sqrt{2}/2) = -3 \tag{6-17}$$

The combination of these two results gives the original number, $\mathbf{W} = (3 - j3)$.

To understand the utility of complex numbers in AC systems, consider a circuit at some specific frequency. At various points in the circuit, voltages will be present with various amplitudes and phases:

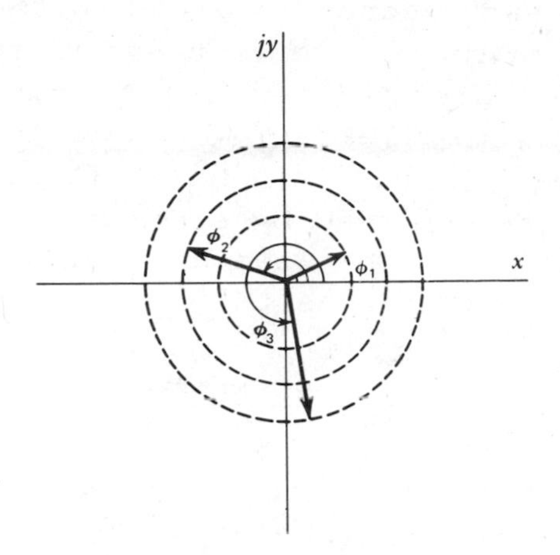

Figure 6-3. A set of rotating vectors at the moment when t = 0.

$$
\begin{aligned}
&A_1 \sin(\omega t + \phi_1) \\
&A_2 \sin(\omega t + \phi_2) \\
&\quad \cdots \\
&A_n \sin(\omega t + \phi_n)
\end{aligned}
\tag{6-18}
$$

Each of these sinusoidal functions can be described by a rotating vector of angular velocity ω. The entire cluster of vectors (Figure 6-3) rotates at the same velocity and with fixed relative orientations as determined by the phase angles. In addition, the vectors describing the currents in the same circuit also rotate with this velocity.

It is advantageous to represent the collection of vectors with their relative orientations as of a certain moment in time, commonly taken as t = 0. It can be regarded as an instantaneous photograph of the cluster. Such stationary vectors are called *phasors*.

In this notation, the angle between the positive abscissa and each vector is given by the phase angle, whereas its length represents the corresponding amplitude. All the rules for manipulation of complex numbers also apply to phasors. The great utility of this representation is that trigonometric computation with sinusoids can be replaced by algebra.

The relationship between the sinusoidal and phasor notations can be stated as

$$A \sin (\omega t + \phi) \longrightarrow A \angle \phi \tag{6-19}$$

in which the arrow indicates that the relationship is one of transformation (the *phasor transform*), rather than of equality, since the right-hand side does not contain time as a variable.* Both voltages and currents can be represented as phasors. Amplitudes are conventionally expressed as the RMS values, which simplifies many computations.

IMPEDANCES AND TRANSFER COEFFICIENTS IN COMPLEX FORM

Since voltages and currents in phasor form are complex numbers, it follows that their ratios, including impedances and transfer coefficients, will also be complex. This feature, far from obscuring numerical manipulations, makes them more convenient and less likely to produce errors.

The impedance of combinations of resistors, capacitors, and inductors can be written directly as follows:

* Although the loss of time as a variable might appear to involve also a loss of information, this is not the case, as phasors are applicable only to steady-state signals, where time is irrelevant.

1. Impedance of resistors:

$$\mathbf{Z}_R = R \angle 0 \tag{6-20}$$

2. Impedance of capacitors:

$$\mathbf{Z}_C = \frac{1}{j\omega C} = \frac{1}{\omega C} \angle -90^\circ \tag{6-21}$$

3. Impedance of inductors:

$$\mathbf{Z}_L = j\omega L = \omega L \angle +90^\circ \tag{6-22}$$

4. Series impedances:

$$\mathbf{Z}_T = \mathbf{Z}_1 + \mathbf{Z}_2 + \mathbf{Z}_3 + \cdots \tag{6-23}$$

5. Parallel impedances:

$$1/\mathbf{Z}_T = 1/\mathbf{Z}_1 + 1/\mathbf{Z}_2 + 1/\mathbf{Z}_3 + \cdots \tag{6-24}$$

The impedance reduces to resistance for nonreactive circuits or for DC signals. It can be considered to be an extension of the notion of resistance to a new dimension, where not only amplitudes but also phases are specified. This extension is necessary since the use of the scalar quantities where vectors should be employed can lead to errors.

To illustrate the types of computations that can be made with complex impedances, consider the circuit shown in Figure 6-4*a*. The impedance of the series combination is given by

$$\mathbf{Z}_T = R + \frac{1}{j\omega C} \tag{6-25}$$

Suppose that a potential of 100 V is applied to the circuit and that the frequency is such that the impedance of the capacitor is 1000 Ω. "Common sense" would suggest that the total impedance would be the sum $R + Z_C = 1000 + 1000 = 2000\ \Omega$, which would predict a current of 100/2000 = 0.050 A. Actually, a meter would show a larger reading, 0.071 A. The error in the calculation results from treatment of $\mathbf{Z}_C$ as if it were a resistance rather than an impedance. The correct computation requires obtaining the complex impedance:

$$\mathbf{Z}_T = R + \frac{1}{j\omega C} = (1000 - j1000) \qquad (6\text{-}26)$$

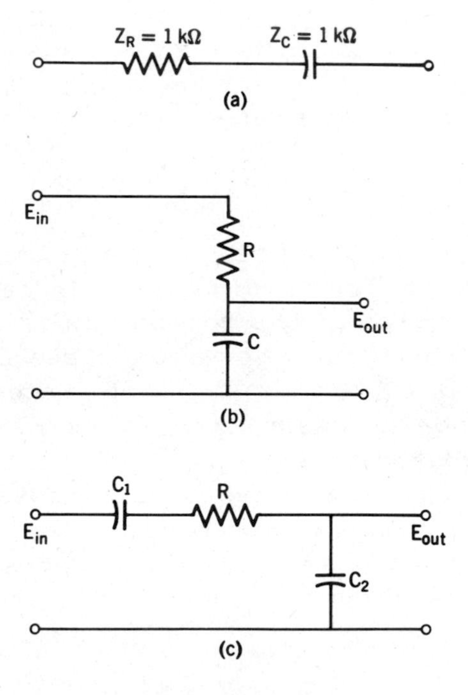

Figure 6-4. *RC* circuits illustrating the phasor method.

which can be rewritten in the polar form as

$$\mathbf{Z}_T = \sqrt{1000^2 + 1000^2} \angle \arctan\left(-\frac{1000}{1000}\right) \tag{6-27}$$
$$= 1414 \angle -45°$$

The current can then be computed as

$$\mathbf{I} = \frac{\mathbf{E}}{\mathbf{Z}_T} = \frac{100 \angle 0}{1414 \angle -45°} = 0.071 \angle +45° \tag{6-28}$$

Since an AC ammeter cannot disinguish phase angles, it will indicate simply 0.071 A, which is quite different from the tentative result, 0.050 A.

Another unexpected result is seen in the voltages across the resistor and capacitor in the same circuit. Common sense would predict that the voltage across the whole circuit would distribute itself equally between the two impedances as 50 and 50 V. If the voltages are measured, however, they turn out to be 71 and 71 V. In other words, two 71 V signals add to give a 100 V sum. This is caused by the fact that the voltages across the two components are 90° out of phase. By trigonometric manipulation, one can show that in fact the two out-of-phase voltages do add to 100 V total, but with the use of phasors the result comes more easily:

$$\mathbf{E}_R = \mathbf{I}\mathbf{Z}_R = (0.071 \angle 45°)(1000 \angle 0°)$$
$$= 71 \angle 45° = (71/\sqrt{2})(1 + j) \tag{6-29}$$

$$\mathbf{E}_C = \mathbf{I}\mathbf{Z}_C = (0.071 \angle 45°)(1000 \angle -90°)$$
$$= 71 \angle -45° = (71/\sqrt{2})(1 - j) \tag{6-30}$$

$$\mathbf{E}_{\text{total}} = \mathbf{E}_C + \mathbf{E}_R$$

$$= (71/\sqrt{2})(1 + j) + (71/\sqrt{2})(1 - j)$$

$$= 142/\sqrt{2} = 100 \text{ V} \tag{6-31}$$

Phasors are also applicable to transfer coefficients and gains. The resulting expressions are complex and hence contain information about both the amplitudes and the phase relationships between input and output of a device. For example, in Figure 6-4*b* the same pair of impedances shown in (*a*) is now connected as a voltage divider. The transfer coefficient $\mathbf{E}_{\text{out}}/\mathbf{E}_{\text{in}}$ is given by the voltage-divider equation:

$$\frac{\mathbf{E}_{\text{out}}}{\mathbf{E}_{\text{in}}} = \frac{\mathbf{Z}_C}{\mathbf{Z}_R + \mathbf{Z}_C} = \frac{1/j\omega C}{(1/j\omega C) + R} = \frac{1}{1 + j\omega RC} \tag{6-32}$$

From this we can deduce the behavior of the circuit at various frequencies. At very high frequency, the transfer coefficient is zero, whereas for DC it becomes unity. Thus the circuit is a low-pass filter.

A slightly more complicated circuit is shown in Figure 6-4*c*. By the same method, the transfer coefficient can be shown to be

$$\frac{\mathbf{E}_{\text{out}}}{\mathbf{E}_{\text{in}}} = \frac{1/j\omega C_2}{1/j\omega C_1 + R + 1/j\omega C_2} \tag{6-33}$$

It is left as an exercise for the reader to rewrite this expression in the conventional format (a + *j*b).

In another application, phasors can be used to determine the frequency response of op amp circuits.

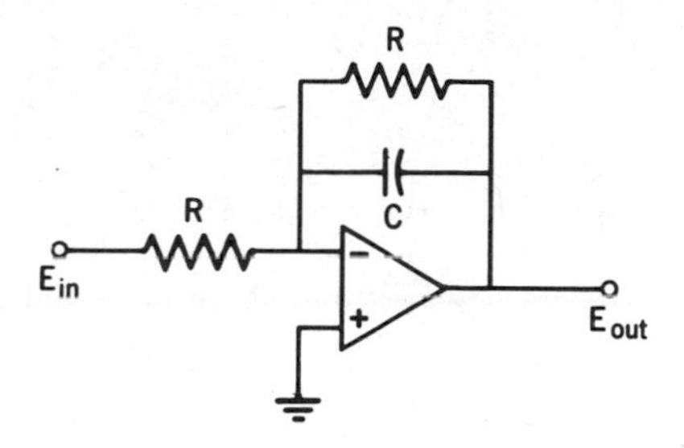

Figure 6-5. An op amp circuit with a complex feedback impedance.

Steady-state sinusoidal voltages are assumed. The familiar relationship $\mathbf{E}_{out}/\mathbf{E}_{in} = -\mathbf{Z}_f/\mathbf{Z}_{in}$ is applicable. As an example, in the circuit shown in Figure 6-5 the complex gain is

$$\frac{\mathbf{E}_{out}}{\mathbf{E}_{in}} = -\left(\frac{1}{1/R + j\omega C}\right)\left(\frac{1}{R}\right) = -\frac{1}{1 + j\omega RC} \qquad (6\text{-}34)$$

which is identical (except for sign inversion) to Eq. 6-32 for a low-pass filter.

EQUIVALENT CIRCUITS

Complicated electrical circuits can generally be represented by simpler equivalent circuits that perform in an identical manner. In principle, all linear circuits, including those containing transistors and op amps, can be reduced to equivalent circuits consisting of no more than five elements: ideal voltage and current sources, resistors, capacitors, and inductors. An ideal voltage source has zero internal impedance, whereas for an ideal current source the internal impedance is infinite.

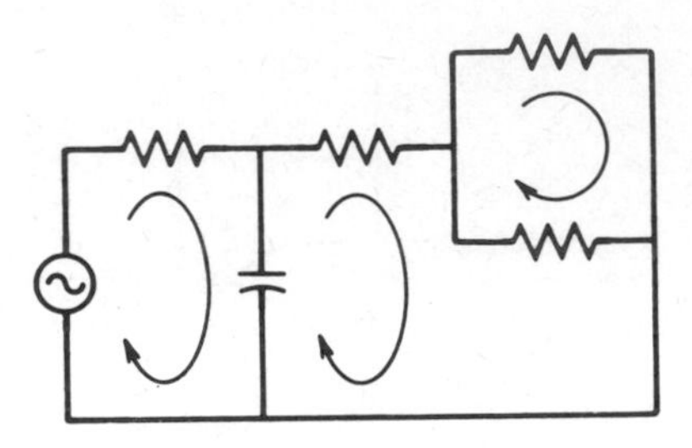

Figure 6-6. A circuit showing three loops (arrows) and four nodes (letters).

A circuit consists of a closed path for energy flow. It can contain one or several subsections called *loops*. The point where two or more loops meet is a *node*. Figure 6-6 can serve to clarify these definitions.

There are several very useful concepts that help in determining the voltages and currents in the circuit. Instead of a complete mathematical analysis, this can be accomplished by establishing an equivalent circuit that allows one to eliminate whatever information is not actually necessary for the purpose at hand. A simpler circuit with identical electrical behavior is used for calculations in place of the original. Occasionally the hardware suggested by the equivalent circuit can be physically substituted for the original, but this seldom can be done without the loss of some important features.

A collection of impedances between any two points in a circuit can be simplified by using the conventional rules of parallel and serial combination. These relationships are equally valid if the impedances are represented by phasors. We shall, for simplicity, discuss them in terms of resistors. An example is given in Figure 6-7.

In a similar way ideal DC voltage sources connected in series can be replaced by a single equivalent source. Its value is the algebraic sum of the component voltages, as seen in Figures 6-7*d* to 6-7*f*.

THÉVENIN AND NORTON EQUIVALENTS

The question arises as to the form of the simplest possible equivalent circuit when both impedances and sources are present. A principle called the *Thévenin theorem* states that a circuit composed of voltage or current sources of a single frequency, together with resistors, capacitors, and inductors, can be reduced to a single complex impedance in series with a voltage source. The Thévenin equivalent circuit so formed has only two terminals; thus only the behavior of the circuit between two points can be described by it. All information about other parts of the circuit is

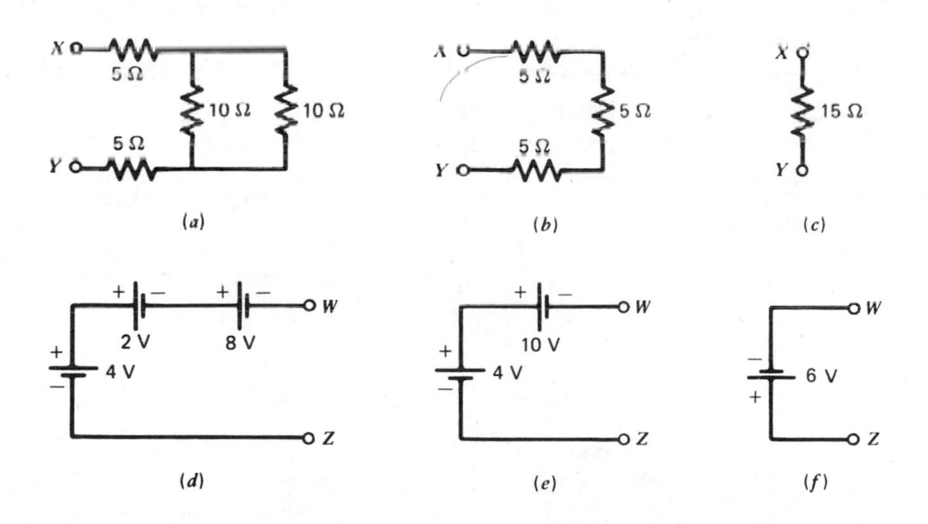

Figure 6-7. Equivalent circuits. An ohmmeter connected to points *X* and *Y* would detect no difference between (*a*), (*b*), and (*c*). Similarly, a voltmeter connected at *W* and *Z* would show identical readings from (*d*), (*e*), and (*f*).

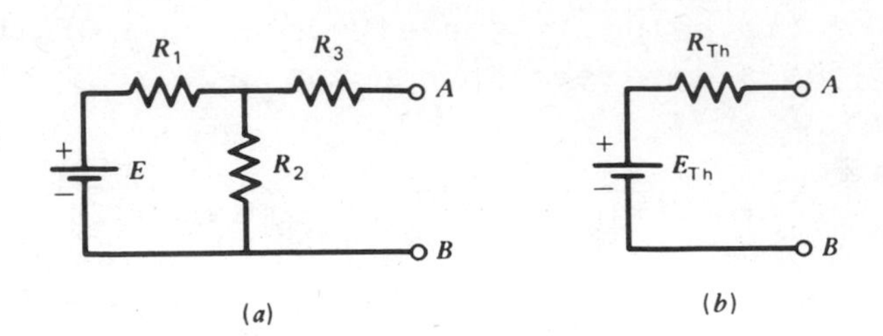

Figure 6-8. The Thévenin equivalent (*b*) of the circuit in (*a*).

lost in the process of simplification.

An example of Thévenin procedure is shown in Figure 6-8, where again only resistors and a voltage source are present. The general method for effecting the simplification consists first in the application of the rules for combination of impedances. In the next step, we can make use of the fact that the Thévenin equivalent must be valid for both extreme conditions, short-circuited and open-circuited.

Take, for instance, the circuit of a Wheatstone bridge, as in Figure 6-9. It is not immediately evident how the Thévenin equivalent can be obtained in this case. The procedure outlined in the preceding

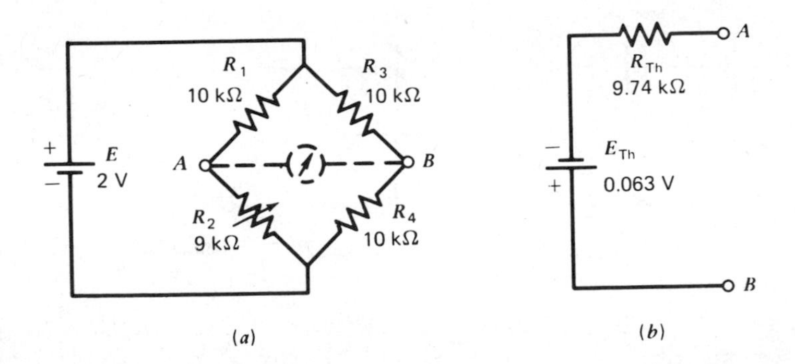

Figure 6-9. A Wheatstone bridge (*a*) and its Thévenin equivalent (*b*).

paragraphs, however, can be of use. In open circuit, the bridge acts as a pair of voltage dividers, generating the voltages

$$V_A = 2\left(\frac{9000}{9000 + 10{,}000}\right) = 0.947\ \text{V} \tag{6-35}$$

$$V_B = 2\left(\frac{10{,}000}{10{,}000 + 10{,}000}\right) = 1.000\ \text{V} \tag{6-36}$$

Consequently, under open-circuit conditions, the voltage between *A* and *B* is given by $V_A - V_B = -0.063$ V. Since the open-circuit current is zero, the drop of voltage across the Thévenin resistance R_{Th} is also zero, which makes $V_A - V_B$ equal to E_{Th}.

The value of R_{Th} can be calculated by considering the bridge with a short circuit between points *A* and *B*. In this case R_1 and R_3 are placed in parallel, giving an equivalent resistance of 5 kΩ. Similarly, R_2 and R_4 give a combined resistance of 4.74 kΩ. The Thévenin resistance is then the sum of these two values, or 9.74 kΩ. The complete equivalent circuit is shown in Figure 6-10*b*. As the bridge is brought to balance by increasing R_2, the Thévenin equivalent tends toward its limit, $E_{Th} = 0$ V, and $R_{Th} = 10$ kΩ.

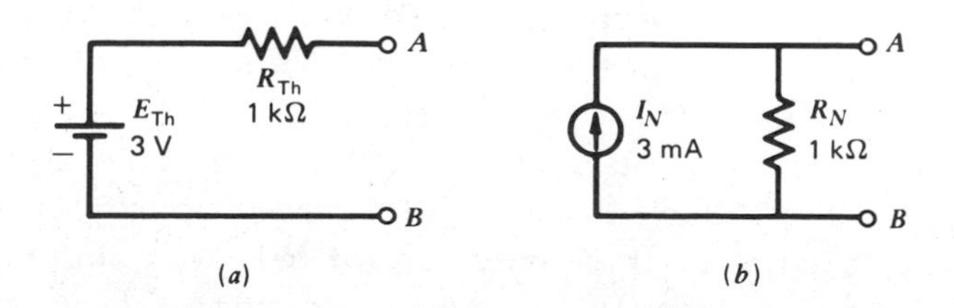

Figure 6-10. A Thévenin circuit (*a*) with its Norton counterpart (*b*).

Thévenin equivalent circuits are often used to describe energy sources such as power supplies and oscillators. The theorem remains valid for an AC source of a single frequency, in which case R_{Th} is replaced by an impedance, $\mathbf{Z}_{Th}$. The dependence of the output voltage on the load is now a simple matter and can be computed by the voltage divider equation:

$$\boldsymbol{E}_{\text{out}} = \frac{\boldsymbol{E}_{\text{Th}} \cdot \boldsymbol{Z}_{\text{load}}}{\boldsymbol{Z}_{\text{load}} + \boldsymbol{Z}_{\text{Th}}} \tag{6-37}$$

For a given network the Thévenin parameters can be measured by observations entirely external to the circuit itself. This is done by (1) measuring the voltage at the terminals under no-load conditions (i.e., without drawing current) and then (2) determining the load resistor that will cause the output voltage to drop to half its no-load value. The voltage measured in step 1 is the Thévenin voltage, whereas the resistance in step 2 is the Thévenin impedance. By another procedure, applicable to known circuits, the Thévenin parameters can be computed by calculation of the no-load output voltage and the short-circuit output current. The Thévenin impedance is the ratio of these two quantities.

Another type of equivalent circuit can be obtained by combining a *current source* with a *parallel* impedance. This is the *Norton equivalent* circuit. The correspondence between Thévenin and Norton circuits can be understood by considering the short- and open-circuit procedures. In the example given in Figure 6-10, the short-circuit current in the Thévenin diagram is evidently 3/1000 = 3 mA. Inspection of the Norton counterpart shows that the short-circuit current must be equal to I_N, namely, 3 mA. On removal of the short circuit, the voltage output must be equal to E_{Th} = 3 V. In the Norton circuit this requires that R_N = 3 V/3 mA

= 1000 Ω. This completes the calculation.

It is interesting to consider how these equivalent circuits can be applied to the op amp. Figure 6-11*a* shows an amplifier in terms of its Thévenin equivalent input and output circuitry. The symbol R_s designates the internal or Thévenin resistance of the source of voltage E_{in}. The potential source E_{Th} is a voltage that is *A* times the input voltage, delivered to the output terminal through a series resistance R_{out}. If we were dealing with a current-to-voltage converter, a Norton equivalent might be more appropriate at the input, as shown in Figure 6-11*b*. These figures suggest that the input and output sections of the amplifier are somehow separated and can be treated independently. This is in agreement with our previous notions about op amps, namely, that we do not need to

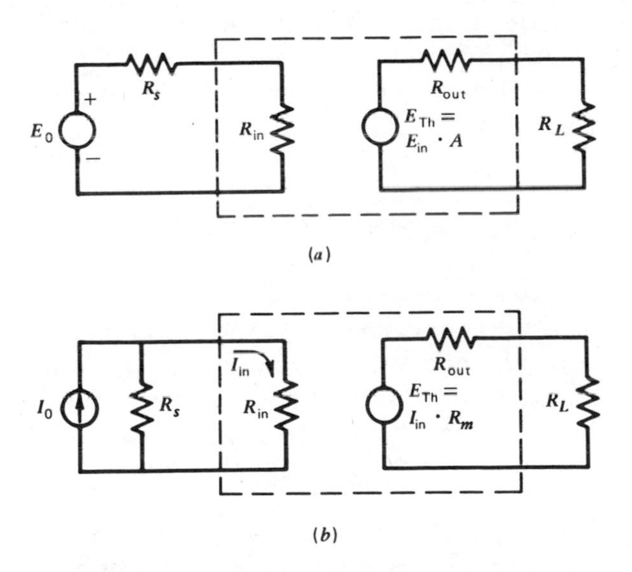

Figure 6-11. Equivalent circuits of op amps: (*a*) an inverter, based on Thévenin equivalents at both input and output; (*b*) a current-to-voltage converter with a Norton input; R_m is the "mutual resistance," namely, the ratio E_{out}/I_{in}.

know about the details of internal circuitry.

The relationship between the Norton and Thévenin circuits illustrates a general proposition in electronics, called the *duality principle*, which states that identical mathematical forms describing circuits can be obtained if we interchange voltage with current, inductance with capacitance, and resistance with conductance.* In the present case, identical behavior can be obtained by either a current source and parallel conductance or a voltage source with a series resistance.

Both types of equivalent circuit are defined equally for DC and for AC of a single frequency. If more than one type of source is present, the situation becomes somewhat more complicated. One must now invoke the *superposition principle* (see also Chapter 14), which states that when a *linear* system contains more than one energy source, the total effect is the sum of the individual actions of the several sources. This principle can be extended to simultaneously include sources at various frequencies.

KIRCHHOFF'S LAWS

The behavior of electronic circuits can be described mathematically by writing conditions associated with the various nodes and loops, as defined in Figure 6-6. The equations are based on two laws enunciated by Kirchhoff, both of which can be derived from the laws of conservation of charge and energy:

1. The algebraic sum of all currents meeting at a node (a junction) is zero. We shall take the currents entering the node as positive and those leaving as negative.

* Compare, for example, the differential equations $E = L(dI/dt)$ and $I = C(dE/dt)$

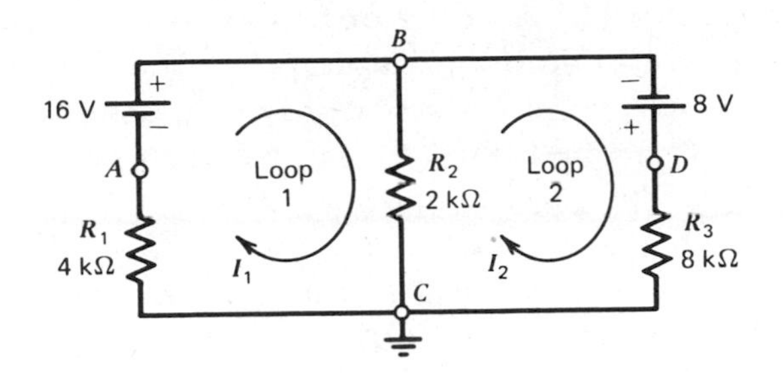

Figure 6-12. A two-loop circuit to illustrate Kirchhoff's laws.

2. The algebraic sum of all the voltages around a loop is zero. A given direction (such as clockwise) must be chosen for the entire circuit. We shall use a positive sign for increasing voltages as we proceed around a loop.

The utility of Kirchhoff's laws lies in providing a general method of calculating voltages and currents in a circuit. It is often desirable first to simplify the problem by constructing equivalent circuits, but this is not always necessary, since the laws can be applied directly even in cases of considerable complexity.

Consider the example given in Figure 6-12, in which there are two unknown currents, I_1 and I_2. Let us select point C to be ground, the point of zero voltage. Three unknown voltages are now present, E_A, E_B, and E_D. From the second law, we can write, starting from point C for both loops

$$-V_{R1} + 16 - V_{R2} = 0 \tag{6-38}$$

$$-V_{R2} + 8 - V_{R3} = 0 \tag{6-39}$$

The negative signs come from the convention that increasing voltages are written as positive. If we follow the positive current (arrows), the voltages decrease across each resistor. The energy sources (batteries) are shown with positive potential since we cross them from (-) to (+). We can now replace the voltages in Eqs. 6-38 and 6-39 with the values from Ohm's law. Note that in the case of R_2, we must consider both currents I_1 and I_2, as required by the superposition principle. This gives us

$$-4000I_1 + 16 - 2000(I_1 - I_2) = 0 \tag{6-40}$$

$$-2000(I_2 - I_1) + 8 - 8000I_2 = 0 \tag{6-41}$$

These equations can be rewritten as

$$6000I_1 - 2000I_2 = 16 \tag{6-42}$$

$$-2000I_1 + 10000I_2 = 8 \tag{6-43}$$

Eqs. 6-42 and 6-43 can be solved simultaneously to give I_1 = 3.14 mA and I_2 = 1.43 mA. The voltages at the three lettered points are V_A = -12.6, V_B = 3.4, and V_D = 11.4 V. The choice of loops is arbitrary; for example, one could use the loop ***ABDCA***. The only restriction is that all components must be included in at least one loop.*

Phasor Form of the Kirchhoff Laws

In systems using AC currents, it is advantageous to express Kirchhoff's laws in phasor notation. In this case (1) the sum of phasor currents at a node must

* By the duality principle, there is a second method of solving for the circuit values, by writing current conditions at the nodes, rather than the potentials around the loops. This makes nodes and loops duals with respect to each other.

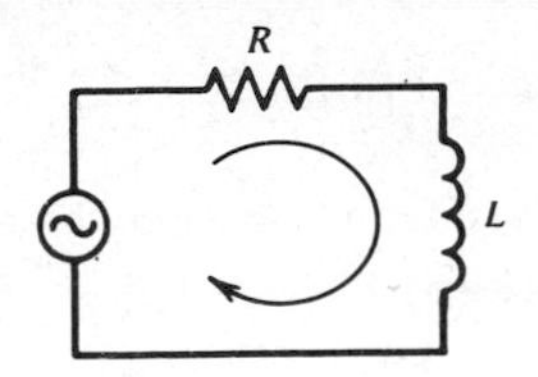

Figure 6-13. A phasor application of Kirchhoff's second law.

be zero and (2) the sum of phasor voltages around a loop must be zero.

Let us consider the circuit shown in Figure 6-13, to illustrate the sign conventions used in this application. By convention, we indicate by (+) and (−) the polarity of the reference voltage at a fixed instant. As before, the components that decrease the potential are said to give negative contributions. For a clockwise scan,

$$\mathbf{V} - \mathbf{I}R - j\omega L\mathbf{I} = 0 \tag{6-44}$$

from which it follows that

$$\mathbf{I} = \frac{\mathbf{V}}{R + j\omega L} = \frac{\mathbf{V}(R - j\omega L)}{R^2 + \omega^2 L^2} \tag{6-45}$$

This can be rewritten in the form

$$\mathbf{I} = \mathbf{V}\left(\frac{R}{R^2 + \omega^2 L^2} - j\frac{\omega L}{R^2 + \omega^2 L^2}\right) \tag{6-46}$$

or, in polar form,

$$\mathbf{I} = \mathbf{V}\left(\frac{1}{\sqrt{R^2 + \omega^2 L^2}} \angle \arctan \frac{-\omega L}{R}\right) \tag{6-47}$$

In other words, the current has the RMS amplitude of $V/\sqrt{R^2 + \omega^2 L^2}$ and a phase angle with respect to the voltage of arctan $(\omega L/R)$, which makes it possible to compute the voltages at various points.

PROBLEMS

6-1. Convert the following to their equivalents in polar coordinates:

(a) $(6 + j6)$
(b) $(-6 + j6)$
(c) $(6 - j6)$
(d) $(-6 - j6)$
(e) $(0 + j10)$
(f) $(10 + j0)$
(g) $(0 + j0)$

6-2. Convert the following to their equivalents in cartesian coordinates:

(a) $(10 \angle 60^\circ)$
(b) $(10 \angle -60^\circ)$
(c) $(0 \angle 60^\circ)$
(d) $(10 \angle 0^\circ)$
(e) $(0 \angle 0^\circ)$

6-3. Perform the following operations:

(a) $(1 + j2) + (6 - j7)(2 + j2)$
(b) $[(1 + j)/(1 - j)](2 - j2)$
(c) $(3 + j3)^3$
(d) $[(1 + j2)/j]^2$

6-4. Convert to polar coordinates the expressions in Problem 6-3 and carry out the indicated operations.

6-5. Write the complex conjugates for each expression in Problem 6-1.

6-6. An AC signal of 150-V RMS at 10,000 rad/s is generated by an oscillator with an output (Thévenin) impedance of 10 Ω. The signal is applied directly to a 100-pF capacitor.

(a) What is the frequency of the signal in hertz?

(b) Compute the current and power dissipation in the capacitor.

6-7. For the circuits shown in Figure 6-14, calculate the voltages across and the currents through the load for R_L = 100, 1000, and 10,000 Ω. Show that the difference between sources of current and voltage is simply a matter of impedance ratios.

6-8. Repeat the procedure followed in Problem 6-7 with the circuits in Figure 6-15.

6-9. Find the Thévenin equivalent of the circuits shown in Figure 6-16.

6-10. Consider the circuit shown in Figure 6-17, consisting of an ideal current source and a parallel resistor R_n. Find its Thévenin equivalent.

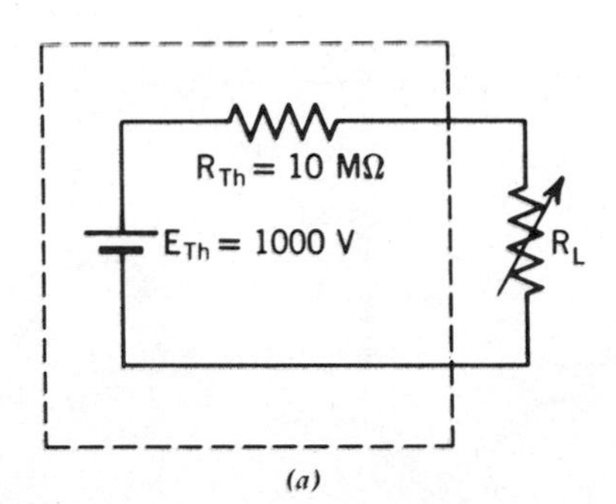

(a)

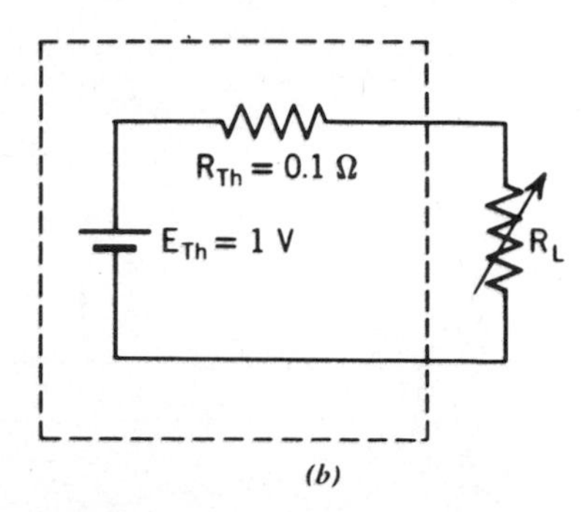

(b)

Figure 6-14. See Problem 6-7.

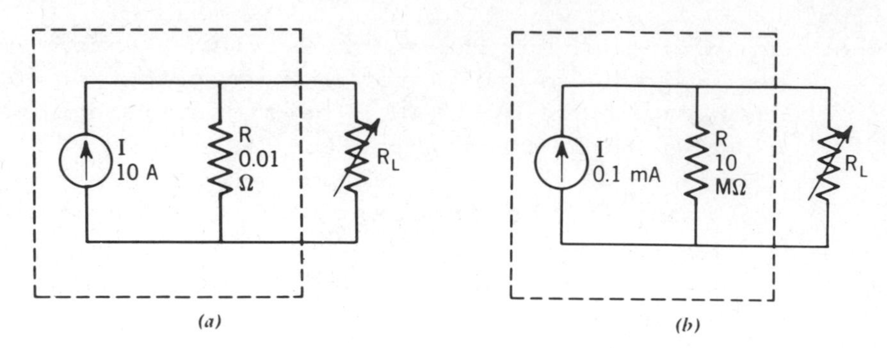

Figure 6-15. See Problem 6-8.

6-11. (a) In the circuit shown in Figure 6-18, with the load connected, compute the input resistance and the voltage, current, and power attenuations (in decibels). (b) What will happen to the input impedance if a number of similar "black boxes" are connected in series between source and load?

6-12. By application of the Kirchhoff laws, calculate the currents and voltages in the circuit shown in Figure 6-8*a*, where $R_1 = R_2 = R_3 = 100\ \Omega$, and $E = 2$ V. The points *A* and *B* are connected together.

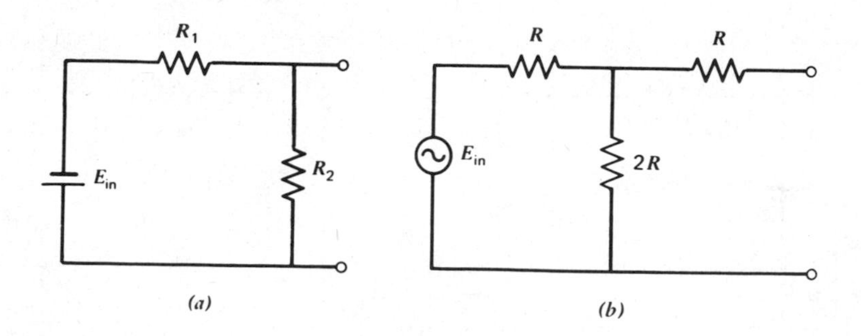

Figure 6-16. See Problem 6-9.

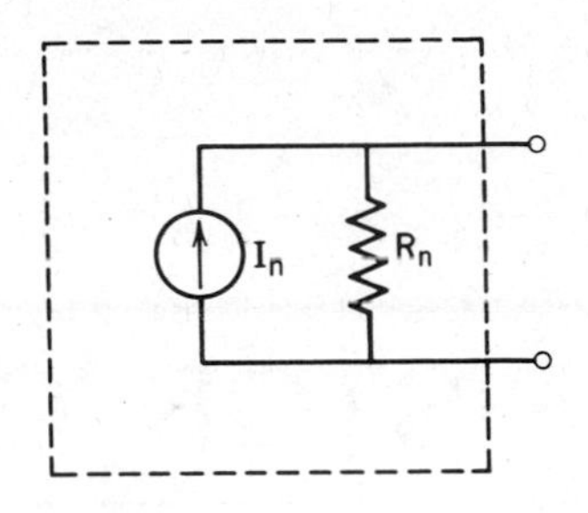

Figure 6-17. See Problem 6-10.

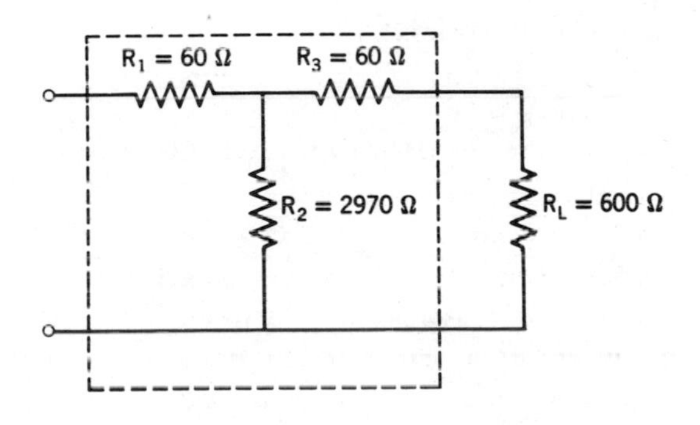

Figure 6-18. See Problem 6-11.

VII
TRANSISTOR CIRCUITS

In Chapter III we examined the structure and the physical principles on which bipolar and field-effect transistors (FETs) operate. Now we will take up transistor circuitry in greater depth, starting with equivalent circuit considerations.

EQUIVALENT CIRCUITS

Any linear amplifier, by definition, produces an output signal functionally related to its input. Hence the operation of an AC transistor amplifier can be analyzed by treating the input and output sections separately, as was done for op amps in Chapter VI. Figure 7-1 shows a basic voltage amplifier and its equivalent circuit based on two embodiments of the Thévenin theorem. The input circuit is represented by a voltage source V_S with its associated internal resistance R_S. This is connected to R_{in}, the input resistance of the amplifier, which is actually formed by R_1 and R_2 in parallel plus a contribution from the transistor itself. The voltage source contained in the output segment is designated as $E_{in} \cdot A_V$, the input voltage multiplied by the voltage gain. The Thévenin resistance R_{Th} constitutes the internal resistance of the transistor. (Note that both R_2 and R_L are treated as though they were going directly to ground, rather

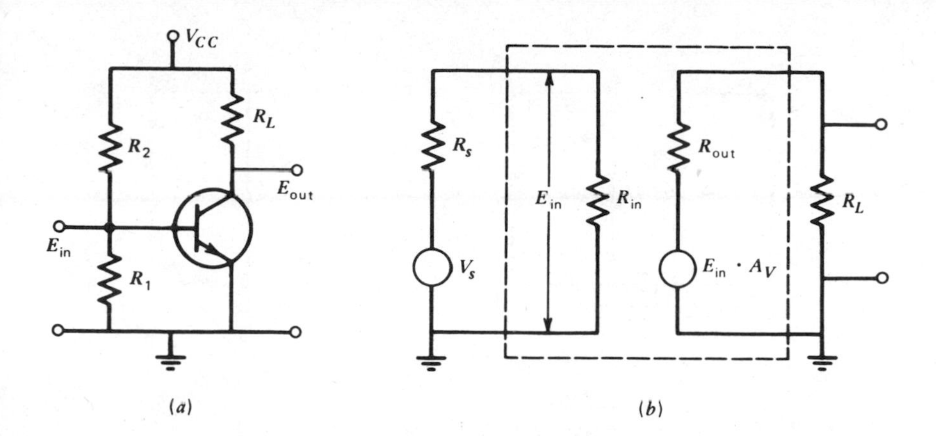

Figure 7-1. A transistor voltage amplifier (*a*) and its Thévenin equivalent (*b*).

than to the power supply. This is permissible since the output impedance of a regulated power supply is very small, often a fraction of an ohm.)

Other transistor circuits can be analyzed similarly, but it is simpler to construct an equivalent circuit that more closely follows the layout of the device itself. The transistor can be represented by three conductors, corresponding to the emitter, base, and collector, meeting at a point, as in Figure 7-2. Note that all three currents are shown flowing *into* the

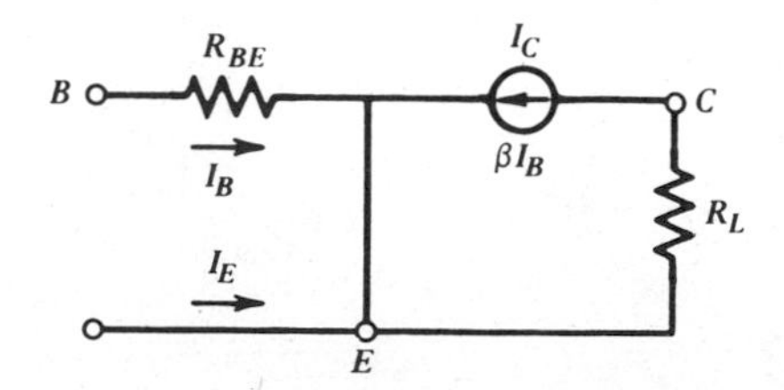

Figure 7-2. An approximate equivalent of a bipolar transistor.

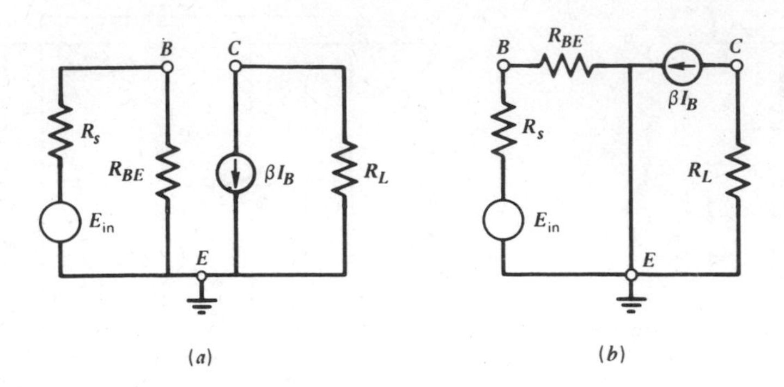

Figure 7-3. The equivalent voltage amplifier corresponding to Figure 7-1a: (a) in the format of Figure 7-2, and (b) redrawn to show the resemblance to Figure 7-1b.

transistor. The collector current I_C is indicated by the symbol for a current generator with the value βI_B. The resistance R_{BE} refers to the thin slice of silicon constituting the base. According to this method, the circuit shown in Figure 7-1 would be represented by the equivalent circuit in Figure 7-3.

In Figure 7-4 is another common configuration of a

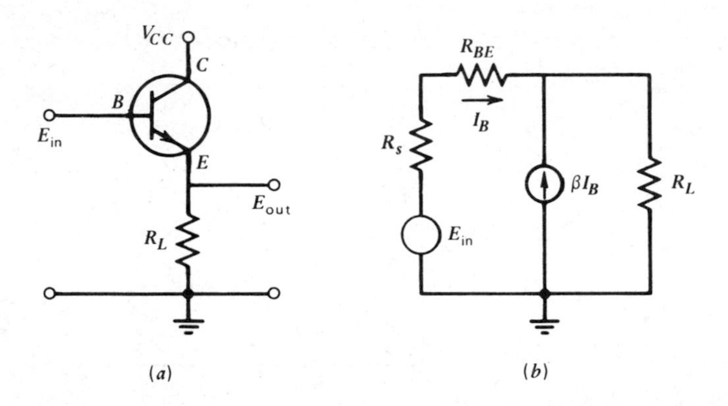

Figure 7-4. An emitter follower and its equivalent circuit.

transistor amplifier. This is called an *emitter follower* because the voltage gain is nearly unity, so the output is essentially equal to the input. (This is comparable to the voltage-follower connection for an op amp and can be used for analogous purposes.) The equivalent circuit is similar to the previous one but oriented differently, since the collector in this circuit is connected directly to the power supply with no intervening resistor. The load is taken from the emitter to ground. The current gain A_I in this amplifier can be found by the application of Kirchhoff's current law at the emitter node *E*. Since the current I_L in the load is always taken as flowing toward ground, this gives

$$I_L = -I_C = I_B + \beta I_B \tag{7-1}$$

and

$$A_I = \frac{I_L}{I_B} = 1 + \beta \tag{7-2}$$

This should be compared to the voltage gain A_V, which we have seen to be practically unity.

AN APPROXIMATE CURRENT REGULATOR

It is often desirable to be able to maintain the current in a circuit at a stable value. This can be done with a single transistor that can be an integral part of the controlled circuit itself. Figure 7-5 gives two circuits, using a bipolar transistor and a FET. In both, R_L is the effective resistance of the

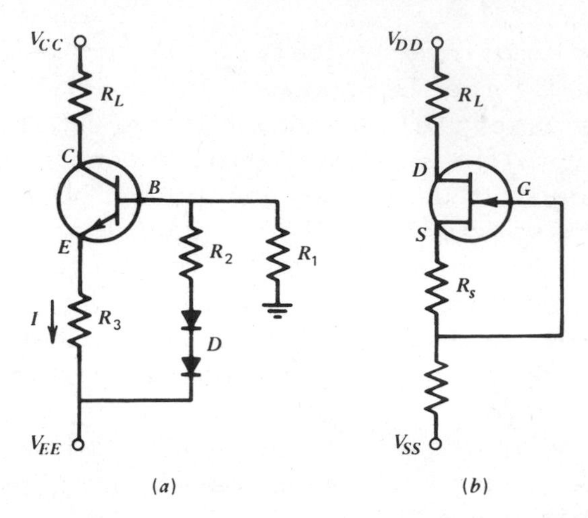

Figure 7-5. Transistors as current regulators.

circuit to be controlled. In Figure 7-5*a*, the application of Kirchhoff's voltage law to the emitter loop gives

$$IR_3 + V_{BE} = V_D + (V_{EE} - V_D)\left(\frac{R_2}{R_1 + R_2}\right) \tag{7-3}$$

where V_{BE} is the base-emitter voltage of the transistor and V_D is the forward drop across the two diodes in series. Solving for I and simplifying, we obtain

$$I = \frac{1}{R_3}\left(\frac{V_{EE}R_2}{R_1 + R_2} + \frac{V_D R_1}{R_1 + R_2} - V_{BE}\right) \tag{7-4}$$

The circuit parameters can be chosen to make

$$\frac{V_D R_1}{R_1 + R_2} = V_{BE} \tag{7-5}$$

This gives a current value of

$$I = \frac{V_{EE}R_2}{R_3(R_1 + R_2)} \qquad (7\text{-}6)$$

which is independent of the properties of the transistor. The two diodes are needed in order to satisfy the condition in Eq. 7-5, since the voltage drop across each is close to V_{BE}. If the diodes and the transistor are kept at the same temperature, the current will be insensitive to changes in this parameter also.

The FET circuit shown in Figure 7-5*b* is similar, but since no current can flow into the gate, the connections can be much simpler. The action is based on the independence of the drain current on source-drain voltage for a given gate potential. Zero temperature coefficient can be attained only for a particular cur-

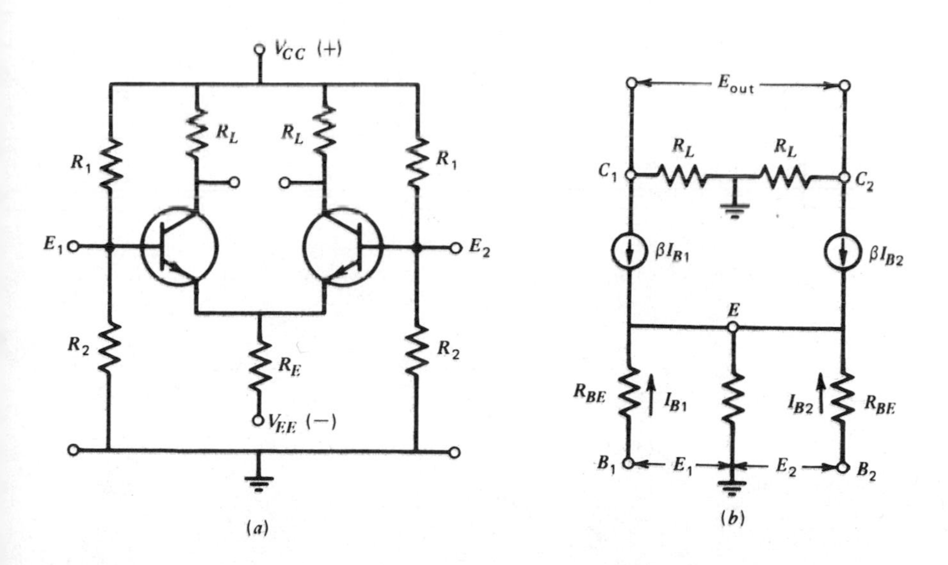

Figure 7-6. A differential amplifier and its equivalent circuit.

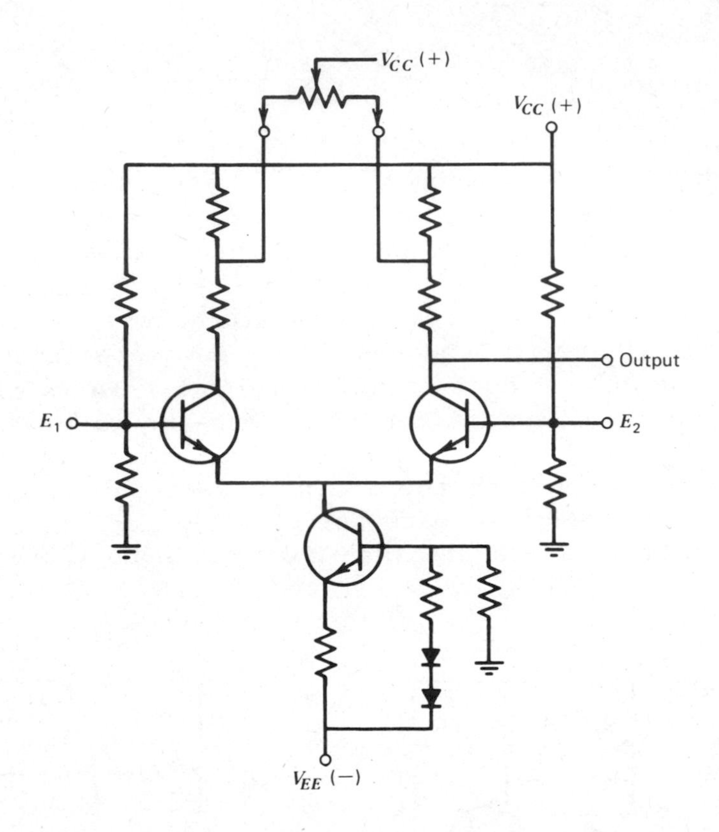

Figure 7-7. A differential amplifier with a current source as emitter impedance and a balance potentiometer in the collector circuit.

rent (cf. Figure 3-36), dependent on the transistor type, but it will be low for any reasonable current.

THE DIFFERENTIAL AMPLIFIER

The two-transistor circuit shown in Figure 7-6 is widely used as the input section of an op amp. Its symmetry ensures that common-mode voltages, temperature effects, and many other factors will cancel out. In addition, it has two equally effective signal connections, which can be used as inverting and noninverting inputs.

The equivalent circuit can be set up as shown in Figure 7-6*b*. Application of the Kirchhoff voltage rule to the loop that includes both bases and emitters gives us

$$E_1 - R_E \cdot I_{B1} + R_E \cdot I_{B2} - E_2 = 0$$

$$E_1 - E_2 = R_E(I_{B1} - I_{B2}) \quad (7\text{-}7)$$

The same rule applied to the emmitter-collector loop gives

$$E_{\text{out}} + \beta I_{B1} \cdot R_L - \beta I_{B2} \cdot R_L = 0$$

$$E_{\text{out}} = \beta(I_{B1} - I_{B2})R_L \quad (7\text{-}8)$$

Combination of these two expressions gives

$$E_{\text{out}} = -\beta\left(\frac{R_{\text{L}}}{R_{\text{BE}}}\right)(E_1 - E_2) \quad (7\text{-}9)$$

The common-mode rejection of an op amp with this type of input circuit is increased by a large value of the emitter resistor R_E and is further improved by the current regulator shown in Figure 7-7. This figure also shows a potentiometer connected across the collectors, which can be used to balance out any residual asymmetry between the two halves of the circuit. Connections for this manual control are often provided in IC op amps; the control can be omitted if not needed but easily added if further correction is necessary.

A corresponding circuit using FETs will provide an even better input section for an op amp, with very high

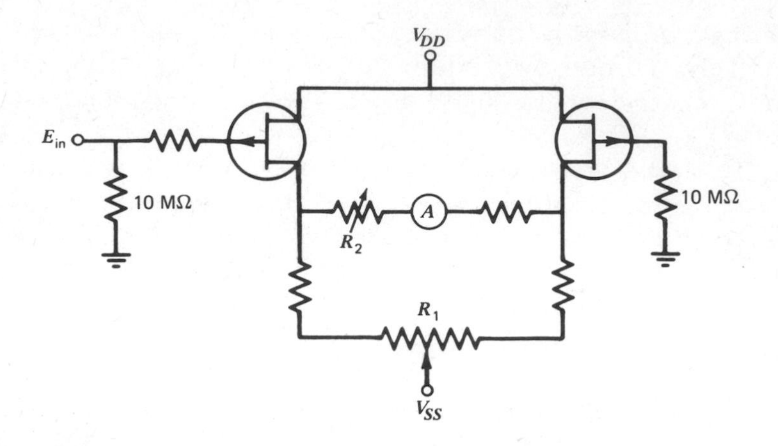

Figure 7-8. A differential amplifier as an input to a moving-coil meter.

input impedance as well as a high CMRR.

A differential circuit especially well suited for use as a meter amplifier is shown in Figure 7-8. The FETs (or transistors) are connected as voltage followers with a microammeter located between the two sources (or emitters). This circuit constitutes a bridge with the transistors as two arms and the source resistors as the others. Two controls are provided: R_1 is used to balance the circuit so that 0 V in gives 0 V out; R_2 is for full-scale calibration of the meter.

CASCADED TRANSISTOR STAGES

The two transistors in the circuits just presented are not considered to be distinct stages since the output of one does not feed into the other. If the gain of a single stage is not adequate, two or more transistors can be connected in series. Figure 7-9 shows two ways in which this can be done. The circuit in (*a*) is the more straightforward of these in that the two stages are identical. The coupling between stages

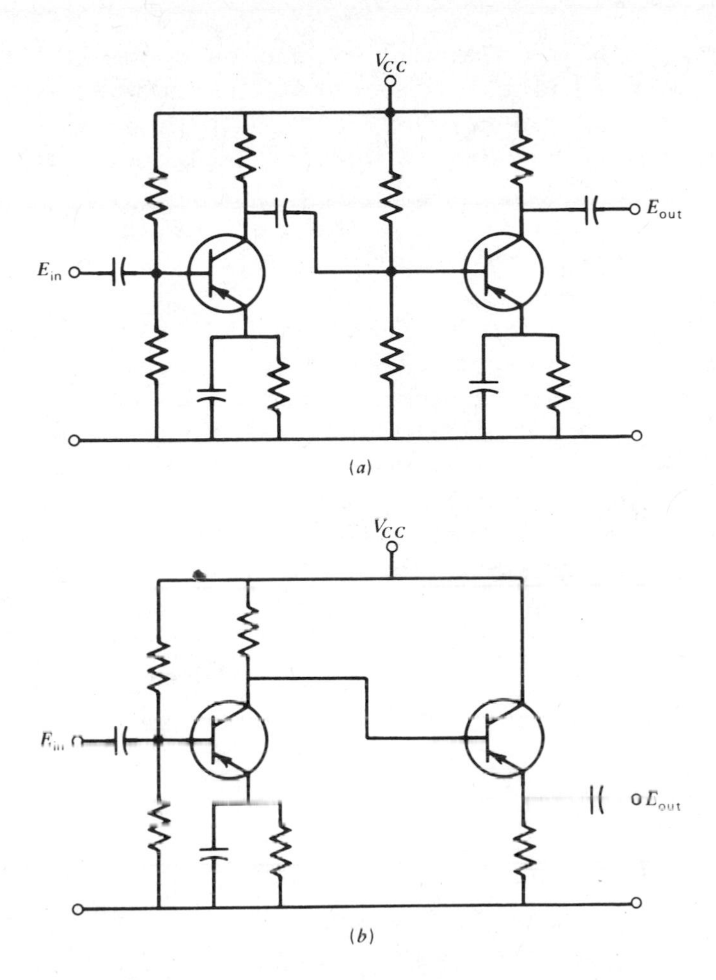

Figure 7-9. Dual transistor amplifiers: (*a*) capacitively coupled, and (*b*) direct copupled.

is capacitive, restricting the signal to AC. Each stage requires its own bias resistors. The circuit in (*b*) uses a voltage follower stage to provide a low-impedance output; this can be coupled directly, and additional bias resistors are not needed. Tasks requiring multistage amplifiers for low frequencies are better implemented with op amps.

Two bipolar transistors can be connected directly as shown in Figure 7-10, so that the emitter current of Q_1 forms the base current of Q_2. This combination can be treated as through it were a single unit, as suggested by the large circle in the figure. This is called a *Darlington transistor*. The β of the composite transistor is equal to the product of the β's of the individual components.

Cascaded stages of FET amplification can be designed in a manner similar to that used with bipolar transistors. Enhancement MOSFETs are particulary convenient for multiple-stage application, as shown in Figure 7-11. The stages are coupled directly rather than through capacitors, and the drain potential of each provides the gate bias for the following stage. This circuit is well adapted for use in ICs.

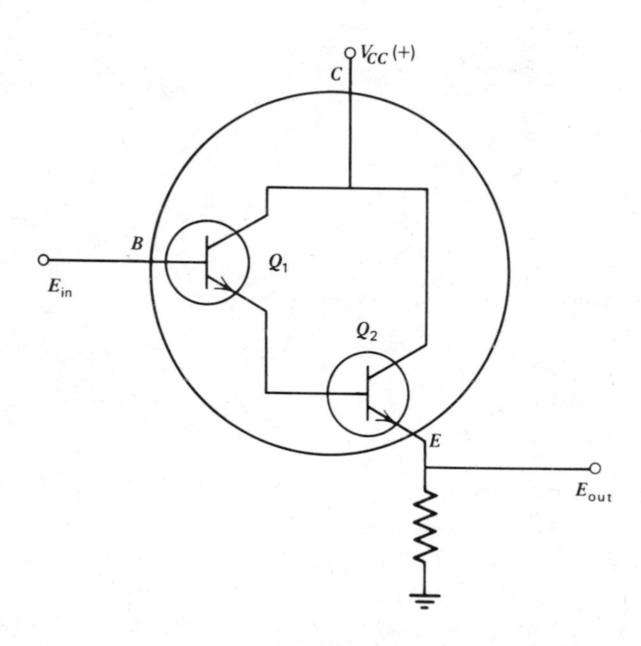

Figure 7-10. The Darlington transistor used as a follower.

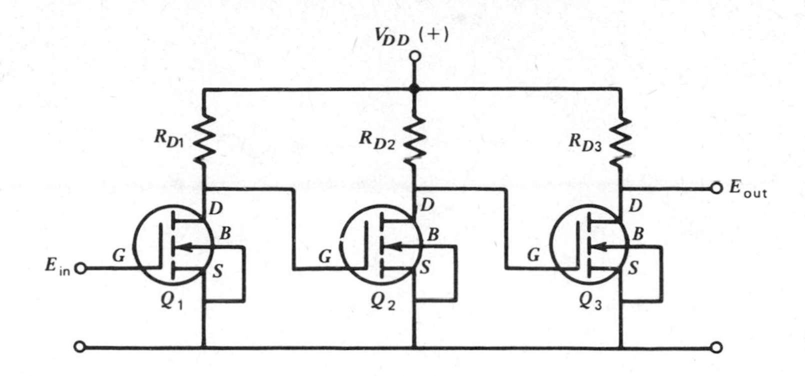

Figure 7-11. A three-stage MOSFET amplifier.

PHASE SPLITTERS AND COMPLEMENTARY SYMMETRY

There are occasions when an AC signal must be converted to two signals 180° out of phase with each other. This can be done with the differential circuit shown in Figure 7-6; the incoming signal is applied to one of the base connections, and the other is left open. The two outputs will then be out of phase, as required.

The same end can be achieved with a single transistor in the circuit shown in Figure 7-12. Equal valued resistors are placed in both emitter and collector leads. An AC signal applied to the base will cause equal excursions above and below the quiescent point. This is reflected in equal AC voltage drops in the two resistors. The phases are opposed since one subtracts from V_{CC}, and the other adds to the ground potential. This type of circuit is used extensively in digital ICs, particularly in the TTL series, as is discussed in Chapter IX.

Another technique for phase splitting depends on the use of *npn* and *pnp* transistors in the same circuit (complementary symmetry). This was encountered in the

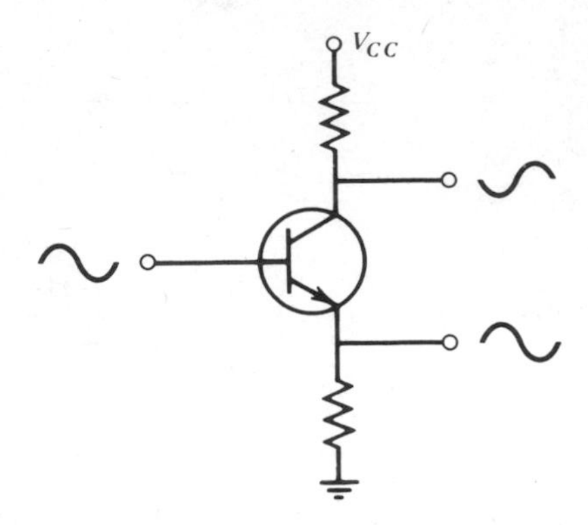

Figure 7-12. A phase-splitting amplifier.

booster depicted in Figure 5-1. Such a circuit requires separate power supplies for V_{CC} and V_{EE}. This is no problem in a booster for op amps, as a dual supply is already present, but in an amplifier contructed with discrete components, it is an inconvenience. The transistors must be accurately matched, and this is difficult to do for the two disparate types.

TRANSISTOR OSCILLATORS

Single transistors are better suited for use as high-frequency oscillators than are op amps. Several designs in addition to the phase-shift and Wien bridge circuits (see Chapter V) are available. Four of these are shown in Figure 7-13. Circuits (*a*), (*b*), and (*c*) utilize a resonant *LC* "tank" as the frequency-determining element. This parallel combination of an inductor and capacitor resonates at the frequency that makes the impedance of the two arms equal:

$$\omega L = \frac{1}{\omega C} \tag{7-10}$$

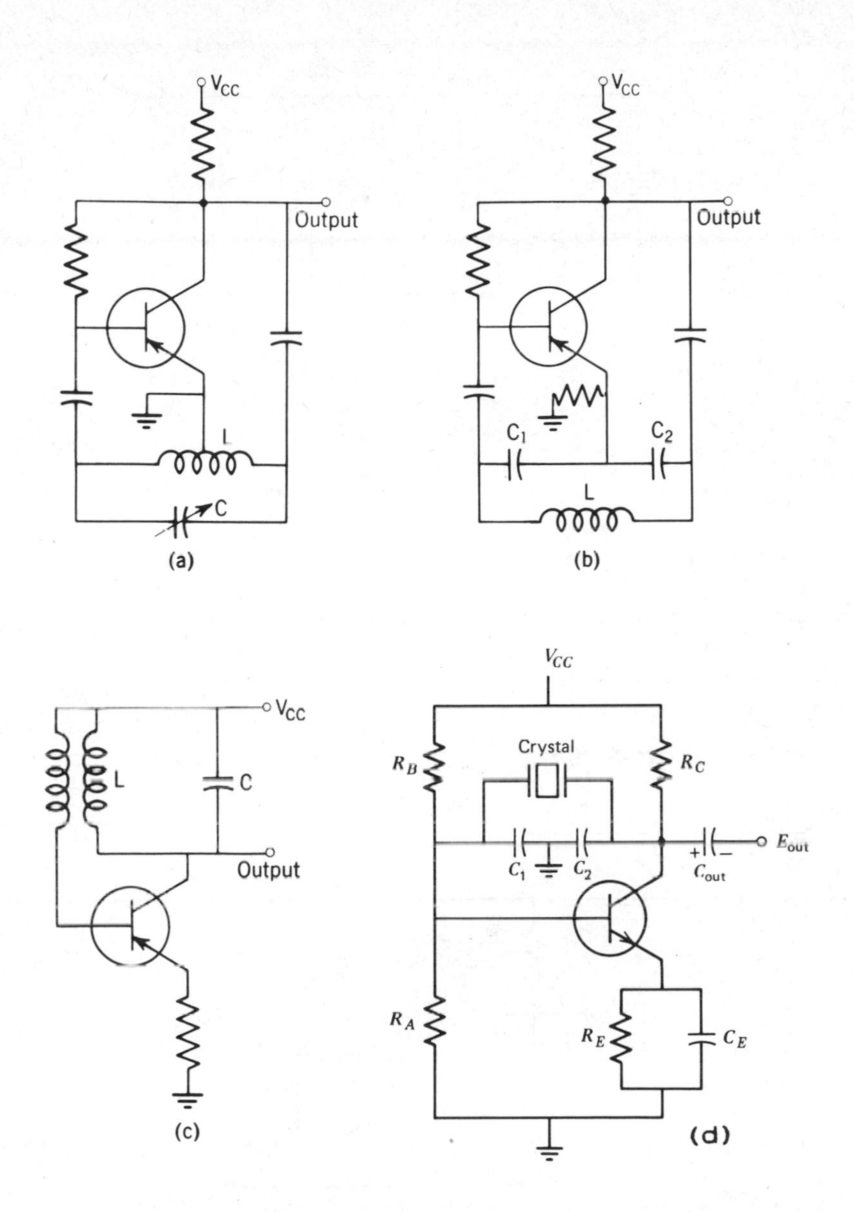

Figure 7-13. Four single-stage oscillators: (*a*) the Hartley circuit, with a tapped inductor; (*b*) the Colpitts circuit, with a divided capacitor; (*c*) a transformer-coupled oscillator; (*d*) the Pierce crystal-controlled circuit.

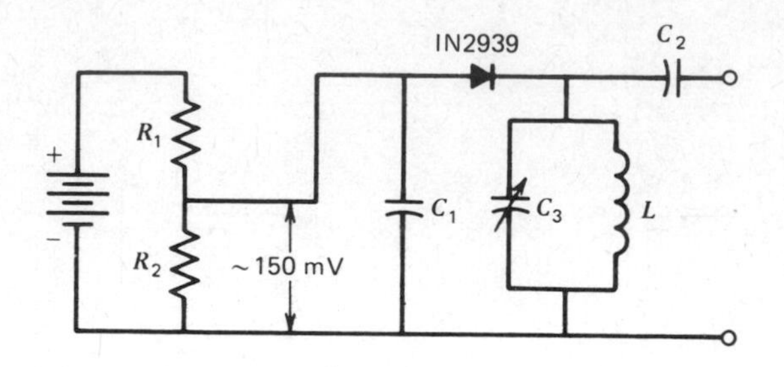

Figure 7-14 A tunnel diode oscillator.

or

$$f = \frac{1}{2\pi\sqrt{LC}} \qquad (7\text{-}11)$$

In (*a*), the *Hartley* oscillator, the inductor is tapped to provide feedback to the emitter of the transistor, whereas in (*b*), the *Colpitts*, feedback is taken between two capacitors. Circuit (*c*) uses a small transformer to provide feedback. In all three the output could be

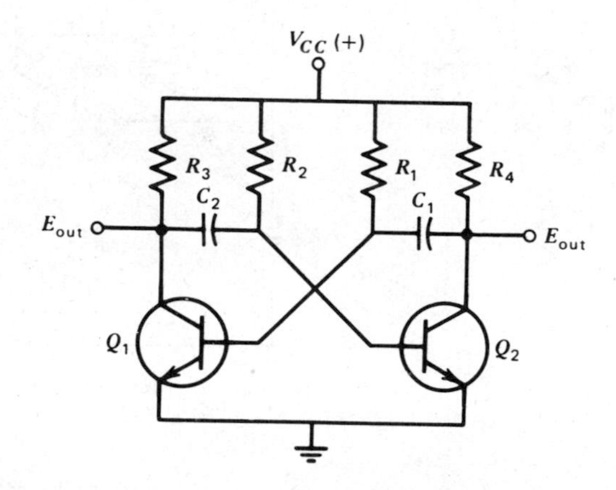

Figure 7-15. A two-transistor astable multivibrator. The two outputs give signals that are 180° out of phase with each other.

taken from an extra winding on the inductor instead of from the collector as indicated.

The oscillator shown in Figure 7-13*d* (the *Pierce* circuit) is built around a quartz crystal that resonates at a frequency dictated by its physical dimensions. The several capacitors are needed to generate appropriate phase relationships. Crystal oscillators are among the most precise devices available. If the crystal is maintained at constant temperature, frequency stability within the sub-parts-per-million range is attainable.

A related oscillator, utilizing an *LC* tank, is designed around the tunnel diode described in Chapter III. A typical circuit is shown in Figure 7-14. The operation depends on the "negative resistance" property of the tunnel diode, that portion of the characteristic curve (Figure 3-25) where the current decreases as the voltage is increased. The I^2R power absorbed by the tank is offset by an equal amount ($-I^2R$) generated by the diode with its negative dynamic resistance. The tunnel diode oscillator is especially valuable at high frequencies (megahertz to gigahertz), a region where *LC* circuits are particularly convenient.

Figure 7-15 gives a circuit for a square-wave generator that is easily assembled from discrete parts; this is called an *astable multivibrator*. The connections are such that both transistors might be expected to operate simultaneously in a heavily conducting mode since the bases of both are forward biased through R_1 and R_2. However, when the power supply is first turned on, inevitably one of the transistors, say Q_1, will fire first as the result of slight inequalities. During the turn-on of Q_1, a negative pulse will be sent through C_2 to the base of Q_2, preventing it from turning on immediately. Because of the capacitive coupling, this blocking is only temporary. The flow of current through R_2, neutralizes the negative charge on C_2, and brings the base of Q_2 to more and more positive potentials, until Q_2 suddenly turns on. This generates a pulse that turns Q_1 off, and the process repeats itself indefinitely. The repetition frequency, for identical transistors, is given approximately by

$$f = \frac{1.5}{C_1R_1 + C_2R_2} \tag{7-12}$$

If $C_1 = C_2$ and $R_1 = R_2$, the square wave will be symmetrical.

NONLINEAR APPLICATIONS OF TRANSISTORS

A transistor can be used as an on-off switch merely by giving it appropriate signals that will cause it to go into cutoff or saturation. Field-effect transistors differ from bipolars in that the controlled current passes only through pure resistance, not through any *pn* junctions. This means that the current can pass in either direction and still be interrupted by switching action corresponding to cutoff or saturation. A FET used this way is called a *transmission gate* or *analog gate*, contrasting with the digital gates used in logic systems.*

Figure 7-16*a* shows the basic transmission gate using an n-channal JFET. If a voltage more negative than either drain or source is applied to the control input, the switch will be turned off, whereas if the gate is left open, the switch will be turned on. (This action can be verified by reference to Figure 3-35.)

A transmission gate composed of a pair of enhancement-mode MOSFETs connected in parallel (Figure 7-16*b*)

* It is unfortunate that the word "gate" is used in so many senses, including the control element of the FET. One must guard against ambiguity.

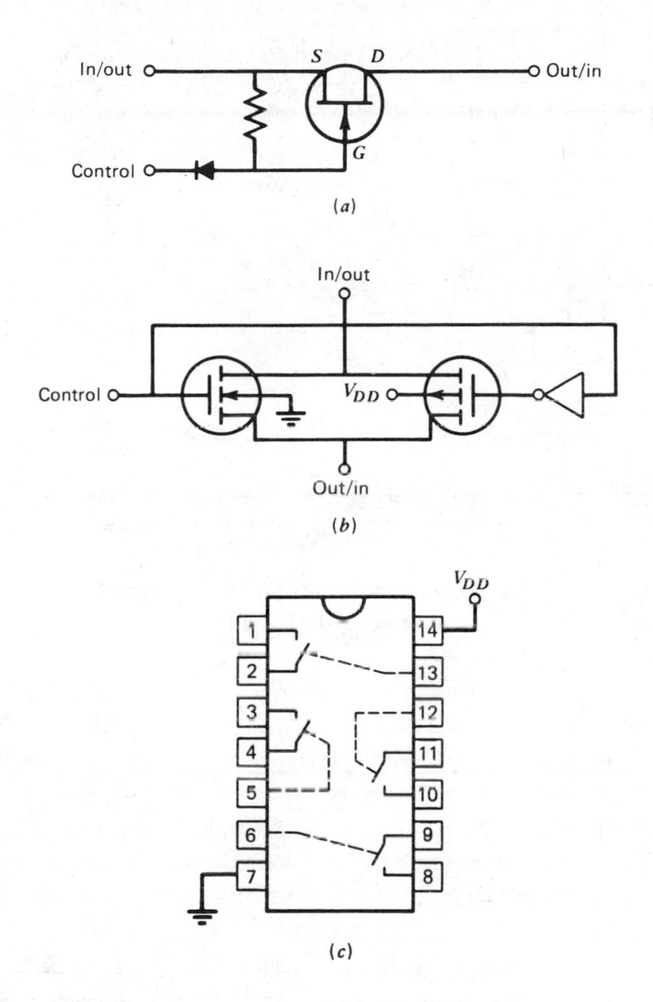

Figure 7-16. Transmission gates: (*a*) using a JFET; (*b*) a gate made from a pair of MOSFETs; (*c*) the CD4016, an IC carrying four gates; the numbers are the pin designations; pin 13 connects to the control for the switch between pins 1 and 2, and so on.

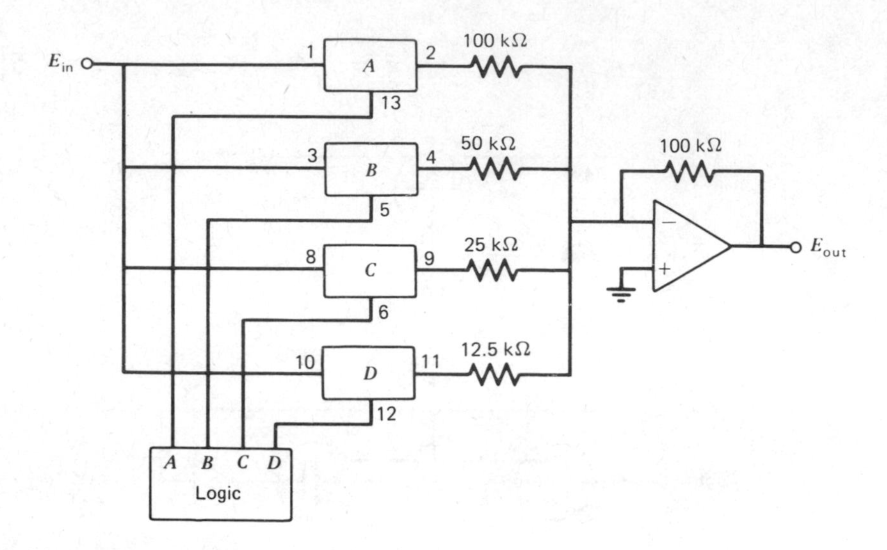

Figure 7-17. A variable gain amplifier using a CD4016. The numbers correspond to those in Figure 7-16.

is often used in ICs. This gives greater speed of operation than a single MOSFET alone because the time constant of each member of the pair is lowered by the presence of the other. Figure 7-16*c* shows the functional diagram of the CD4016, containing four identical switches that can be used independently of each other. Each switch consists of a MOSFET pair similar to that just described. Similar ICs are available with switches in various connections, such as single-pole double-throw, or with all segments responding to the same command signal.

An application for a quad switch such as the CD4016 in a programmable amplifier is shown in Figure 7-17. Combinations of the four different input resistors for the op amp can be selected by closing one or more switches. A little arithmetic will show that with the resistance values given, the gain of the amplifier can assume integral values from -1 to -15.

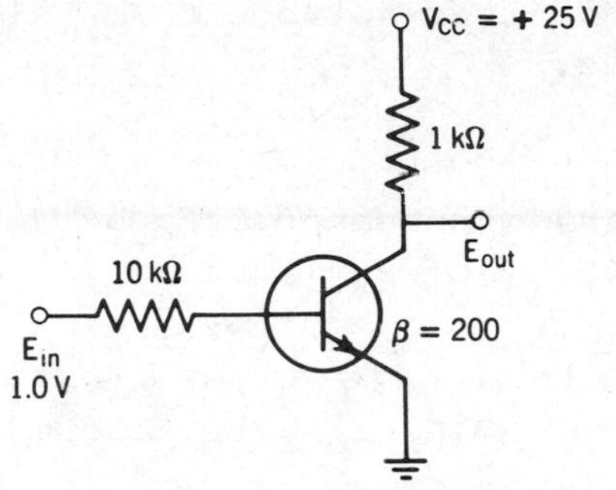

Figure 7-18. See Problem 7-1.

PROBLEMS

7-1. For the circuit shown in Figure 7-18, compute the following, neglecting the base-emitter voltage drop:

(a) I_B
(b) I_C
(c) E_{out}
(d) A_V

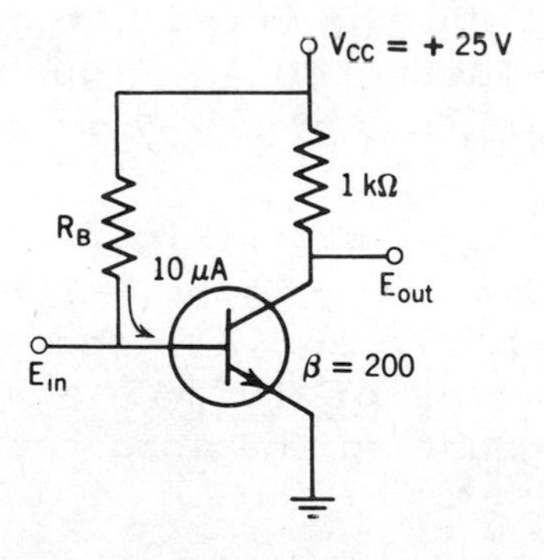

Figure 7-19. See Problem 7-2.

7-2. For the circuit shown in Figure 7-19, neglecting the base-emitter voltage drop, find the following quantities:

(a) E_{out}
(b) the transimpedance, dE_{out}/dI_{in}
(c) the value of R_B

7-3. For the circuit given in Figure 7-1, calculate the voltage gain, E_{out}/V_S, assuming that $R_{out} << R_L$.

7-4. Calculate the output voltage E_C in the circuit shown in Figure 7-3.

7-5. An emitter follower, a source follower, and an op amp connected as a voltage follower all perform similar functions. Compare them with respect to the precision with which the output and input voltages are equal.

7-6. Given a bipolar transistor with $\beta = 200$ in the emitter-follower configuration, calculate the ratio E_{out}/E_{in}. Do the same for a source follower and for an op amp with $A = 10^4$.

7-7. What would be the effect of connecting the two outputs illustrated in Figure 7-12 to the two inputs shown in Figure 7-7 or 7-8?

7-8. A current regulator (Figure 7-5a) is assembled using the 2N4074 (Figure 3-31). What resistor values (R_1, R_2, and R_3) should be selected to give a current of 10 mA? What would be the effect of doubling the load resistance R_L?

7-9. The FET for which the characteristics are given in Figure 3-35 is connected in the circuit shown in Figure 7-5*b* as a current regulator. What value of resistor R_S should be selected to produce a current of 4 mA? What variation in current could be expected to result from doubling the load resistance R_L?

7-10. In the differential circuit shown in Figure 7-7, determine the effect on the output resulting from:

(a) raising E_1 while holding E_2 constant
(b) raising E_2 while holding E_1 constant
(c) raising E_1 while decreasing E_2 by the same amount
(d) raising both E_1 and E_2 by the same amount
(e) holding both E_1 and E_2 constant and varying the balance potentiometer

7-11. Show that the circuits shown in Figures 7-6 and 7-8 can be combined to give a meter of greater sensitivity. Select suitable resistance values and compute the voltage sensitivity possible with a 0- to 10-μA meter movement.

7-12. Show how FETs can be used to control the "integrate," "hold," and "reset" modes of an integrator. What is the distinction between an integrator with these controls and a sample-and-hold amplifier?

7-13. Sketch an equivalent circuit for the Darlington amplifier shown in Figure 7-10.

7-14. Devise a circuit whereby a phase-shift oscillator can be rendered voltage-tunable by the addition of a pair of matched FETs.

7-15. Calculate the natural period of oscillation of the *LC* tank shown in Figure 7-13*b*, if $L = 300$ mH, $C_1 = 0.033$ µF, and $C_2 = 0.022$ µF.

7-16. Using an integrator, a comparator, and a FET switch, design a signal generator to give saw tooth waves.

7-17. Suppose that in the precision rectifier shown in Figure 4-29*b*, the diodes are replaced by FET switches driven by alternate half-cycles of a square wave at the same frequency as the input signal. (a) Can this be used as a rectifier? (b) Show that it can be applied to the measurement of phase angles.

7-18. Sketch the output of the circuit shown in Figure 7-20 for each of the following inputs:

(a) 4 V, DC
(b) 4 V, RMS sine wave at 50 Hz

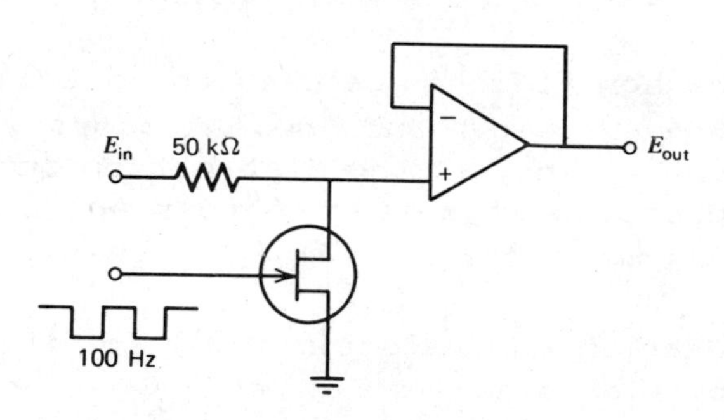

Figure 7-20. See Problem 7-18.

VIII
POWER DEVICES

Circuits and devices based on operational amplifiers are limited in power to the amount needed to transport and process small signals. Thus the highly precise and reliable op amp is limited in practice by what we might call the rule of the three tens: 10 V, 10 mA, and 10 kHz. Although each of these can be exceeded for some models, this rule is a convenient guide.

In this chapter we discuss semiconductor devices involving higher voltages and currents and thus higher powers. They require additional design considerations because of the heat produced, which must be dissipated to avoid self destruction. Actually these devices also have a wider frequency range (into the megahertz region). Because the mathematical description of the behavior of such circuits is complex, approximation methods are usually required.

POWER CALCULATIONS

The Power Factor

The general statement describing power, as shown in Chapter II, is

$$P = EI \cos \phi \qquad (8\text{-}1)$$

where cos ϕ is the *power factor*, which becomes unity for DC. Power can also be defined as the flow of energy per unit time.

The power in a circuit must be dissipated in some form. It may be transformed into some other domain, such as mechanical or electrochemical, or it may appear as heat generated in various circuit components. We can write for the heat energy in joules:

$$\text{Heat} = EIt \cos \phi - \text{(useful energy)} \qquad (8\text{-}2)$$

in which we combine under the term "useful energy" energy that is converted to mechanical, luminous, or other forms. Thus for an incandescent lamp (in which the power factor is unity), the heat to be dissipated is EIt minus the luminous energy.

As a rule, in our context, the accumulation of heat is undesirable, and the excess energy must be absorbed by the surroundings. Let us now consider the dissipation of heat by various electronic components.

Power in Resistors

Here the power EI is completely transformed into heat. The power dissipated by a resistor can be calculated by the formulas

$$P = EI = \frac{E^2}{R} = I^2R \qquad (8\text{-}3)$$

For example, at 30 V, a 1000-Ω resistor will produce heat at the rate of $(30)^2/1000 = 0.9$ W. A resistor with a rating at least as large as 1 W should be selected.

At this point we include a few comments about safety factors. In scientific instruments the cost of a malfunction far outweighs the cost of more reliable components. Therefore, one should avoid the operation of components at their maximum heat dissipation rating. The resistor mentioned above will run quite hot, and its useful life will be much shorter than that of a 2- or 5-W resistor in the same circuit. The 5-W resistor will generate exactly the same amount of heat as the 1-W unit, but it has larger dimensions and so will lose heat more readily, resulting in a lower temperature. The life expectancy of electronic equipment doubles for a decrease of only about 10°C in operating temperature. Thus it is recommended that a safety factor of at least five times be allowed. This principle applies even more strongly to transistors and other semiconductors.

Power in Capacitors and Inductors

In these reactive devices the power factor approaches zero. Only to the extent that resistance is present will heat be produced. Hence, as a rule, there is no need to consider heat dissipation in reactive components. The chief exception is in power transformers, which have considerable resistance and may carry high currents.

Power in Diodes

In the conventional silicon diode the forward voltage drop is only about 0.7 V and the power dissipated is small. The rating of a diode is often given in terms of current rather than power (since $P = 0.7 I$). Diodes are also specified in terms of the maximum peak inverse voltage (PIV) they can withstand.

For high currents (over about 10 A), diodes require heat sinks to provide thermal contact with a body that is able to dissipate the heat. This body might be a metal plate on which the device is mounted. If the diode is correctly placed on the heat sink, with a layer of heat-conducting grease on the mounting surface, the allowable flow of heat depends primarily on the area of the heat sink. As a simple rule, if the heat sink is vertical and free of obstructions, one should allow at least 5 cm^2 area per watt of power dissipated. The same criteria apply to heat sinking of power transistors and other devices.

Zener Diodes: The requirements of a Zener diode present a special case, since variable currents are accepted at a fixed, relatively large, voltage. Consider the circuit shown in Figure 8-1. The design objective is to accommodate a variable-output current (0 to 10 mA) at 10 V. Since the input is 15 V and the output only 10 V, clearly the drop across the resistor must be 5 V. If the load is so great that the output current is essentially zero, then the 500-Ω resistor is the only impedance seen by the incoming current, and the entire 10 mA will pass through the Zener. If, however, significant current is drawn through the load, the drop across the resistor will remain the same, and the current will divide between the load and the diode. Thus a load of 1.5 kΩ will draw 10/1.5 = 6.67 mA, leaving 3.33 mA to pass through the Zener. The maximum power dissipated in the Zener will be 10 V • .01 A = 0.1 W, while that in the resistor is $5^2/0.5$ = 0.05 W. Both components should be rated at 0.25 W or larger. (Zener diodes can have a rather high temperature ocefficient dV_Z/dt, so it is especially desirable to keep them from overheating.)

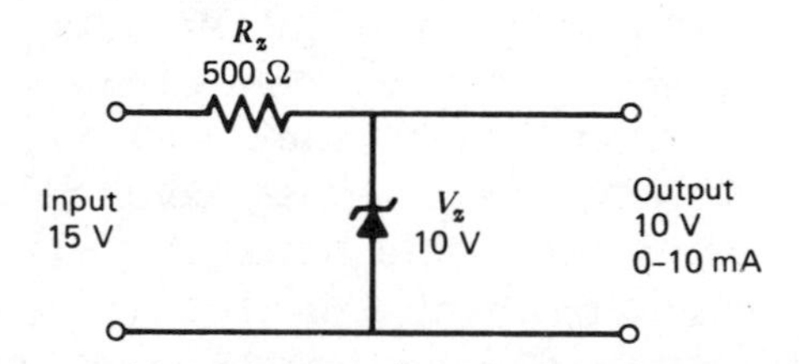

Figure 8-1. A Zener regulator. (Cf. Figure 3-26).

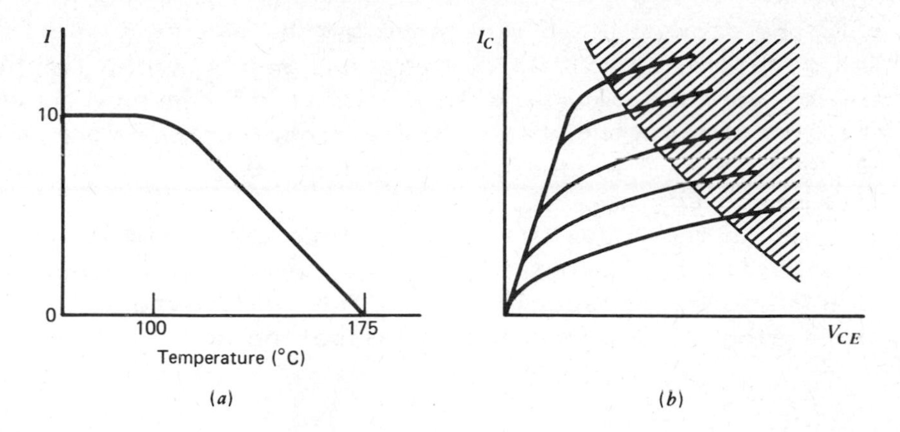

Figure 8-2. Transistor derating curves: (*a*) the maximum permissible current for a power transistor as a function of temperature; (*b*) power limitations of a transistor--the dashed line must not be exceeded.

Power in Transistors

Transistors must be treated with care from the point of view of heat dissipation since they are more prone to thermal failure than other devices. It is the junction temperature of a semiconductor that must be kept within limits, but this is not a directly measurable quantity. Consequently a set of empirical restrictions in current and power must be employed. The maximum current limitation for a typical power transistor as a function of the easily measured case temperature is indicated in Figure 8-2*a*. Up to about 100°C, no derating is necessary.

The power limitation on transistors can be calculated by a thermal equation reminiscent of Ohm's law:

$$\Delta T = P\theta \tag{8-4}$$

where ΔT is the difference of temperature between the transistor junctions and the ambient air ($T_j - T_a$), and

P is the flow of heat generated by the device in watts. The constant θ, called the *thermal resistance*, is the sum of various barriers in the path of the heat from the junction to the air. These contributions to θ can be labeled as θ_{JC} (junction to case), θ_{CH} (case to heat sink), and θ_{HA} (heat sink to air).

It is thus possible to calculate the maximum power dissipation by taking for T_j the permissible junction temperature, as specified by the manufacturer, and expressing the maximum heat dissipation as

$$P = \frac{T_j - T_a}{\theta_{JC} + \theta_{CH} + \theta_{HA}} \tag{8-5}$$

A reasonable value of θ_{JC} is about 0.3 K/W (kelvin per watt), and of θ_{CH} about 2 K/W. The dominant factor here is θ_{HA}, which depends on the area of the heat sink; this can be taken as about 200 K/W for a 1 cm^2 sink and proportionately less for larger areas.

If T_j is allowed to go only to 100°C, and not to its ultimate safety limit of 175° or so, the power dissipation can be calculated for a typical (5-cm^2) heat sink, assuming that the temperature of the air in the instrument is 50°C, as follows:

$$P = \frac{100 - 50}{0.3 + 2 + 40} = 1.2 \text{ W} \tag{8-6}$$

Once the maximum power is established, the output characteristic curve of the transistor can be modified, as in Figure 8-2*b*, to include the maximum permitted combination of currents and voltages. The transistor can be safely operated if the load line does not enter the forbidden area.

POWER SUPPLIES

All electronic circuits require a source of power, usually DC voltages in the 5- to 30-V region. Many analog ICs require a dual supply of ±12 or 15 V, whereas digital devices usually need only +5 V. Power can be provided by batteries, which are particularly useful for portable instruments. In laboratory apparatus, however, it is better economy to use a *power supply*, a device that converts AC line power into appropriate DC voltages. Modular plug-in power supplies are convenient and moderately priced, but sometimes it is advantageous to assemble one's own unit.

The chief requirements of a power supply, in addition to its current and voltage capabilities, are (1) absence of residual AC ripple in the output, which implies good filtration; (2) an output voltage that is independent of variations in the line voltage and in the load current, in other words, good regulation; and (3) a means of limiting the current to a safe value.

The major parts of a power supply are shown in the block diagram in Figure 8-3. Simpler units may lack the regulation and protection features shown. We shall discuss the various component parts in turn.

Transformers

A transformer, in its simplest form, is a device

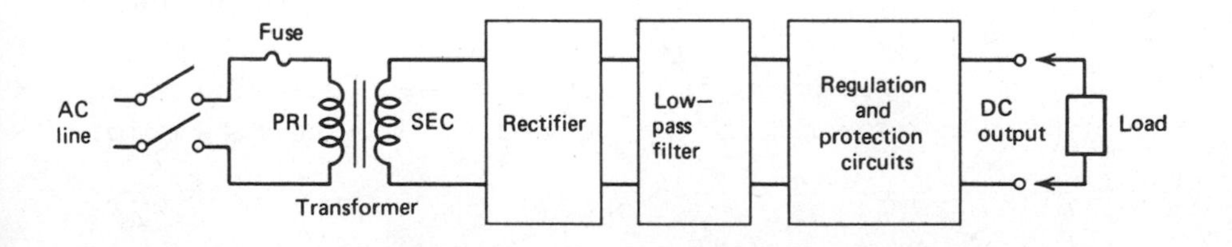

Figure 8-3. Block diagram of a typical power supply.

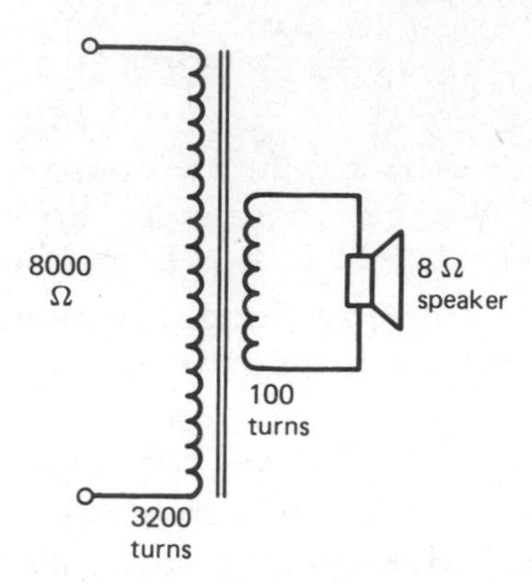

Figure 8-4. A transformer used for matching impedances.

consisting of two inductors coupled through their magnetic fields. If the primary coil is energized by AC, the secondary will generate a voltage depending only on the ratio of the number of turns n_1/n_2 in the two windings. One can thus write

$$\frac{n_1}{n_2} = \frac{E_1}{E_2} = \frac{I_2}{I_1} \tag{8-7}$$

indicating that the voltages in the two windings are proportional to the corresponding number of turns. The currents, on the other hand, bear an inverse relationship to the number of turns. Equation 8-7 is strictly valid only if the efficiency of the transformer is 100%, which implies zero internal resistance.

The impedance of a circuit as seen through a transformer depends on the *square* of the turns ratio,

$$\frac{Z_1}{Z_2} = \left(\frac{n_1}{n_2}\right)^2 \tag{8-8}$$

Consequently, transformers can be used to match impedances. Thus if it is desired to have the 8-Ω impedance of a loudspeaker appear as an 8000-Ω impedance to its driving circuit, a coupling transformer with a square ratio $(n_2/n_1)^2 = 1000$, or $n_2/n_1 = 32$, could be used, as shown in Figure 8-4.

Rectification and Filtration

Rectification is the first step in the conversion of AC into DC. This can be done by inserting diodes between the transformer and load, as in Figure 8-5. This limitation to one direction of flow results in a pulsating current that contains both AC and DC components. The AC ripple is very pronounced in half-wave (*a* and *b*) but less so in full-wave rectifiers (*c* and *d*). These circuits are not ordinarily useful as such,

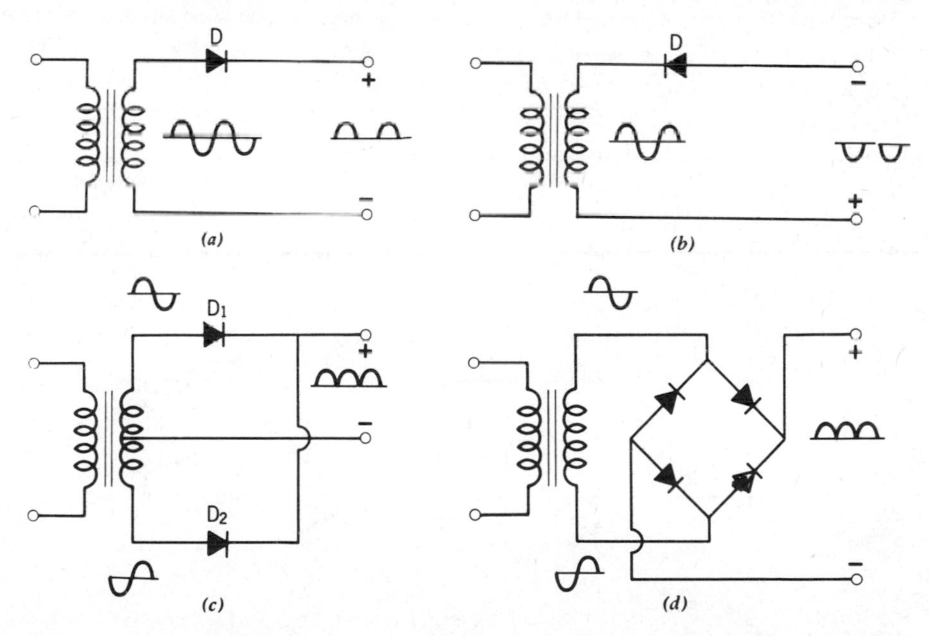

Figure 8-5. Rectifier circuits; (*a, b*) half wave; (*c, d*) full wave. The bridge (*d*) is available commercially as a four-terminal module.

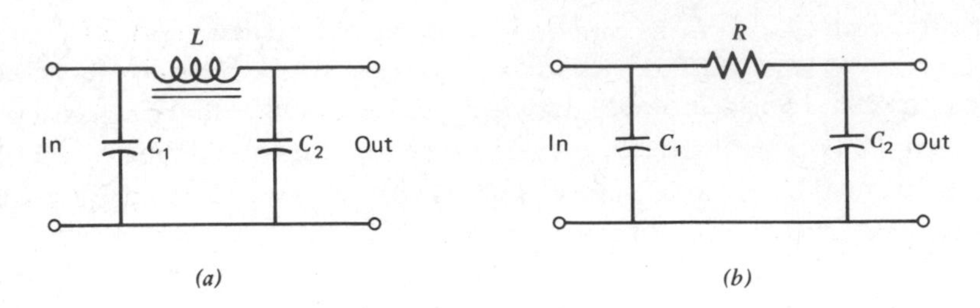

Figure 8-6. Two types of power-supply filters: (*a*) *LC*; (*b*) *RC*. Both are called "π-sections" because of the shape of their diagrams.

and filtering must be added to reduce the ripple.

The output from the rectifier can be fed into a low-pass filter to attenuate the ripple, as shown in Figure 8-6. The *LC* circuit in (*a*) is very effective and useful in high-power systems but requires a bulky and relatively expensive inductor, called a *choke*. For low-power applications it is more convenient to use circuit (*b*) in connection with an IC regulator. The regulator eliminates ripple as well as stabilizing the voltage.

A typical unregulated power supply using an *RC* filter is shown in Figure 8-7. With the component values indicated, the residual ripple will be about 0.1% of the DC voltage, which is usually satisfactory.

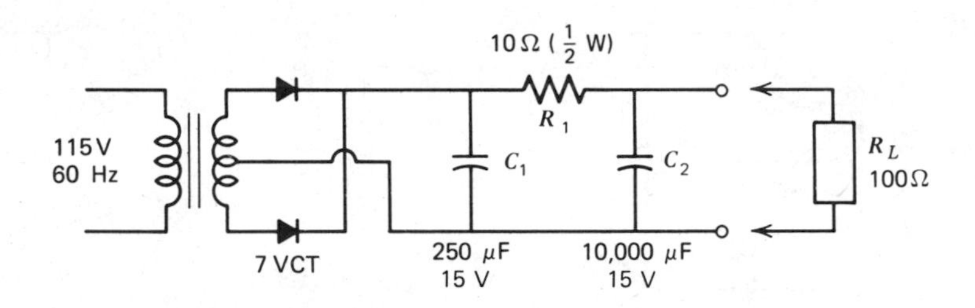

Figure 8-7. A nonregulated power supply to give 5 V at 50 mA. Care must be taken to select components with sufficiently high voltage and power ratings. Capacitor C_1 is relatively small, to avoid overloading the transformer.

The filtering improves in proportion to the product $R_1 R_L C_1 C_2$, so large values of R_1 give better ripple rejection. On the other hand, R_1 and R_L form a voltage divider, so that variations in the load resistance will have less effect on the voltage if R_1 is small. Hence a compromise value for R_1 must be selected. In the example shown, under no-load conditions E_{out} = 5 V, whereas with a 100-Ω load, E_{out} can be calculated to be about 4.5 V, a drop of only 10% for a change of current from zero to full load.

Zener Regulation

A considerable improvement for small power supplies is obtained by placing a Zener diode in parallel with the load (*Zener shunt regulation*), as in Figure 8-8. It now becomes possible to use a larger value for R_1 and to reduce the capacitors. The Zener maintains a reasonably constant 5 V across the load, regardless of variations in line voltage or load current. This is valid as long as R_L is greater than about 100 Ω, below which the diode drops out of conduction.

One factor to be taken into account in Zener regulation, as mentioned previously, is the power dissipated as heat. If the load is disconnected, the diode current is limited only by R_1, in this case to 50 mA,

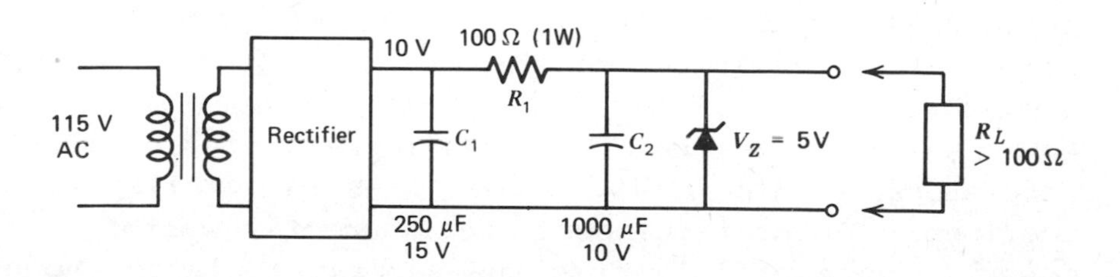

Figure 8-8. A power supply with a shunt Zener diode as regulator, designed to give 5 V at 50 mA.

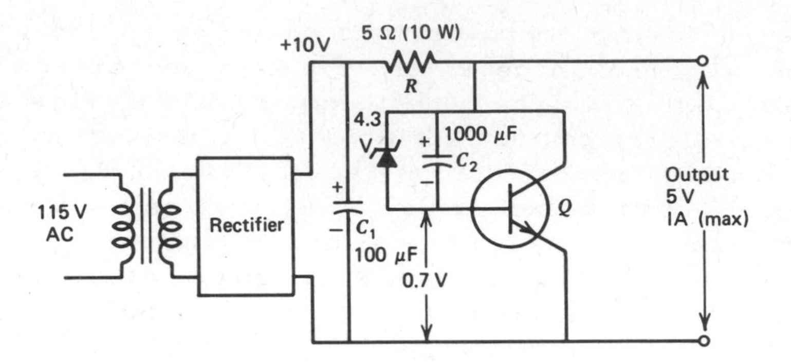

Figure 8-9. An amplified Zener shunt regulator. The combination of Z, C_2, and Q can be regarded as a single, two-terminal unit that will carry a maximum current of 1 A. The power dissipation in R is 5 W, and in Q is also 5 W at its maximum. Proper heat dissipation precautions must be taken.

and the power dissipated is 250 mW; a diode rated at 750 mW should be used.

Transistor-Regulated Power Supplies

A very effective, inexpensive regulator can be implemented by a combination of a Zener diode and a transistor, as shown in Figure 8-9. The Zener current is amplified by the β of the transistor (assumed to be 50), so that a 1-A current can be controlled by a 20-mA Zener. An additional merit is that the AC current through the capacitor C_2 is also amplified by β. The circuit behaves as though the combination of C_2 and Q were a capacitor of $\beta C_2 = 50 \times 1000 = 50{,}000$ μF.

The circuits just described are examples of *shunt* regulators, as the active elements is in parallel with the load. Shunt regulators are inherently wasteful of power. More efficient is the *series* regulator shown in Figure 8-10*a*, in which the load current flows *through* the controlling element. The transistor is operating as an emitter-follower, reproducing the voltage of the Zener at the output. The circuit can be regarded as a three-terminal regulator, as in Figure 8-10*b*. Regulators of this type, available commercially, may contain short-circuit protective circuitry. We highly recommend these units, as they are moderately priced yet give a precision within about 0.1% in terms of regulation, drift, and ripple rejection. A typical power supply in which one of these regulators is used is shown in Figure 8-11.

The circuit shown in Figure 8-12 uses an op amp and a booster transistor to generate a dual ±15-V supply suitable for powering op amp circuits. The feedback guarantees that the midpoint between the resistors, and hence between positive and negative outputs, is at ground potential.

Current Supplies

In the power supplies presented so far, the output voltage is held constant regardless of the current drawn by the load up to the point of saturation. In other words, by Ohm's law, since *E* is maintained con-

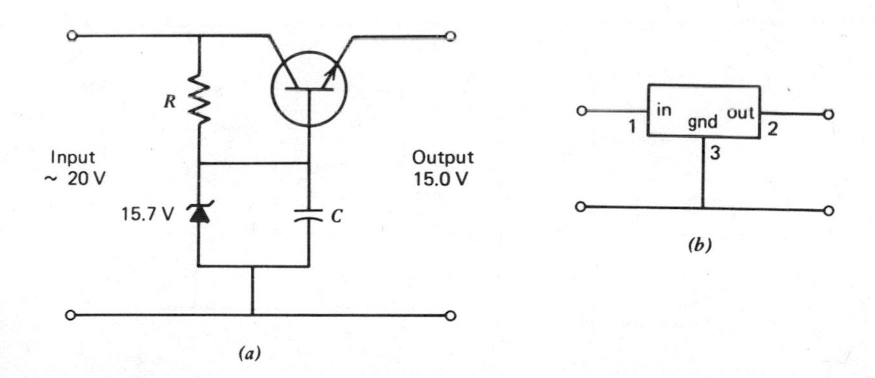

Figure 8-10. A series regulator (*a*) and its black-box representation (*b*). The capacitance of *C* is effectively multiplied by the β of the transistor.

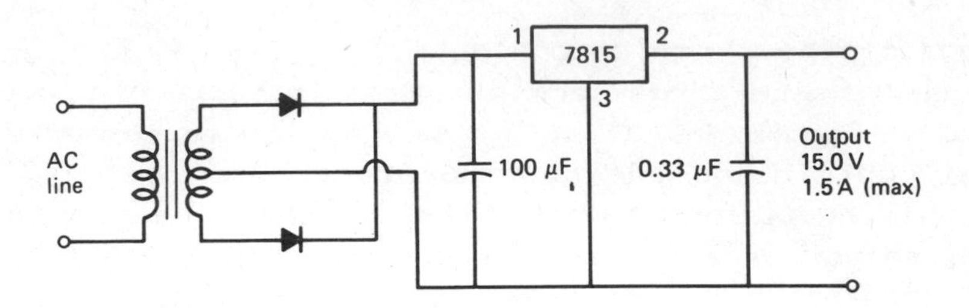

Figure 8-11. A regulated power supply using a 7800-series regulator. The last two digits of the 7800 number give the output voltage.

stant, it must follow that the current is determined by the resistance of the load. This is valid up to a maximum called the *current compliance* of the supply. Another possibility is to keep the current constant. In this case it is the voltage that is determined by the resistance of the load, and the maximum available voltage is the *voltage compliance*.

The Thévenin equivalent of a good current source consists of a high E_{Th} coupled to a high R_{Th}, such that $I_{out} = E_{Th}/R_{Th}$. There are several ways of implementing a current source, one of which involves the actual construction of a "Thévenin" unit. For exam-

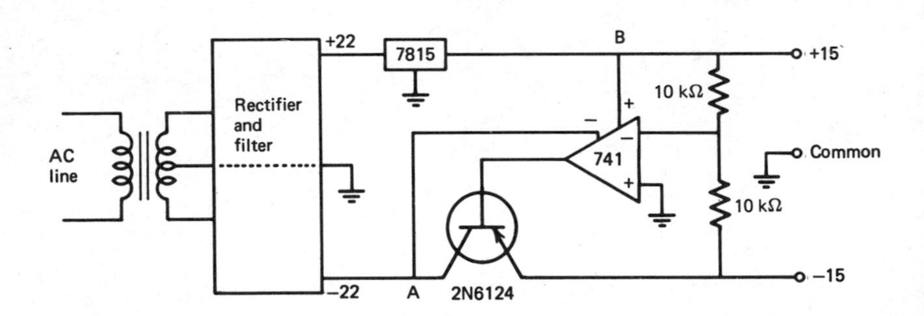

Figure 8-12. A dual-polarity "tracking" power supply. The op amp is itself powered from −22 V at point *A* and +15 V at point *B*. The 10-kΩ resistors must be precisely matched.

ple, a 100-V battery and 100 kΩ resistor form a nearly ideal 1-mA source, with a 100-V compliance.

Figure 8-13*a* shows an op amp supplying current to a load n the feedback loop. A booster *B* permits greater current to flow than the op amp alone could handle. A simple current regulator with small compliance can be made with a modular voltage regulator, as in Figure 8-13*b*.

SWITCHING CIRCUITS

A frequently encountered power device is the switch, which either permits or interrupts the flow of energy in a circuit. The most important types are based on mechanical operations, transistors, or *pnpn* devices called ***thyristors***.

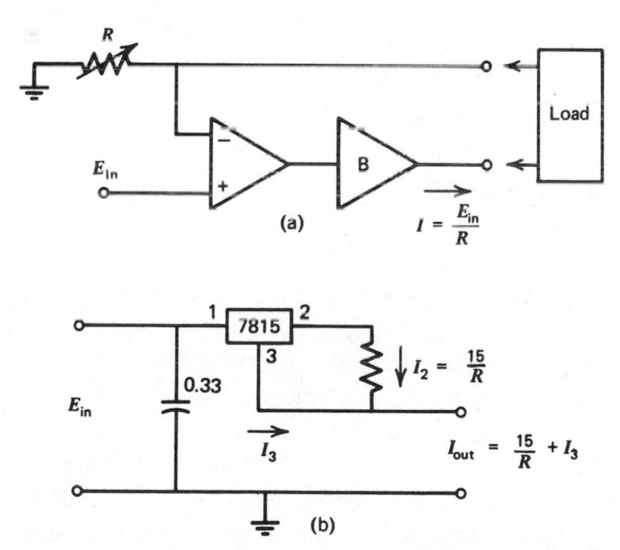

Figure 8-13. Current regulators: (*a*) a convenient circuit if the load need not be grounded; (*b*) a circuit for a grounded load, using the 7815 IC; I_3 is a constant current component, typically 5 mA.

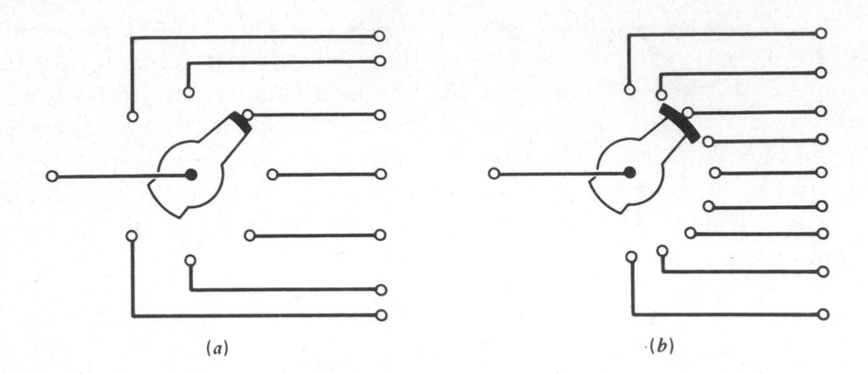

Figure 8-14. Multiple-contact rotary switches: (*a*) seven-position, non-short-circuiting; (*b*) nine-position, short-circuiting. The wiper contact in (*b*) will bridge the gap between adjacent fixed contacts, whereas that in (*a*) will not.

Mechanical Switches and Relays

These familiar types have many contact arrangements, from simple on-off devices to multiposition units that can break several circuits simultaneously. Figure 8-14 shows two of these. A distinction is made as to whether a new connection is made before the previous contact is broken (a short-circuiting or make-before-break switch), or a momentary interruption occurs on switching (nonshort-circuiting or break-before-make switch). It was mentioned in connection with Figure 3-7 that a multipoint switch used to select ammeter shunts *must* be of the short-circuiting type, as the meter would otherwise be subjected to a probably fatal momentary current during switching action.

Relays are also mechanical switches, but they are electrically actuated (Figure 8-15). The position of the contacts in the absence of energizing is referred

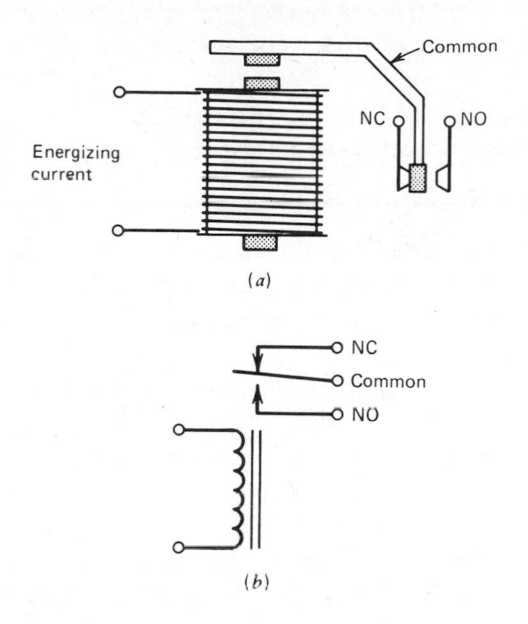

Figure 8-15. A typical relay: (*a*) one mode of construction; (*b*) the circuit symbol.

to as "normal," in the sense that a contact can be "normally open" (NO) or "normally closed" (NC). The characteristics of relays of importance to us are the voltage and current needed to energize the magnet and the maximum voltage and current ratings of the contacts. A good relay can complete a switching operation in as little as 10 ms or even less.

Relays can seldom be driven directly by op amps, since most require at least 0.5 W to operate, whereas an op amp might be able to deliver only about 0.2 W. The usual solution to this problem is to make use of a transistor booster, as illustrated in Figure 8-16. The op amp is configured as a comparator, since the rapid transition from one state to the other is desirable to give a fast snap action to the relay. The diode shunted across the relay coil provides a low-impedance path to ground for the inductive current produced by the coil on interruption of the energizing current.

For highly reliable, fast-acting relays, *mercury-wetted* types are advisable, even though they are rather costly. They do not tend to "chatter" or bounce on closure, as some other kinds may do. *Circuit breakers* are relays that are set to open automatically in case a circuit draws excessive current; they must be reset manually after the short circuit has been removed.

Transistor Switches

A transistor can be used as an on-off switch by supplying it with the appropriate signal to drive it into saturation or cutoff, respectively. The two states are marked *A* and *B* in Figure 8-17. Note that the load line crosses the forbidden area of excessive heat dissipation. This is permitted for a switch since it will not operate at any points other than *A* and *B*. A circuit operating as described is shown in Figure 8-18, where the input is a logical signal of 0 or +5 V, whereas the output is 25 or 0 V.

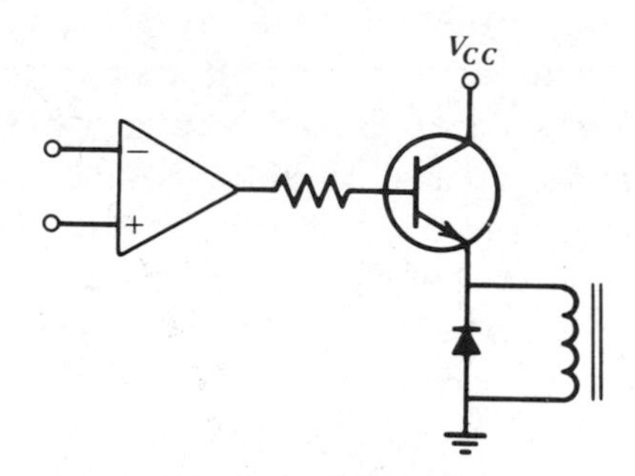

Figure 8-16. A relay coil operated by a comparator with a booster transistor. A relay with a winding of several hundred ohms DC resistance is appropriate for this connection.

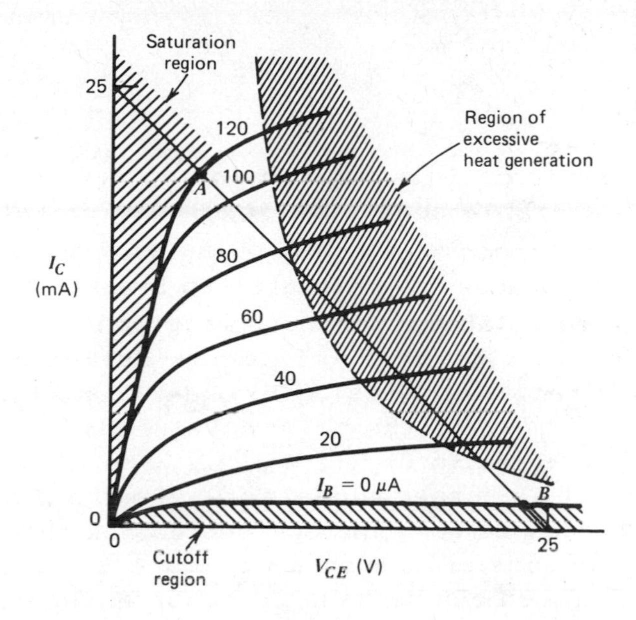

Figure 8-17. Typical output characteristic curve of an *npn* transistor. The distances between points *A* and *B* and their respective axes are exaggerated for clarity. The load line corresponds to a load resistance of 1000 Ω if V_{CC} = 25 V.

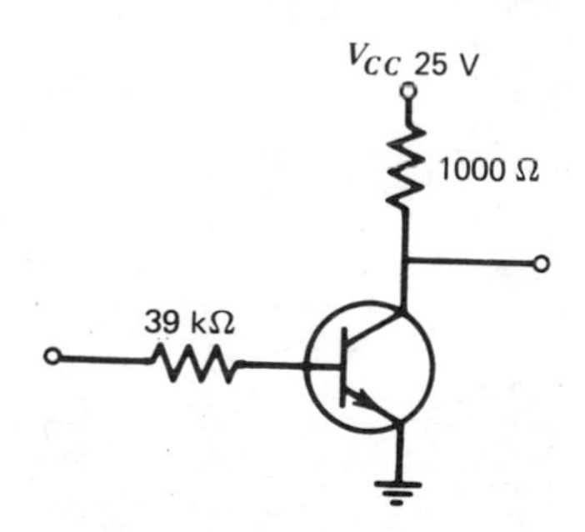

Figure 8-18. A switch using the transistor described in Figure 8-17.

THYRISTORS

The thyristor is a power-handling device that can remain in a stable condition in either of two modes, conducting or nonconducting. A pulse applied to a gate terminal will cause the thyristor to break into conduction and remain latched in that condition.

The basic type of thyristor is called a *silicon controlled rectifier* (SCR). To understand its operation, refer first to the half-wave rectifier using a standard diode, as in Figure 8-19*a*. The diode acts as a switch that is turned repeatedly on and off as the AC voltage crosses zero. The SCR represents an extension of the same mode of operation (Figure 8-19*b*). Here the switching on can be made to occur anywhere in the positive half period, but the switching off occurs only on crossing zero.

Physically the SCR is a combination of four semi-

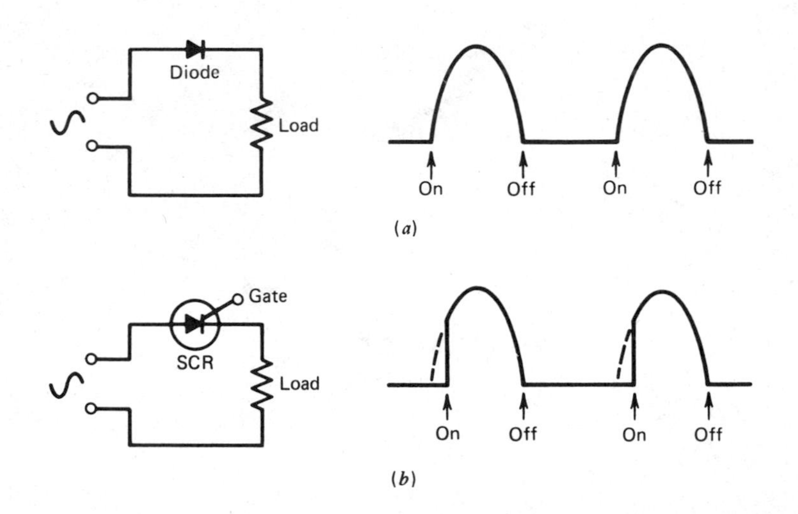

Figure 8-19. Comparison of (*a*) a diode, and (*b*) an SCR, as rectifiers.

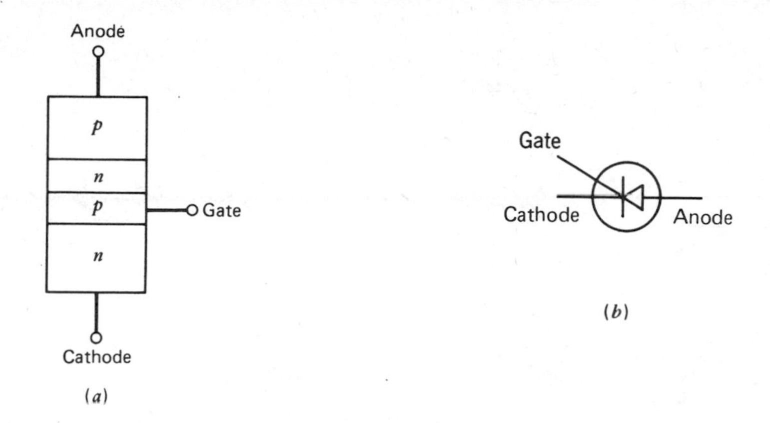

Figure 8-20. An SCR: (*a*) construction (schematic) and (*b*) the symbolic representation.

conductor layers in the sequence *pnpn* (Figure 8-20). The current-voltage characteristic of such a device is shown in Figure 8-21*a*. The general behavior is that of a conventional silicon diode, except for the region corresponding to small forward currents. Let us assume first that no connection is made to the gate electrode. If the anode-cathode potential is now increased in the forward direction, the current will initially assume a very small value (a few milliamperes in an SCR capable of carrying tens of amperes). It will follow the gently sloping curve for a considerable voltage range and then turn upward. At this point, labeled V_{BRO}, a forward "breakover" occurs (dotted line), and the device reverts to normal diode operation, with the current limited only by the resistance of the external circuit. At the same time, the potential drop across the device falls to perhaps 1 V. This abrupt decrease in impedance is called "firing" or "turning on" of the thyristor. Once fired, it cannot be turned off again except by interruption of the anode current. The breakover voltage can be decreased by supplying a positive potential to the gate relative to

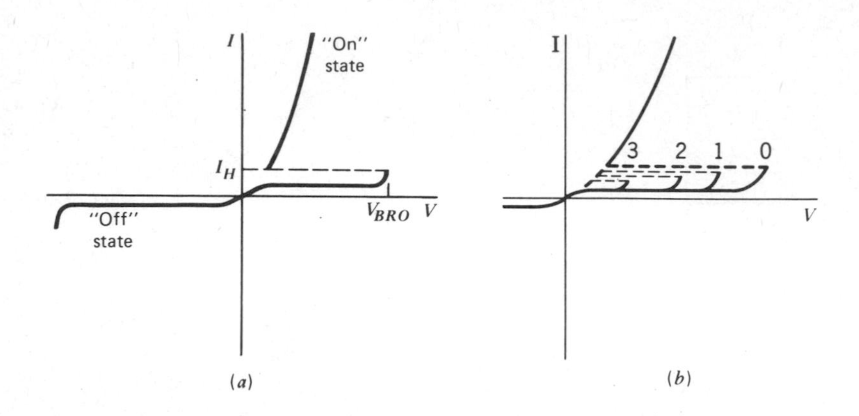

Figure 8-21. Characteristics of an SCR: (*a*) with zero gate current, where I_H designates the holding current, below which the device turns off; (*b*) the effect of varying the gate current I_G.

the cathode, as seen in Figure 8-21*b*.

The switching-on process can be better understood if we refer to the circuit and load line shown in Figure 8-22. The unusual feature of the graph is that there are *three* intersections of the load line with the characteristic curve. Point *A* is no different from an operating point in any conducting diode. The voltage is about 1 V, and the current, depending on the load, may be several amperes or tens of amperes. The fact that the voltage drop in the conducting mode is so small keeps the heat dissipation from being excessive. Point *C* corresponds to an off condition, with the voltage across the diode nearly equal to V_{AA}. In this case the anode current is very low, so that, again, the heat dissipation is small. Point *B* is unique in that it does not correspond to any stable state.

When first connected, the SCR normally goes to the off state.* Firing can be brought about by applying a

* If the voltage V_{AA} is applied too suddenly, the thyristor may self-trigger and go directly into conduction.

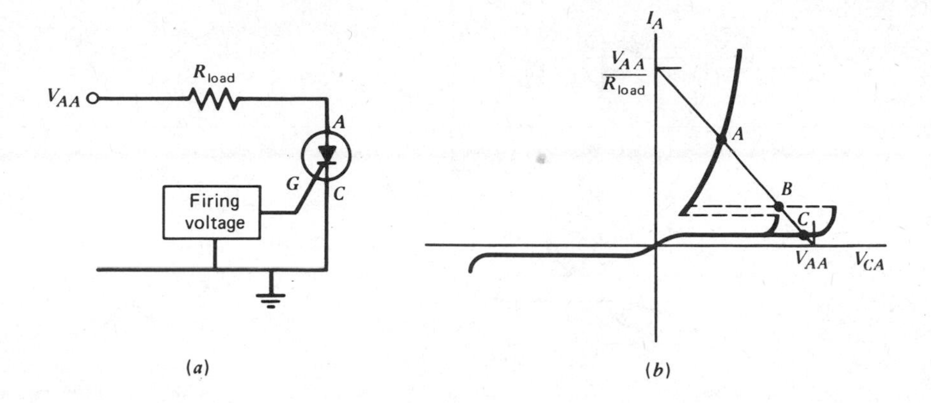

Figure 8-22. (*a*) The basic SCR circuit and (*b*) a load line superimposed on the curve shown in Figure 8-21.

positive pulse to the gate electrode. This changes the characteristic curve produced by the device, which now follows the dotted line in Figure 8-22*b*. At this time points *B* and *C* disappear and only *A* remains, so that the state of the SCR is forced to assume the current and voltage corresponding to this point, and the device is successfully fired.

The SCR cannot be turned off by any kind of signal at the gate electrode but will continue to conduct until the potential V_{CA} is brought momentarily close to zero (or negative). When the SCR is acting as a rectifier, the current is automatically interrupted once in every AC cycle as the waveform crosses zero.

There are several types of thyristors in addition to the SCR. One of these is the *triac*, which is a bidirectional modification of the SCR. The characteristic curve of the triac is the same as for the SCR in the first quadrant and repeats that curve symmetrically in the third quadrant (Figure 8-23).

In principle, a thyristor can be triggered by a DC current, but heat dissipation considerations require that only pulses be used. One device that can give an appropriate pulse is the *unijunction transistor* (UJT),

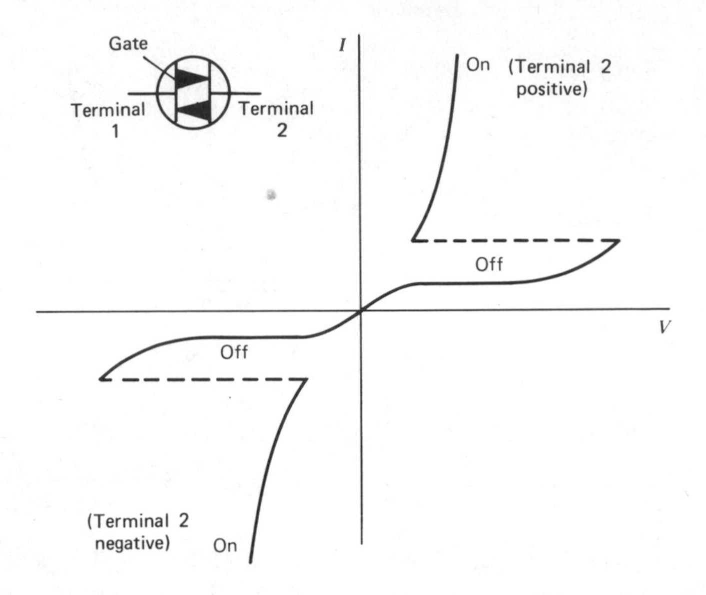

Figure 8-23. The triac thyristor and its symbol.

a three-terminal device shown in a typical circuit in Figure 8-24. The important characteristic is that the resistance between the emitter E and first base B_1 suddenly decreases from its normal level of several kilohms to nearly zero when the potential of the emitter is raised above a critical value, thus permitting sizable currents to pass. This property allows the design of a simple circuit for generating a train of pulses. When the power supply is connected, the capacitor begins to charge through R_T. When its potential equals the critical value, the resistance between E and B_1 drops abruptly and the capacitor discharges through the UJT and R_1. When it reaches some low level, depending on the size of R_1, the base-emitter resistance returns to its original high level, the capacitor starts to recharge, and the cycle repeats.

Figure 8-25 shows an application of a UJT to the control of an SCR in a temperature regulator. This circuit contains neither a transformer nor a relay,

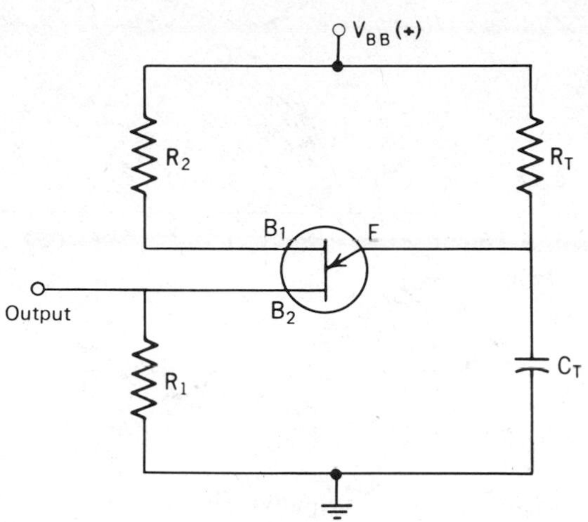

Figure 8-24. A pulse generator using a unijunction transistor.

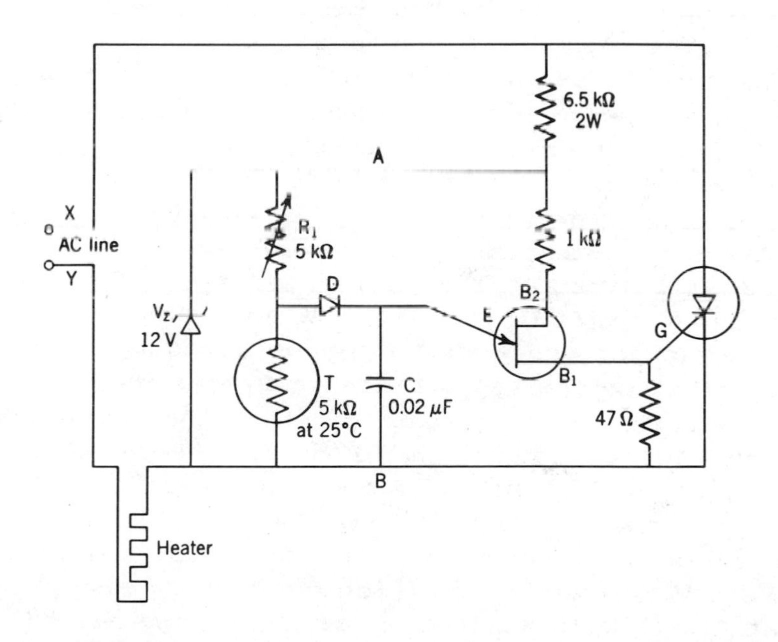

Figure 8-25. A temperature-control circuit using a UJT to drive an SCR.

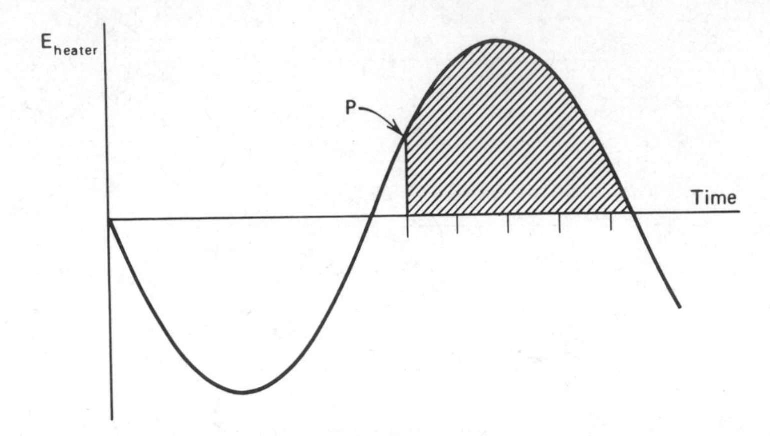

Figure 8-26. Operating principles of the circuit shown in Figure 8-25. The heater is *on* during the shaded portions of the cycle. The tick marks correspond to successive firings of the UJT.

hence can be constructed in a conveniently compact form. The condition of the circuit must be studied separately for successive half-cycles of the AC power. When the line connection marked Y is positive, the SCR cannot conduct, regardless of what happens to its gate, whereas the Zener diode conducts freely in its forward direction. Hence points A and B will be at essentially the same potential, and the UJT cannot operate. On the next half cycle the power lead X is positive while the Zener maintains A at 12 V more positive than B. The capacitor proceeds to charge through R_1 at a rate determined by the resistances of R_1 and of the *thermistor* T* immersed in the bath of which the temperature is to be controlled. The lower the temperature, the larger the resistance of T and the greater the fraction of the current through R_1 available to charge C. This increases the frequency of production of pulses by the UJT. The direct connection between B_1 and G ensures that the SCR is turned on at that time within each

* A thermistor is a resistor made of metal oxides sintered together. It shows a characteristically high negative temperature coefficient.

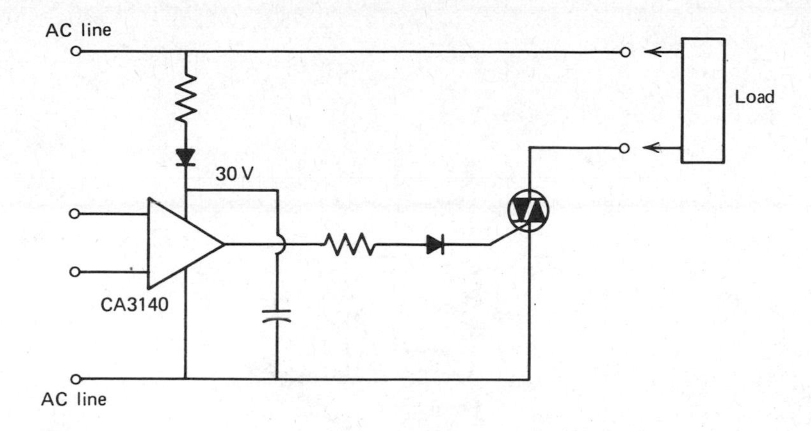

Figure 8-27. A triac triggered by an op amp. Note that the amplifier is powered by half-wave rectified AC. The entire circuit can be considered to be a solid-state relay.

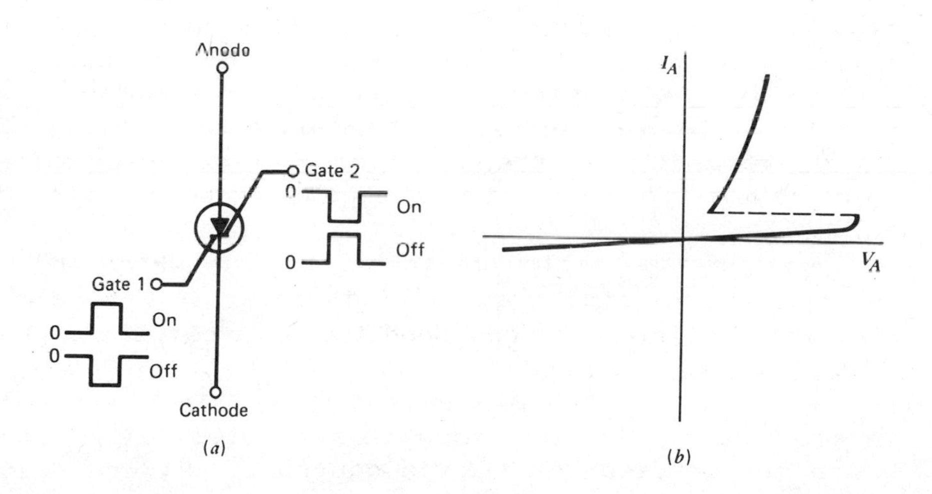

Figure 8-28. (*a*) A silicon-controlled switch and (*b*) its current-voltage characteristic, resembling that of an SCR.

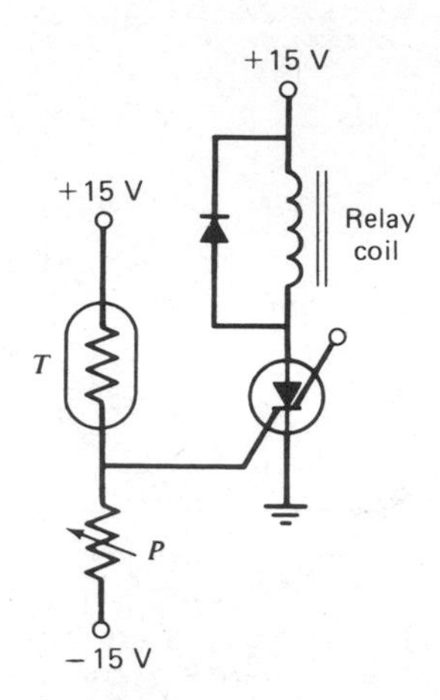

Figure 8-29. An SCS used as an overheat sensor. The potentiometer sets the gate voltage at zero when the temperature is at its normal value. The gate current is about 1 μA and the anode current, perhaps 50 mA.

cycle when the first pulse from the UJT appears (Figure 8-26). The SCR turns off automatically when the voltage crosses zero. The energy supplied to the heater, indicated by the shaded portion of the diagram, is greater when the pulse rate is faster, corresponding to a temperature lower than desired. The temperature level can be set by adjustment of the variable resistor R_1.

Another application of thyristors, in which an op amp is used to trigger a triac, is shown in Figure 8-27. The amplifier can be connected as a comparator to trigger at a given input level.

A *silicon controlled switch* (SCS) is a low-power version of an SCR provided with two gates, either of which can be used for firing control. Figure 8-28 shows the symbol and characteristic curve, and Figure 8-29 gives a typical application. A *diac* is a two-terminal device that can be used to control the firing of a triac or SCR (Figure 8-30).

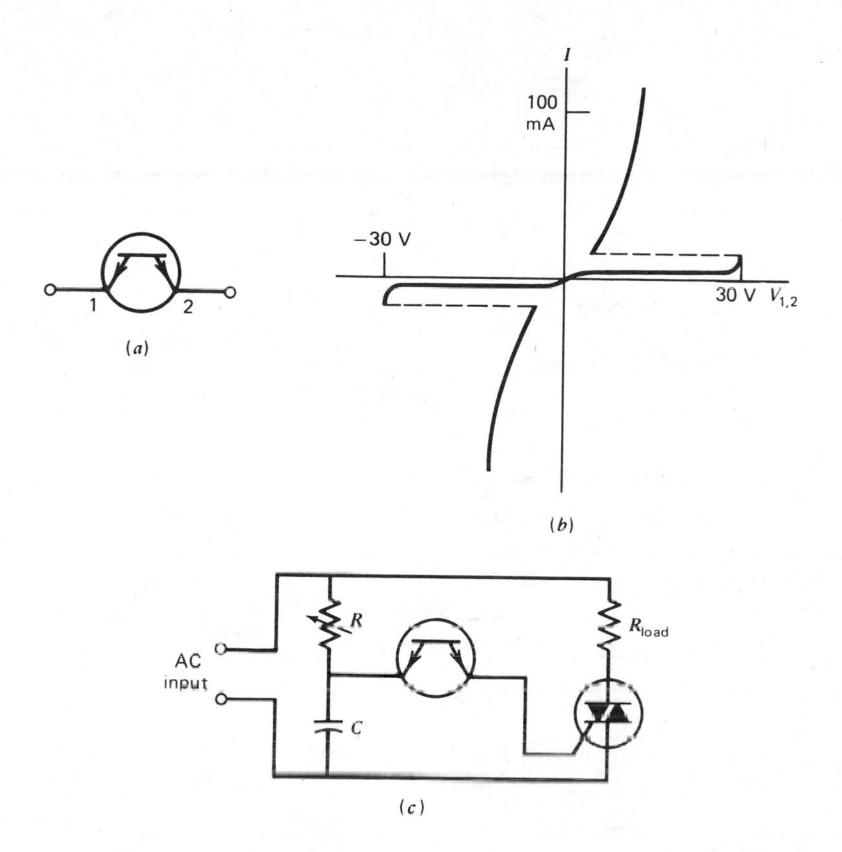

Figure 8-30. (*a*) Symbol for a DIAC; (*b*) its characteristic curve; (*c*) a circuit used to vary the power in the load by varying the fraction of each half period of the AC current during which the TRIAC is on.

PROBLEMS

8-1. Calculate the average and RMS output voltages for the circuits shown in Figure 8-5, assuming that the transformer gives 20 V (RMS) across its secondary. Neglect the forward voltage drops of the diodes.

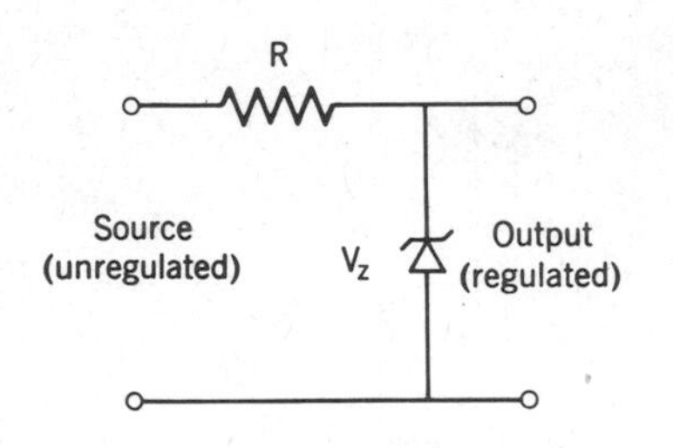

Figure 8-31.
See Problem 8-2.

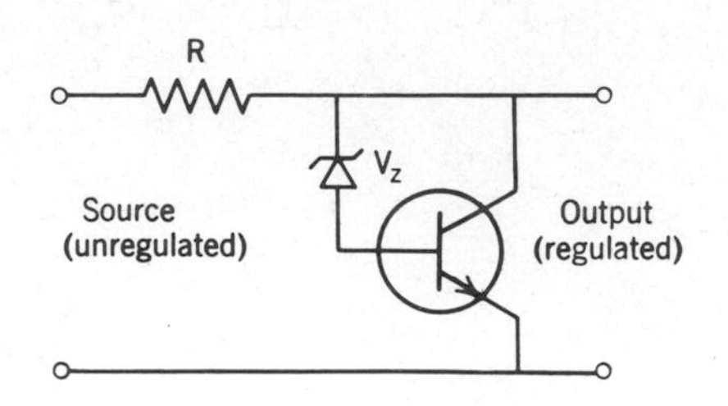

Figure 8-32
See Problem 8-3.

8-2. In the half-wave rectifier shown in Figure 8-31 the diode and resistor pair constitute a voltage divider. Calculate the ratio E_{out}/E_{in} for both the positive and negative half-cycles of the AC. Use the following data: $R = 1\ k\Omega$, $R_{fwd} = 1\ \Omega$, $R_{rev} = 1\ M\Omega$. (The terms R_{fwd} and R_{rev} denote the forward and reverse resistances of the diode. Neglect the forward voltage drop.)

8-3. What should the power ratings of the three components be in the circuit shown in Figure 8-32 with $R = 100\ \Omega$, $V_Z = 10$ V, and $E_{in} = 20$ V? Assume that $V_{BE} = 0.5$ V.

8-4. In the Zener regulator circuit shown in Figure 8-33 with $V_Z = 12$ V, compute the resistance and necessary power rating of R, and the maximum power dissipation in the diode, if the load is

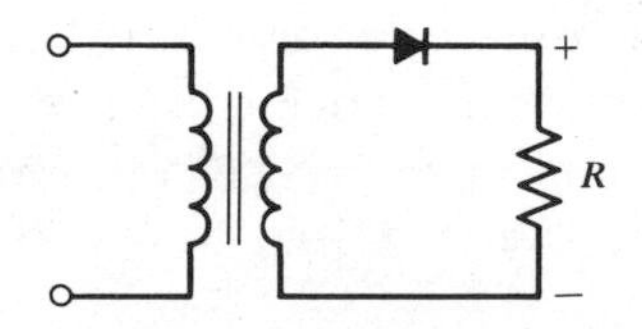

Figure 8-33. See Problem 8-4.

allowed to vary from 100 to 2000 Ω. (It may be assumed that V_Z is constant over the required range of currents. Take E_{in} as +15 V.)

8-5. Consider a constant-current source consisting of a 510-V battery and a 2 MΩ resistor.

(a) What is the current output into a 1 kΩ load? A 10 kΩ load?
(b) What is the regulation, defined as 100 × $(\Delta I/\Delta R)$? What are its units?
(c) What is the voltage compliance?

8-6. Design a variable-voltage power supply to cover the range 0 to 10 V, capable of at least 50 mA, using a regulator of the 7800 series.

8-7. Explain the nature of the three shaded areas in Figure 8-17.

8-8. Make a report on the use of the RCA model CA3059 IC used in connection with triacs.

8-9. What is meant by the term "negative resistance"? In what types of application is it useful?

8-10. If a triac were substituted for the SCR shown in Figure 8-25, what changes would be needed in the circuit? How would it change the diagram in Figure 8-26?

IX
LOGIC SYSTEMS

The major thrust in the previous chapters has been the treatment of signals in the analog domain. We now turn to the digital realm, where in place of amplifiers and function generators we encounter such devices as logic gates, digitizers, and counters.

The interrelationship between the two areas is illustrated in the design of an autoranging voltmeter (Figure 9-1). The analog input, after buffering, is fed into a programmable amplifier, the gain of which can be controlled by a digital signal. Whenever the

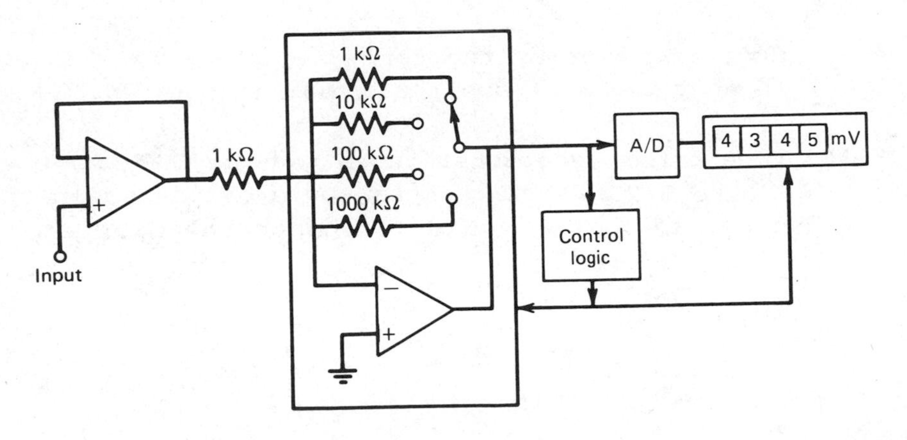

Figure 9-1. An autoranging millivolt meter.

voltage from the amplifier is outside the proper range, the control logic will shift the gain in the required direction to keep the voltage on scale. This permits the measurement of any voltage between 10 mV and 10 V with four-digit resolution. An analog-to-digital converter (A/D in Figure 9-1) is a device that generates the actual digits for display. The location of the decimal point is set by a signal from the logic system.

Before proceeding to further discussion of such instruments and their components, let us devote some attention to fundamentals.

LOGIC STATES

Every logic signal must exist in one of only two states. Hence each logic device must give an output at either of two specified voltages, but nowhere in between. Similarly, devices that receive digital signals must be able to respond appropriately to each of the two levels, but need not produce a defined response to intermediate voltages.

The two levels are designated by the terms "high" and "low" or "one" and "zero." We will use "1" and "0," but many data sheets from manufacturers use high and low (or H and L). The "1" and "0" notation is consistent with the binary system of numeration, introduced in a later paragraph. Logic systems must have established voltage levels corresponding to these two states. These voltages could, in principle, be of any desired values, but in the interests of standardization many manufacturers have agreed on zero volts for "0" and +5 V for "1." There is some latitude in these voltages, as indicated in Figure 9-2.

The smallest portion of information that can be represented in a digital system is the *bit*, a logical "1" or "0" at a given point in a circuit at a given instant. Digital information is usually organized in

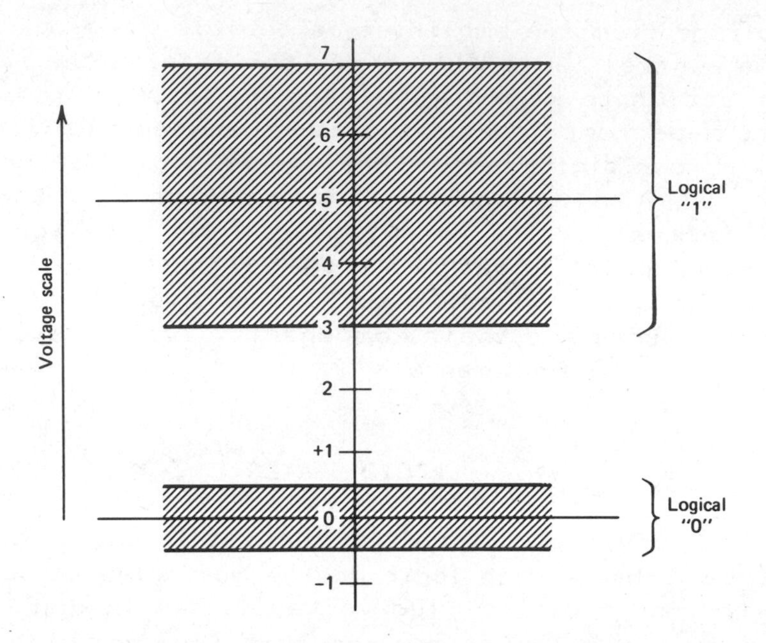

Figure 9-2. Typical logic levels (shaded areas). Nominal values are 0 and +5 V. The region between +0.5 and +3.0 is logically undefined.

groups of 8 bits, called *bytes*. A *word* of information may consist of one or more bytes. These 8 bits can be transmitted from one unit to another either in a parallel arrangement using 8 wires, or sequentially with a single conductor. A group of 8 bits can handle numbers from 0 to 255 (which is $2^8 - 1$), whereas 16-bit logic (2 bytes) can extend to 65535 ($2^{16} - 1$).

Operations on Signals

The fundamental unit in handling digital signals is the *logic gate*. This is a device in which a specific output state is generated by each combination of input signals. In a typical logic system, there may be from a few up to hundreds or thousands of individual gates, each of which can be physically small, as they

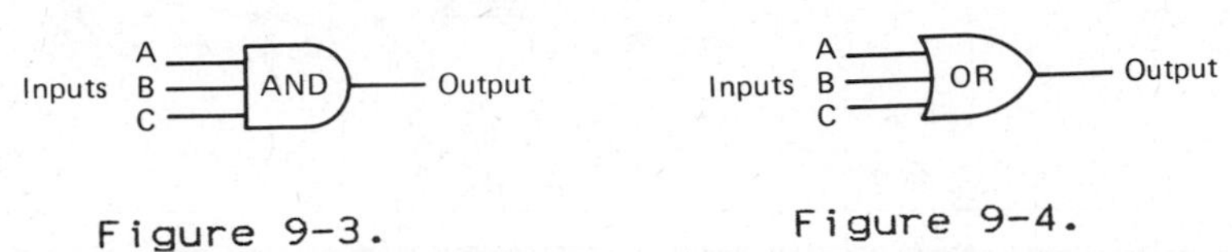

Figure 9-3.

The logical **AND** gate.

Figure 9-4.

The logical **OR** gate.

are never called on to handle appreciable power. Modular ICs are available that combine many gates into a single component to implement complex logic operations.

An example of a unit logic element is the **AND** gate shown in Figure 9-3. The output is a signal taking the value "1" only if *A* **and** *B* **and** *C* are all at logic "1." The **AND** gate answers the query as to whether all inputs are HIGH, giving the response of "1" for "yes" and "0" for "no."

Figure 9-4 shows a related gate, called the **OR** gate. This gate answers the question as to whether any of the inputs *A* **or** *B* **or** *C* are at logic 1. The answer is "1" unless *A*, *B*, and *C* are all zero. There are two other basic gates, called **NAND** and **NOR**, the reverse, respectively, of **AND** and **OR**. Each of these can be made up by a combination of its prototype with an *inverter* (Figure 9-5), a gate that converts a "1" to a "0" and vice versa. The inverter can have only a

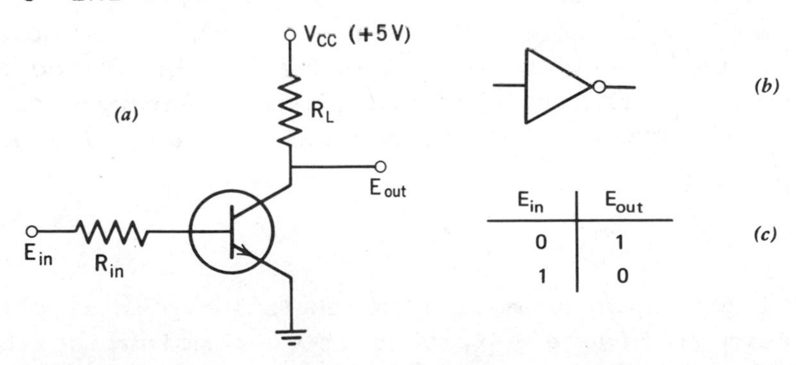

E_{in}	E_{out}
0	1
1	0

Figure 9-5. The logic inverter: (*a*) circuit; (*b*) symbol (not to be confused with an op amp); (*c*) truth table.

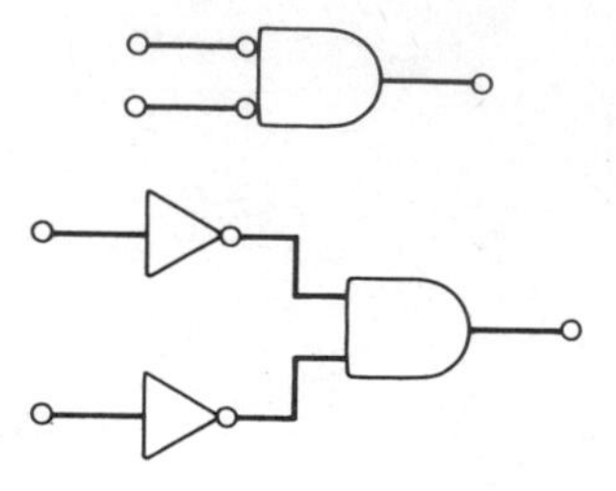

Figure 9-6. These two symbols have identical significance.

single input, as shown in the schematics, but the others may be provided with any number from 2 to 8 or even more.

Also useful is one additional gate, the exclusive-OR, often designated as **XOR**, that can be built up by a combination of basic gates, as will be shown presently. All of these gates are displayed in Table 9-1. Also given is the *truth table* for each, a term borrowed from symbolic logic. This describes exhaustively the outputs for *all* possible combinations of input states.

Note that the symbols denoting the **NAND** and **NOR** gates are the same as the **AND** and **OR**, but with a small circle at the output, indicating inversion. A gate symbol is sometimes seen with one or more circles at the inputs, also showing inversion, as demonstrated in Figure 9-6.

There are a number of families of gates available, of which three will be discussed here, designated as *transistor-transistor logic* (TTL), *complementary metal-oxide semiconductor* (CMOS), and *emitter-coupled logic* (ECL).

TTL Logic

TTL gates, an example of which is shown in simplified form in Figure 9-7, consist of combinations of transistors and resistors. Of the four transistors, the two at the right constitute the output stage, commonly

TABLE 9-1

Truth Tables for Several Gates

Inputs							
A	*B*	*C*	**AND**	**NAND**	**OR**	**NOR**	**XOR**
0	0	0	0	1	0	1	0
0	0	1	0	1	1	0	1
0	1	0	0	1	1	0	1
0	1	1	0	1	1	0	0
1	0	0	0	1	1	0	1
1	0	1	0	1	1	0	0
1	1	0	0	1	1	0	0
1	1	1	1	0	1	0	0

known as a "totem-pole" arrangement, with one transistor sitting on top of another. The preceding stage is a phase splitter: if transistor Q_2 is turned on (i.e., conducting), the voltage drops in the associated resistors will lower the voltage on the base of Q_3, turning it off, and at the same time will raise the voltage to Q_4, turning it on. The result of this action is that the output will be nearly at ground potential, within the permissible limits for logical "0." On the other hand, if Q_2 is off (nonconducting), the base of Q_3 will be almost at the +5 V potential of the power supply, and that of Q_4 will be at ground, so Q_3 will be on and Q_4 off, and the output will be essentially +5 V.

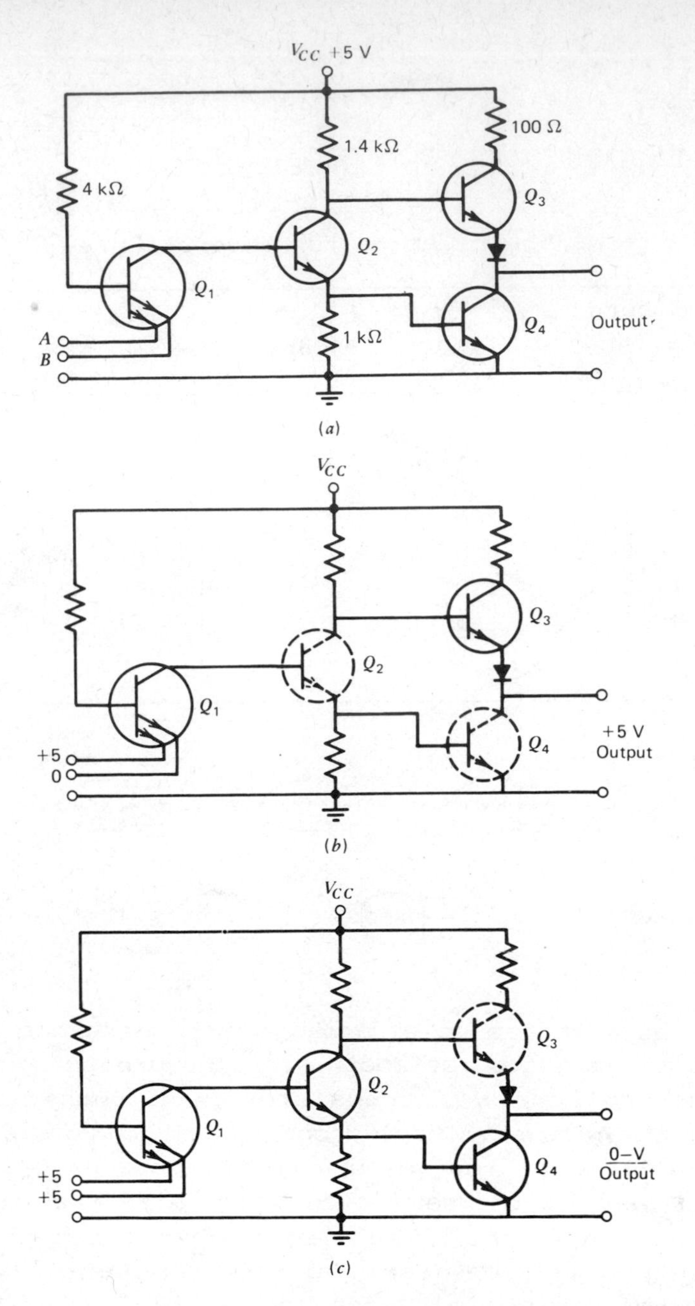

Figure 9-7. A TTL two-input **NAND** gate: (*a*) the basic circuit; (*b*) one input high, the other low; (*c*) both inputs high. Transistors shown dashed are *off* and can be considered to be removed from the circuit.

The difference between various gate types lies in the input section. In the **NAND** gate shown in Figure 19-7 the input uses a two-emitter *npn* transistor as driver for Q_2. If either (or both) of the emitters are grounded ("0" signal; Figure 9-7*b*), the voltage presented to the base of Q_2 will be above ground by only about 0.6 V (the forward voltage drop of one *pn* junction), which cuts it off. This results in the final output being high, following the reasoning outlined previously. However, if both emitters are high, Q_2 will be turned on, and the output will go low (Figure 9-7*c*).

TTL units are available in a large variety of standardized forms (the 7400 series). Figure 9-8 gives the pin-out diagrams for a selection of them. They are low priced and robust, and their rather high input impedance guarantees that one gate can feed signal to as many as 10 inputs of other gates at the same time ("fan-out" of 10). Most TTL gates have counterparts that are designated as *open-collector* types. In these the upper transistor of the totem pole is omitted, and the collector of the lower one has no internal connection to the power supply. Hence such a connection must be made externally, permitting added flexibility in some types of circuitry.

There are several subseries modified with respect to the speed with which each gate responds to its inputs (their *transition times*) and to the average amount of power dissipated as heat:

74XX	Standard	10 ns, 10 mW
74LXX	Low-power	33 ns, 1 mW
74HXX	High-speed	6 ns, 22 mW
74SXX	Schottky high-speed	3 ns, 19 mW
74LSXX	Schottky low-power	9 ns, 2 mW

The power dissipation, although small for each gate, could add up to a rather high value in a large instrument. The transition time of any series is sufficiently fast for most laboratory applications but is a limiting feature in computers. The two Schottky series are built with a special type of metal semiconductor diode that prevents the transistors from going into saturation. They are somewhat more expensive than the standard types. TTL units with the prefix 54 instead of 74 are identical except that they are guaranteed for use from -55 to +125°C, whereas the 74 series are rated only for service from 0 to +70°C.

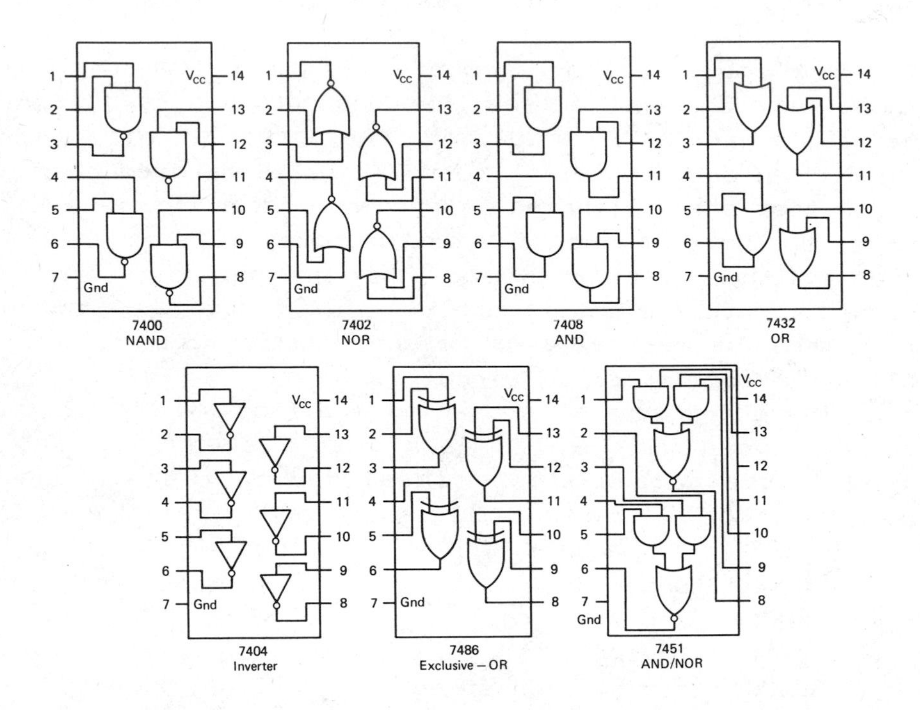

Figure 9-8. A selection of TTL integrated circuits.

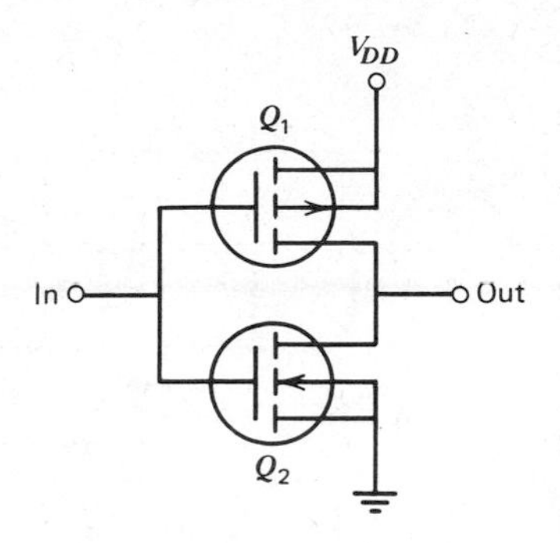

Figure 9-9. A CMOS inverter.

CMOS Logic

The second logic system to be discussed is CMOS. In this type, the bipolar transistors used in TTL are replaced with depletion-mode MOSFETs, an example of which is shown in Figure 9-9. The totem pole now consists of two transistors, respectively *p*- and *n*-channel types, connected between the power source and ground. If the gates of these two transistors are given the same voltage, one will be turned on as the other is turned off, an ideal arrangement for the output of a logic gate. For the inverter, as shown in Figure 9-9, these two transistors are the only essential parts.

Figure 9-10 gives the diagram of a **NAND** gate in the CMOS series, consisting of four transistors, two of each polarity. The function can be understood if the transistors are regarded as switches. When input *A* is high, Q_2 is opened and Q_3 is closed; likewise, when *B* is high, Q_1 is opened and Q_4 closed, thus disconnecting the output from the positive voltage supply and connecting it to ground. Reversal of either or both inputs connects the output to +5 V, simultaneously removing it from ground. A little thought will show that this is

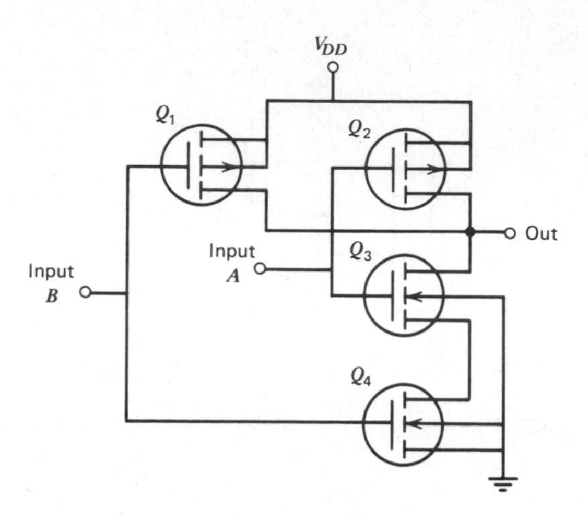

Figure 9-10. A CMOS two-input **NAND** gate.

the requirement for a **NAND** gate.

CMOS gates can operate at supply voltages up to +18 V. They can conveniently be combined with op amps if required, using the +15 V (but not the −15 V) supply. On the other hand they can be interfaced with TTL at +5 V, subject to certain limitations: the CMOS gate may not have enough power to operate more than one TTL gate directly. The TTL output will usually require a resistor (2.2 kΩ is about right) from its output to the +5 V line to permit reliable operation of a CMOS gate (Figure 9-11). (This is called a "pull-up resistor.") It is usually preferable to stay within either system for any particular project.

CMOS has two advantages over TTL: Because of the higher permissible voltage, the power supply rarely needs to be closely regulated. More significantly, CMOS is capable of operating at greater speed than the standard TTL line. However, the 74LSXX series of TTL is nearly as fast as CMOS. Standard TTL gates are somewhat less expensive than CMOS, although the difference is not great.

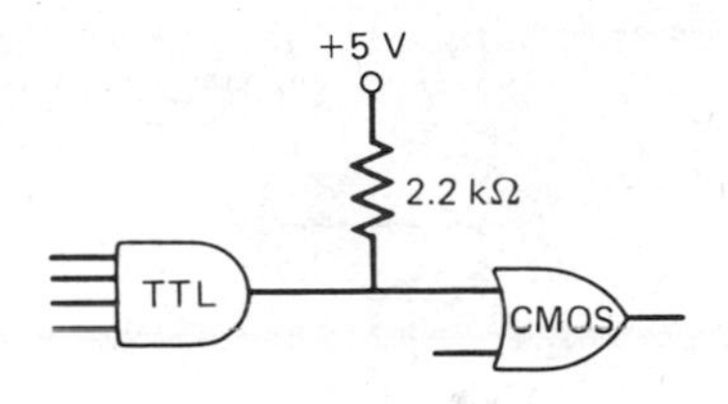

Figure 9-11. Coupling a TTL gate to a CMOS gate.

ECL Logic

Of particular importance for high-speed circuitry is the family designated as ECL. This is based on the two-transistor differential amplifier, which was described in its linear applications in Chapter VII. Consider the circuit shown in Figure 9-12*a*, in which it is assumed that the two transistors are closely matched, and that no current is taken from the output. Clearly, from the symmetry of the circuit, if E_{in} = E_{ref}, the two collector currents will also be equal. If E_{in} exceeds E_{ref}, however, even by as little as 100 mV, Q_1 will become more conductive, so that all the current through R_E will flow through it and Q_2 will be cut off. Resistance R_E is made large enough (several kilohms) that it becomes the chief factor determining the current. The total current drawn from the power supply is essentially the same in the "0" and "1" output states; hence the power-line voltage spikes (*glitches*) often generated by the change of state in TTL devices are almost totally absent in ECL. The high-speed operation permissible with ECL is due to the fact that the transistors never go into saturation, as they do in other logic systems. Saturation causes charge storage, which takes a significant time to dissipate.

Figure 9-12*c* is a diagram of a typical gate in ECL

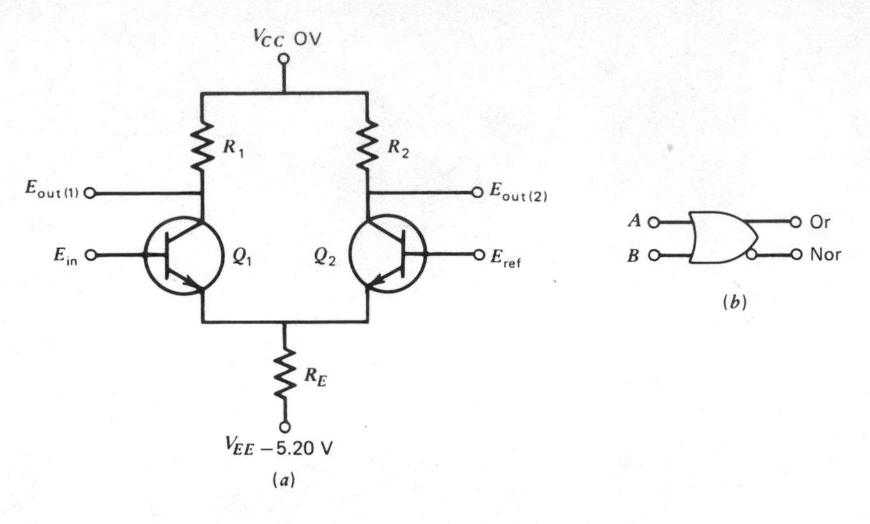

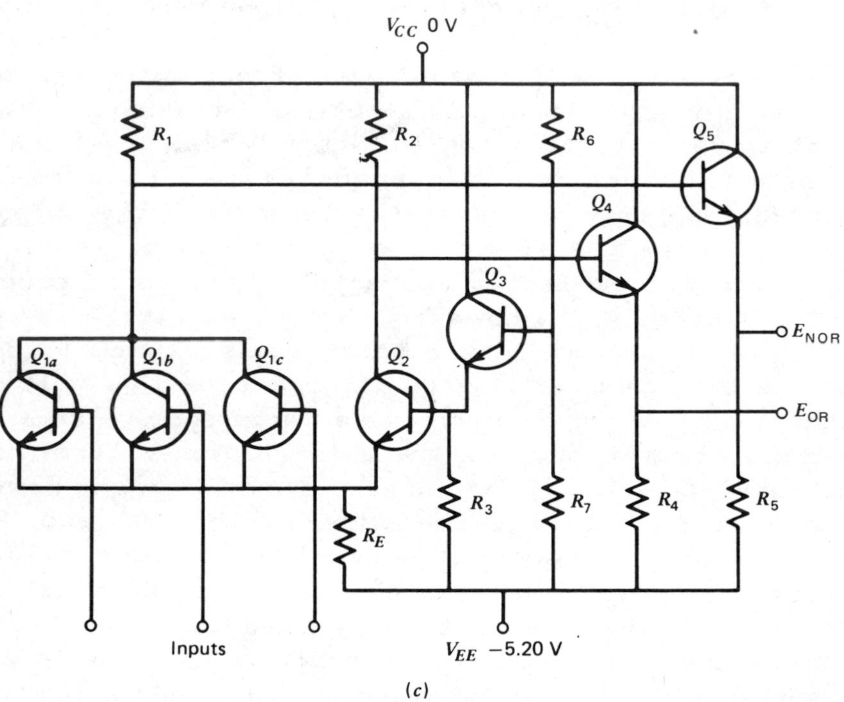

Figure 9-12. ECL gates: (*a*) the basic differential amplifier; (*b*) the symbol for an **OR/NOR** gate; (*c*) the detailed circuit of an OR/NOR gate.

logic. This has three points of difference from the simple circuit discussed earlier. Transistor Q_1 is replaced by several similar transistors in parallel, thus providing for a multiple-input gate. Input E_{ref} is now given a fixed bias through transistor Q_3.

A pair of emitter-follower transistors, Q_4 and Q_5, serve to provide the needed output voltage levels. The voltage levels for ECL logic are different from those we have seen for other systems. For high-speed operation a "high" is nominally -0.9 V, whereas a "low" is -1.75 V. The reference level (the base of Q_2) is set at -1.3 V since it is necessary for the input voltage to vary significantly both above and below this point. For lower-speed applications, the input voltages can be as high as 0 V for the high and as low as -5 V for low. The rise time is somewhat greater with these voltages. It must be noted that the immunity to noise interference is rather less in ECL than in TTL or CMOS. The reason for grounding the high level rather than the low, is to reduce interference from noise.

The output from Q_4 gives the response of an **OR** gate. Since the symmetry of the circuit permits its easy implementation, a second output is provided through Q_5, which gives a **NOR** response. This gate, called an **OR/NOR** gate, is symbolized as shown in Figure 9-12*b*.

BOOLEAN ALGEBRA

The action of logic gates can be interpreted conveniently in terms of *Boolean algebra*, a mathematical discipline for handling two-state variables. According to Boolean notation, the variables *A*, *B*, *C*, • • •, each of which can have only two values (1 and 0), can be related to each other by a series of operators:

1. The dot operator (•):
   $A \cdot B \cdot C$ means A **and** B **and** C
2. The plus operator (+):
   $A + B + C$ means A **or** B **or** C
3. The encircled plus operator ($\oplus$):
   $A \oplus B \oplus C$ means any *one* of A or B or C
4. The bar operator (¯):
   $\bar{A}$ means "**not**-A" or "complement of A"

These notations have a direct correspondence with various logic gates. Thus the outputs of the gates list in Table 9-1 can be expressed as follows:

AND	$F = A \cdot B \cdot C$
NAND	$F = \overline{A \cdot B \cdot C}$
OR	$F = A + B + C$
NOR	$F = \overline{A + B + C}$
XOR	$F = A \oplus B \oplus C$

These functions can be combined in various ways, much as can functions in ordinary algebra. Thus $A + B + C$ is the same as $B + C + A$. However, $A \cdot (B + C)$ is *not* the same as $(A \cdot B) + C$; the operation within parentheses must be made prior to combination with other variables.

A well-established principle in symbolic logic, called *De Morgan's theorem*, states that (**not**-A **AND not**-B) is equal to **not**-(A **OR** B) and that (**not**-A **OR not**-B) equals **not**-(A **AND** B), or in Boolean notation

$$\bar{A} \cdot \bar{B} = \overline{A + B} \qquad \bar{A} + \bar{B} = \overline{A \cdot B} \tag{9-1}$$

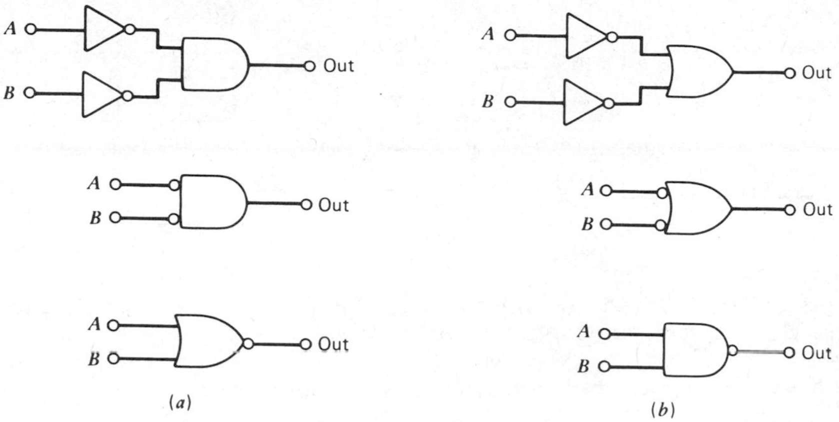

Figure 9-13. Combination of gates to illustrate equivalence of functions: (*a*) three symbols for a **NOR** gate; (*b*) three symbols for a **NAND** gate.

These statements can be related to simple gate structures, as shown in Figure 9-13. (Remember that Boolean "•" means **AND** and "+" means **OR**, and that a circle at an input to a gate inverts the input, just as a circle at the output inverts the output.) This equivalence of two gate assemblies is often convenient in circuit design, as it may permit more efficient use of multigate components such as some shown in Figure 9-8.

SYNTHESIS OF GATE SYSTEMS

It is often necessary to assemble logic systems by combining individual gates in a suitable manner. As an example, consider the exclusive-**OR** gate. This can be built up from **AND**, **OR**, and inverter components according to the diagram in Figure 9-14. To show the correctness of this assemblage, one can make a truth table, also shown in the figure, by including columns corresponding to the signals present at points P and Q

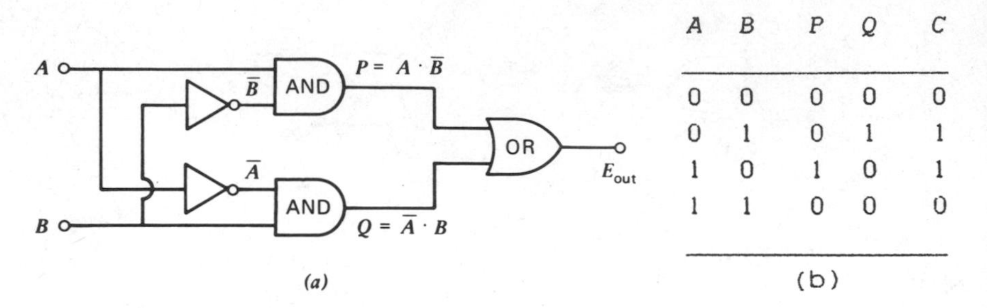

A	B	P	Q	C
0	0	0	0	0
0	1	0	1	1
1	0	1	0	1
1	1	0	0	0

(b)

Figure 9-14. (*a*) An exclusive-**OR** gate with two inputs; (*b*) the corresponding truth table.

as well as at inputs A and B and output C. The values in the P and Q columns are determined by analyzing the effect of the **AND** gates and inverters on the various possible combinations of input values. The P and Q values, in turn, produce the output signal by the action of the **OR** gate. The reader should verify that this result is consistent with that predicted in Table 9-1.

As an example of an industrial application of a multigate system, consider the following hypothetical case. A given mechanism is to be actuated only for the particular combinations of binary signals from four sensors that give OUT = "1" in the truth table shown in Figure 9-14*b*. To implement such a table, it is always possible, but often not optimal, to use a combination of **AND**, **OR**, and **NOT** gates, simply **OR**-ing all the cases in which the output is desired to be "1." Such an implementation for the present example is shown in Figure 9-14*a*.

It is often possible to simplify a logic circuit of this kind. Examination of the truth table in Figure 9-15 will show that the logic can be expressed as "(A **OR** B) **AND** C **AND** **not**-D," or in Boolean notation as "$(A + B) \cdot C \cdot \overline{D}$." This leads to the diagram in Figure 9-16, in which three gates are used rather than nine. Boolean algebra can serve as a much more power-

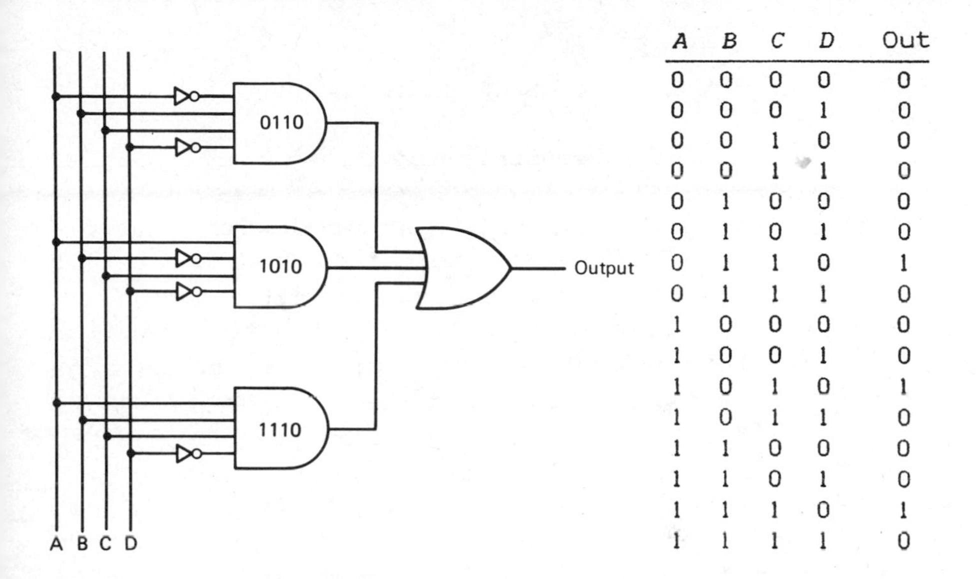

A	B	C	D	Out
0	0	0	0	0
0	0	0	1	0
0	0	1	0	0
0	0	1	1	0
0	1	0	0	0
0	1	0	1	0
0	1	1	0	1
0	1	1	1	0
1	0	0	0	0
1	0	0	1	0
1	0	1	0	1
1	0	1	1	0
1	1	0	0	0
1	1	0	1	0
1	1	1	0	1
1	1	1	1	0

Figure 9-15. A "brute force" implementation of a truth table.

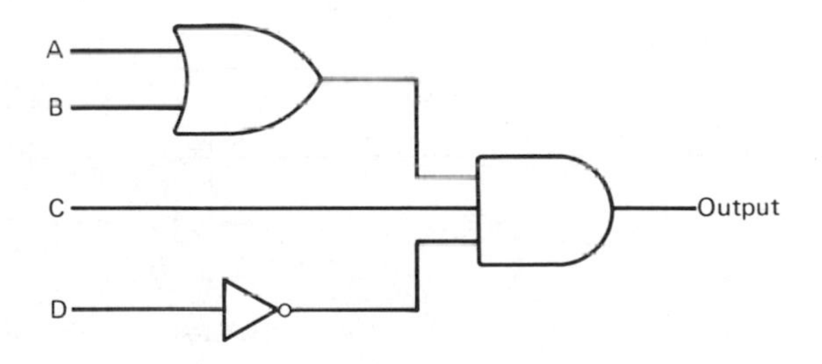

Figure 9-16. Simplified logic circuit implementing the truth table given in Figure 9-15.

ful tool than illustrated here in the design and optimization of logic systems; this is described in more detail in Chapter XIV.

APPLICATIONS OF LOGIC GATES

Three-State Logic

Logic gates are typically provided with multiple inputs and a single output, indicating that connection of several outputs together is not normally permitted. If this is attempted, each gate will try to establish its own desired output. It is possible that large currents may flow between the gates, perhaps resulting in their destruction. Such situations can be avoided by the use of open collector TTL gates. These are constructed so that they can enforce logic "0," but not enforce logic "1." The latter is generated by connecting a pull-up resistor from the output to +5 V, as shown in Figure 9-17*a*. The logical operation performed is $\overline{X}$ **OR** $\overline{Y}$, the so-called *negative-logic wired* ***OR***. Any number of gates can be connected in this way

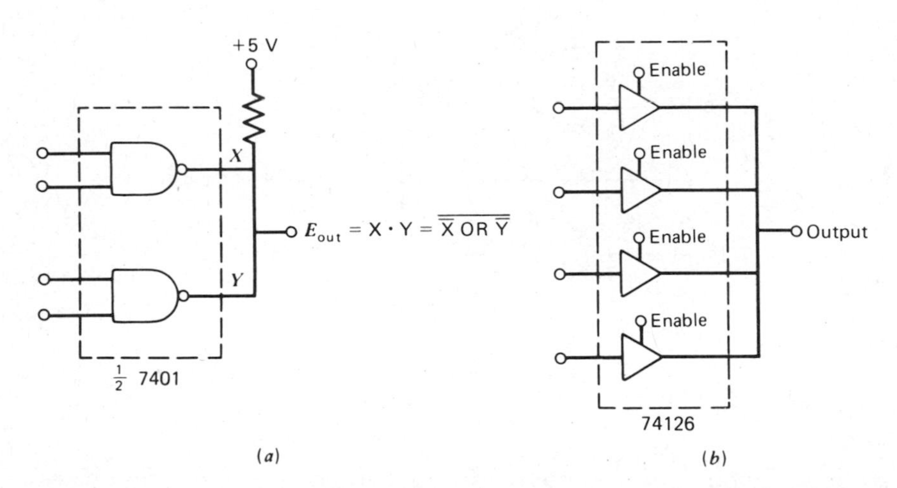

Figure 9-17. (*a*) An example of open collector gates connected in a "wired-**OR**" mode. (*b*) Four three-state buffers leading into a single line. Only one "enable" can be activated at a given time, opening the path for the corresponding signal flow.

to a given point. Such ICs have been extensively used to enter information from different sources onto a common line (or collection of lines, i.e., a *bus*).

The wired-OR system has the disadvantage that logic "1" cannot be actively implemented. A refinement of the principle was introduced in 1972 by National Semiconductor that allows the connection of any number of ICs to a given point, provided all but one are in a special OFF state of high impedance. Such units that have active "0," active "1," and OFF modes are called "three-state" devices.* These modes are illustrated in Figure 9-17*b*.

Pattern Recogition

As indicated previously, a three-input **NOR** gate is able to single out the 000 input combination, and an **AND** gate can identify the 111 state. In other words, of the eight possible combinations, either gate recognizes only a specific combination of ones and zeros. This capability is enhanced when combinations of gates and inverters are used to identify and single out any particular input combination. Consider, for example, the symbol for "carriage return" in the ASCII code, which is specified by the sequence 0001101. To identify this symbol we need a gate circuit that will ignore any pattern other than the command for carriage return where action is to be taken. Figure 9-18 shows a circuit by which such an identification can be made.

* The equivalent term "Tri-State" is a trademark of National Semiconductor.

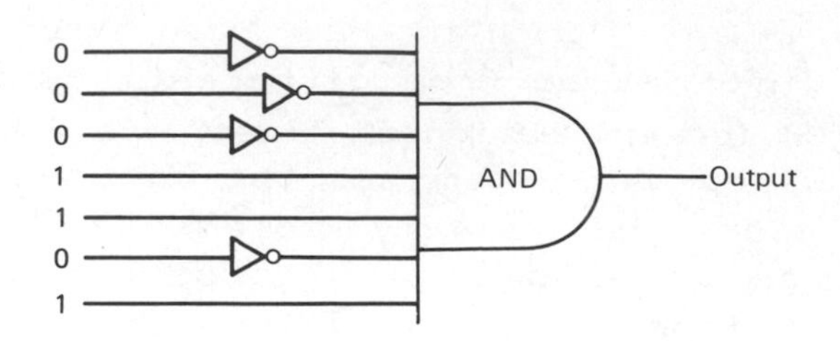

Figure 9-18. A seven-input **AND** gate used to recognize the binary sequence 0001101. The output is at logic "1" if this array of ones and zeros is present at the inputs, and not otherwise.

Coincidence and Anticoincidence Circuits

This type of circuit can be applied in low-level radioactivity measurements, where external radiation is apt to interfere. The system shown in Figure 9-19 contains an array of radiation counters surrounding the measurement area. Whenever a photon of external origin (such as a cosmic ray) penetrates into the instrument chamber, one or more of the counters in the array produces a signal pulse. After level conversion to 5 V (not shown), these pulses are brought together into a six-input **OR** gate, which gives a logical "1" at point *X*. In the logic system shown, the **AND** gate will produce an output pulse only for the case "*A* **AND** *B* but **not**-*X*", equivalent to Boolean $(A \cdot B \cdot X)$. This means that both internal counters A and B have sensed radiation, but none of the outer ring of counters has done so. Cosmic rays have a far greater penetrating power than the radiation from the internal source under measurement. The latter type of radiation can never travel as far as to the outer sensors because of the shielding interposed. Hence the required logical condition can be met only by radiation from the sample, essentially eliminating the background due to cosmic rays.

BINARY NOTATION

The sequence of combinations of ones and zeros that has been used repeatedly in this chapter in describing the action of gates (see Table 9-1, for instance) was not chosen by accident. It represents the equivalent of the numbers 0 to 15 written in the binary system of enumeration. In the familiar decimal system 10 symbols are available (0 through 9), whereas in binary only two can be used (0 and 1). The principle of counting is the same, however. In both cases the

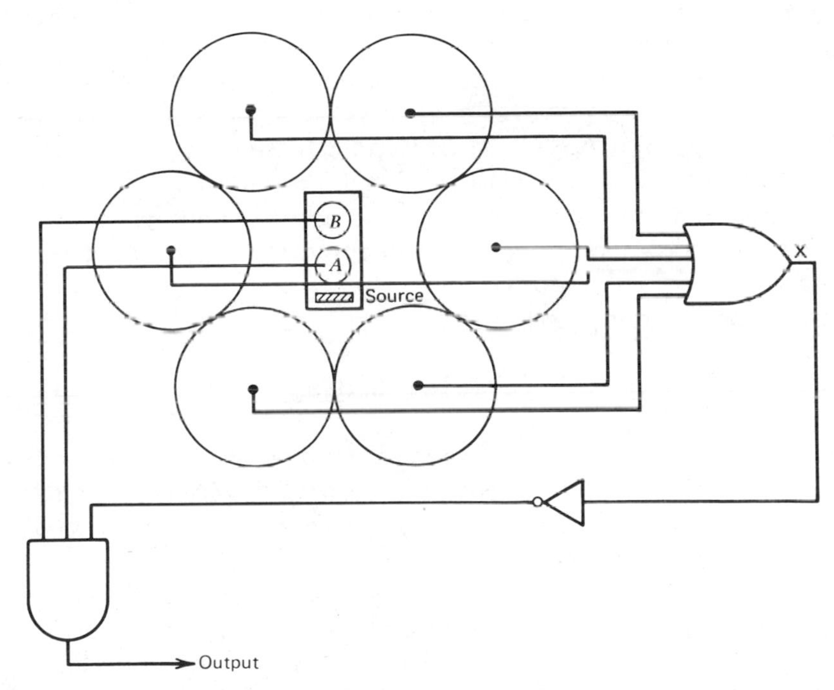

Figure 9-19. A coincidence-anticoincidence circuit. Necessary high-voltage circuitry not shown. The signals from *A* and *B* are said to be in coincidence, and they are in anticoincidence with any signals from the outer ring of counter tubes.

symbols are written down in increasing sequence until they are all used up, when a "1" is added in the next place to the left and the symbols written again in sequence, and so on. Table 9-2 gives the first 32 numbers in both systems.

In using the binary system in electronic work it is customary to include leading zeros to show how many digital spaces are being used; thus decimal 6 is written as "0110" rather than simply "110." The digit in the right-hand column is designated as the *least significant digit*, whereas that in the left-hand column is the *most significant digit*. In the binary system, these are denoted as the LSB and MSB, respectively, the "B" standing for "bit," since each binary digit constitutes a single bit.

In addition to these, the *octal* system, based on 8 digits (0 through 7), and the *hexadecimal* system, using 16 digits, both have important places in electronics, particularly in connection with computers. Appendix V includes details of all these systems and gives arithmetic methods for their interconversion. In the present chapter we need to use the binary system, but the others can wait.

Special Codes

There are several special-purpose systems of encoding numerical information in binary form in addition to the simple numeration described above. One that we will have occasion to use later is called *binary-coded decimal* (BCD). This follows binary notation from 0 to 9 (1001) but then goes directly back to 0000, adding a 1 in the next column. This is often written in the form of two (or more) four-digit columns, each representing one decimal digit. This system is displayed in the right-hand column in Table 9-2. This notation is particularly useful in computer interpretation of

TABLE 9-2
Numeration Systems

Decimal	Binary	Binary-Coded Decimal
0	0	0000 0000
1	1	0000 0001
2	10	0000 0010
3	11	0000 0011
4	100	0000 0100
5	101	0000 0101
6	110	0000 0110
7	111	0000 0111
8	1000	0000 1000
9	1001	0000 1001
10	1010	0001 0000
11	1011	0001 0001
12	1100	0001 0010
13	1101	0001 0011
14	1110	0001 0100
15	1111	0001 0101
16	10000	0001 0110
17	10001	0001 0111
18	10010	0001 1000
19	10011	0001 1001
20	10100	0010 0000
21	10101	0010 0001
22	10110	0010 0010
23	10111	0010 0011
24	11000	0010 0100
25	11001	0010 0101
26	11010	0010 0110
27	11011	0010 0111
28	11100	0010 1000
29	11101	0010 1001
30	11110	0011 0000
31	11111	0011 0001
32	100000	0011 0010

information presented in the form of decimal numbers.

A few other codes will be mentioned here without detailed exposition. *Complementary binary* code is produced from straight binary by replacing every "1" by a "0" and every "0" by a "1." The *two's complement* code replaces every binary number by its complement to which 1 has been added; this code makes binary addition easier. The *Gray code* is designed so that each increment of one causes only one digit to change its value; the first several numbers in one version are:

0	0000
1	0001
2	0011
3	0010
4	0110
5	0111
6	0101
7	0100
8	1100
9	1101

The advantage to Gray codes appears when a sequence of numbers is to be produced by a mechanical device of some sort; errors are less likely to occur than would be the case if several digits changed value simultaneously.

The interconversion between different codes is routinely handled by ICs.

Encoding and Decoding

It is sometimes advantageous to be able to condense the information transmitted along several parallel lines so that the system can be reduced to only a few lines. This can be done by a process known as

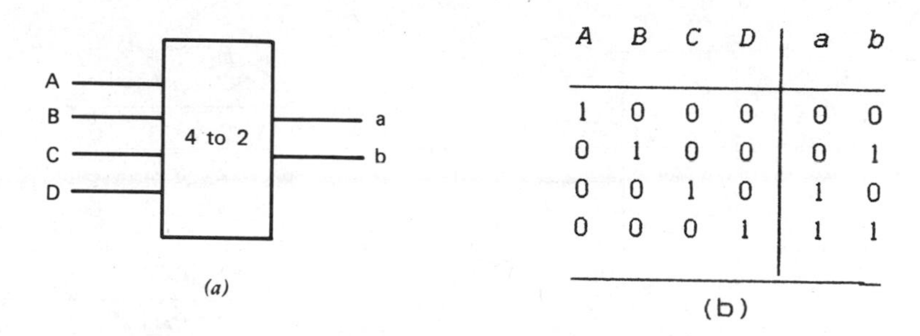

A	*B*	*C*	*D*	*a*	*b*
1	0	0	0	0	0
0	1	0	0	0	1
0	0	1	0	1	0
0	0	0	1	1	1

(b)

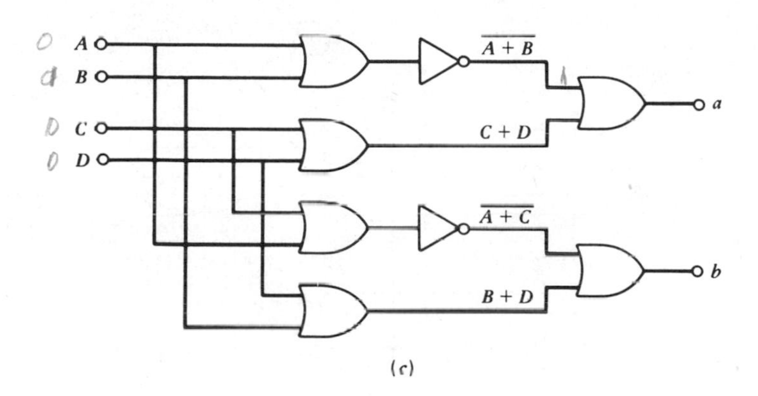

Figure 9-20. A four-to-two encoder: (*a*) symbolic representation; (*b*) truth table, (*c*) logic diagram. Note that only one input is permitted to be high at any given moment.

encoding. Figure 9-20 shows a block diagram and a possible schematic for a four-to-two encoder. The information comes originally on the four wires *A*, *B*, *C*, and *D* and is condensed onto lines *a* and *b*. Because of the bottleneck at the output, it is evident that the flow of information is limited to 2 bits. In the present example, the inputs are subject to the restriction that only one of the four lines is permitted to be high at a given time. This would be the case, for example, in an experiment where the location of a single object is to be monitored, such as an animal in a cage, sensed by four electrical contacts. For any

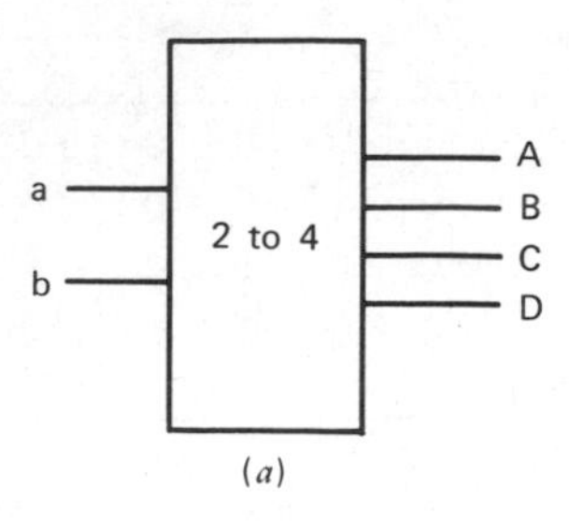

(*a*)

a	*b*	*A*	*B*	*C*	*D*
0	0	1	0	0	0
0	1	0	1	0	0
1	0	0	0	1	0
1	1	0	0	0	1

(b)

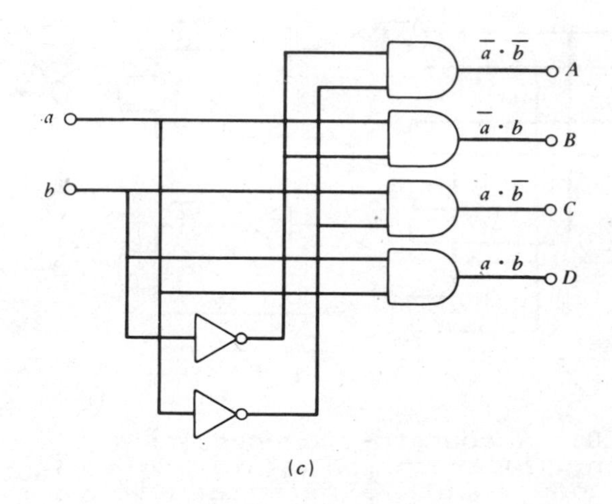

(*c*)

Figure 9-21. The two-to-four decoder (*a*) and its truth table (*b*); (*c*) shows one implementation.

one input high, corresponding to the animal being in one of four compartments, a particular digital code will be given to the output leads, as indicated in the truth table. The reader should verify that the circuit shown in Figure 9-20*c* will actually give this result. An analogous circuit with 14 **AND** gates, 3 **NOR**s, and 12 inverters can be constructed to take digital information from eight lines and produce a three-digit binary code; such an encoder is already available in integrated form as the 74148, at about one dollar.

The inverse process is *decoding*, whereby the digital code on few lines is translated into signals on many lines. Such a circuit is shown in Figure 9-21. The advantage of the coding-decoding process is enhanced for large numbers of bits. Thus eight input lines will decode into 256 outputs, which could, for example, control 256 valves in an oil refinery. Since only one output can be high at a time, a decoder makes a convenient actuator for three-state logic gates.

MULTIPLEXERS

The digital multiplexer is a special type of encoder with two sets of inputs: one for an address and one for data. The address lines serve to select which of the inputs will be steered to the output. Figure 9-22 shows a block diagram and truth table for such a unit with two address lines (S_1 and S_2) and four data lines (*A* through *D*). The logical diagram (Figure 9-23) shows how this works. If the address code stands at "00," the **AND** gate marked *a* will be "enabled"; that

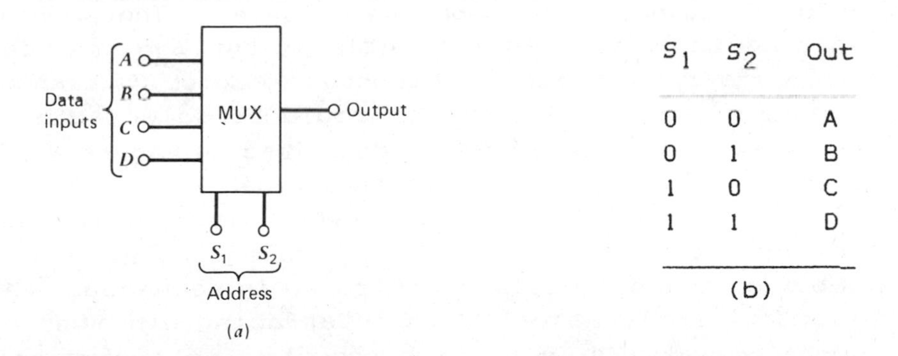

S_1	S_2	Out
0	0	A
0	1	B
1	0	C
1	1	D

Figure 9-22. A four-line multiplexer (*a*) and its truth table (*b*).

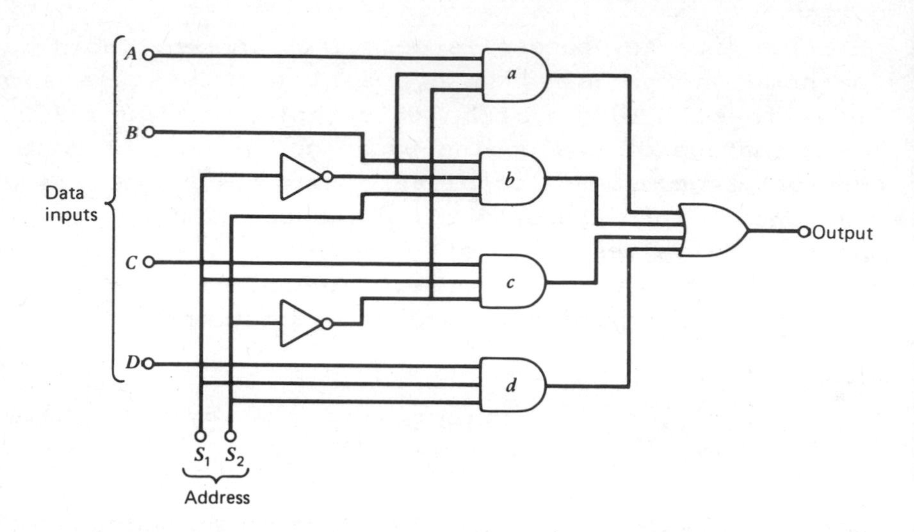

Figure 9-23. Logic diagram of a four-line multiplexer. Note that any given address disables three out of the four signal paths.

is, it is alerted to respond to the *A* input, giving a high or low output according to whether the signal at *A* is high or low. At the same time gates *b*, *c*, and *d* are disabled, so that signals at *B*, *C*, and *D* cannot affect the ultimate output through the **OR** gate. The address lines could be stepped sequentially through the four binary states by means of a counter (Chapter X) responding to a chain of pulses or could be selected by a manual switch. The HA-2400, described in Figure 4-27, makes use of a multiplexer of this type.

Multiplexers with various numbers of input lines are available as ICs. They are useful whenever a number of signals must be monitored in sequence. An important application is in interfacing a number of remote transducers (e.g., thermocouples) to a computer. Each thermocouple produces an analog signal that must be converted to its digital equivalent by some type of digitizer (to be discussed in a later chapter). The digital signals are sent over transmission lines from which the multiplexer feeds them one at a time to the computer.

Another example of the use of address lines to exercise control over data transmission is shown in Figure 9-24, a portion of a bus structure widely used in computer systems. Three lines (*A*, *B*, and *C*) can carry any three-digit binary address (0 to 111), corresponding to a total of eight possibilities. Each "station" connected to the bus, two of which are shown in the diagram, will respond to its own address and no other. The device that is addressed can then receive from the data line, 1 bit at a time, the information destined for it, and transmit it to its output. The bus consists of the set of four conductors (*A*, *B*, *C*, and *DATA*). One additional conductor would permit 16 separately addressed stations to be connected to the same bus. In a computer, much more extensive bus structures are used for the propagation of information.

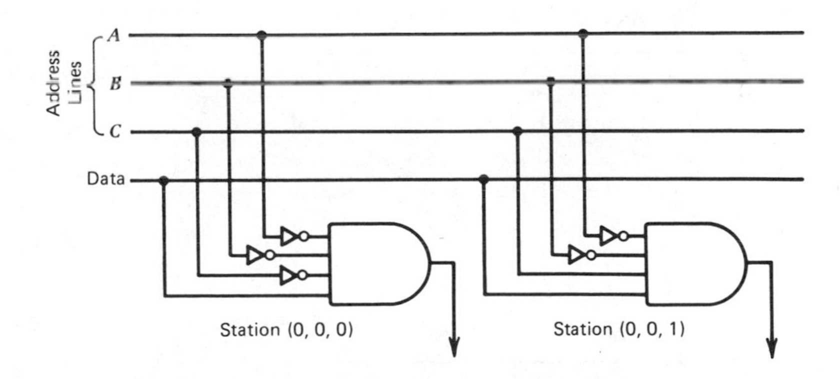

Figure 9-24. Example of remote control by means of a bus. The gates give highs only if the proper address appears on lines *A*, *B*, and *C*, and the DATA line is high.

PROBLEMS

9-1. Design a gate circuit to implement the table:

A	*B*	*C*	*Out*
0	0	0	0
0	0	1	0
0	1	0	1
0	1	1	0
1	0	0	1
1	0	1	0
1	1	0	1
1	1	1	0

9-2. Write the truth table for the circuit shown in Figure 9-25.

9-3. Design a circuit to perform the operations indicated in the following table. Use the minimum number of gates.

A	*B*	*C*	Out_1	Out_2
0	0	0	0	1
0	0	1	1	0
0	1	0	1	0
0	1	1	0	1
1	0	0	1	0
1	0	1	1	0
1	1	0	1	0
1	1	1	0	0

9-4. Implement the function:

$$(A \text{ OR } B) \textbf{ AND } \overline{(C \text{ OR } D)}.$$

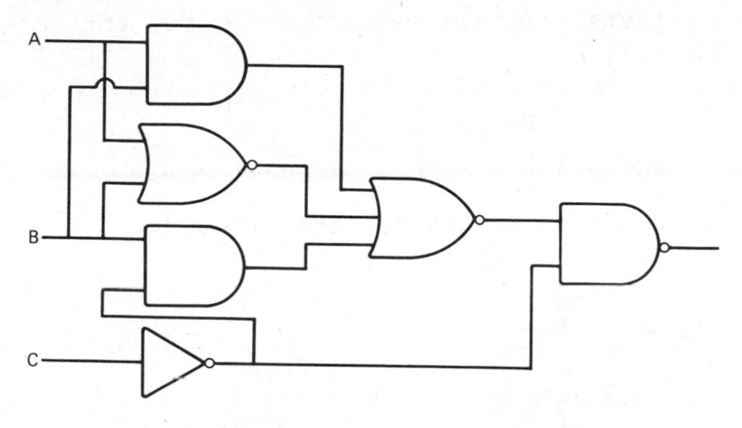

Figure 9-25. See Problem 9-2.

9-5. Show that the gates corresponding to the symbols below perform identical operations. Include the pertinent Boolean notation.

9-6. Prove the following statement to be true:

$$(AB) + (BC) + (\overline{C}A) = (AB) + (\overline{C}A)$$

9-7. Consider the expression $A\overline{BC} + AB$.

(a) Evaluate the expression for the case where $A = B = C = 0$.

(b) Design a logic circuit to implement this expression.

9-8. Evaluate the expression $\overline{AB + C}$ for the following cases:

(a) $A = 1,\ B = 0,\ C = 0$
(b) $A = 1,\ B = 1,\ C = 1$

9-9. Write truth tables for the following:

(a) $A + BC + \overline{B}C$
(b) $A \cdot (\overline{B} + \overline{C})$

9-10. Construct a truth table for an interlock system that prevents a car from starting if the driver's seat is occupied but the seat belt is not fastened. Consider the case for two occupants.

9-11. Devise logic circuits corresponding to the following truth tables:

A	B	E_{out}
0	0	0
0	1	0
1	0	1
1	1	0

(*a*)

A	B	E_{out}
0	0	1
0	1	0
1	0	1
1	1	0

(*b*)

A	B	C	E_{out}
0	0	0	0
0	0	1	1
0	1	0	0
0	1	1	1
1	0	0	0
1	0	1	1
1	1	0	0
1	1	1	0

(*c*)

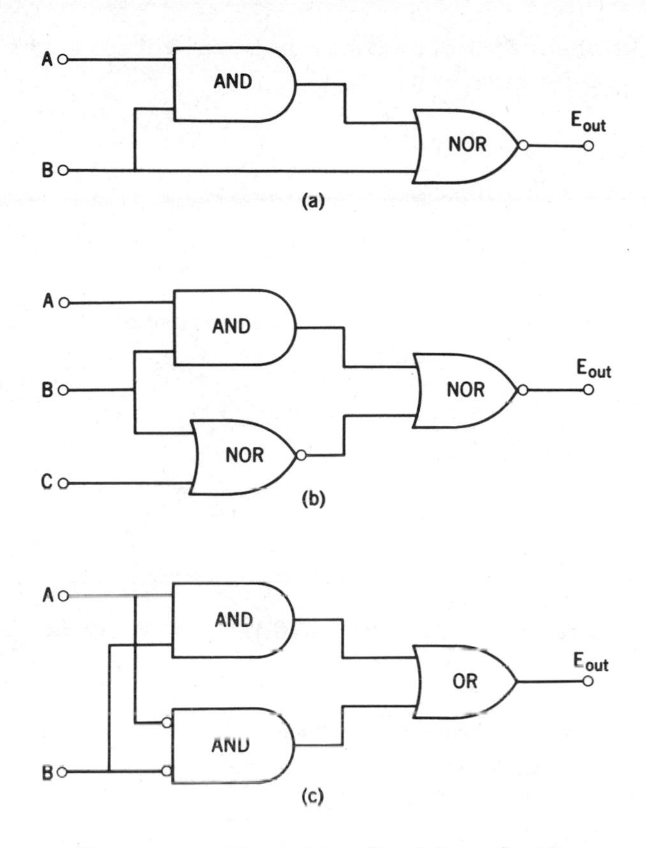

Figuro 9-26. See Problem 9-13.

9-12. Construct the equivalent of a 16-input **NAND** gate, using four 7420 4-input **NAND** gates and 7404 inverters if needed.

9-13. Establish truth tables for each of the circuits of Figure 9-26.

9-14. Explain the function of the diode in the totem-pole TTL output (Figure 9-7).

9-15. Express π (3.1416) in binary and hexadecimal notation.

9-16. What are the decimal equivalents of the following binary numbers?

(a) 1101101
(b) 0101000
(c) 1110
(d) 0101010

9-17. What are the binary equivalents of the following decimal numbers?

(a) 39
(b) 65
(c) 119
(d) 250
(e) 1011

9-18. Write the BCD equivalents of each of the numbers in Problem 9-17.

9-19. Design a logic system for the conversion of the Gray code given in the text to BCD.

9-20. The circuit shown in Figure 9-27 can be used to generate a four-digit code from one of three digits. Write its truth table.

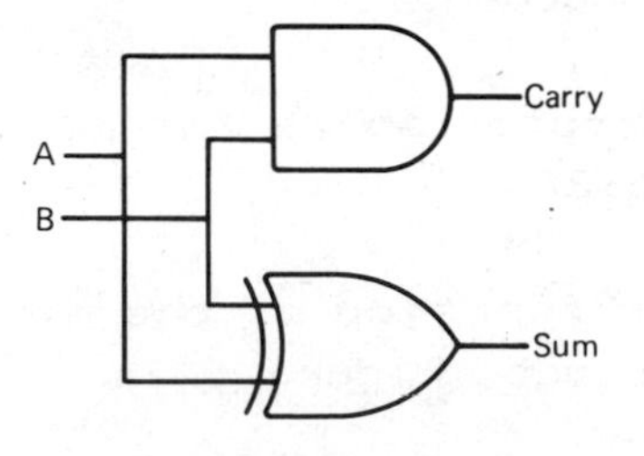

Figure 9-27. See Problem 9-20.

9-21. Give the truth table for Figure 9-28, and explain why the two outputs are called "sum" and "carry."

9-22. A rather frequent occurrence in machine-language computer programming is the need to verify that a certain pattern of ones and zeros exists in a particular eight-digit word. For example, it may be essential that bits 0, 2, and 8 (counting from the right) are each "1", while it doesn't matter what the others bits are. This desired pattern can be represented by the word 1xxxx1x1, where each x can be either "0" or "1." The approach usually taken requires two steps: (1) the digits 10000101 are ANDed to the existing word, (requiring eight AND gates), and then (2) the result from that step is XORed with the same 10000101. If the result is now equal to 10000101, it is established that the original word met the requirements. Explain the logic behind this.

9-23. In the multiplexer shown in Figure 9-23, what would be the effect of replacing the OR gate with an XOR?

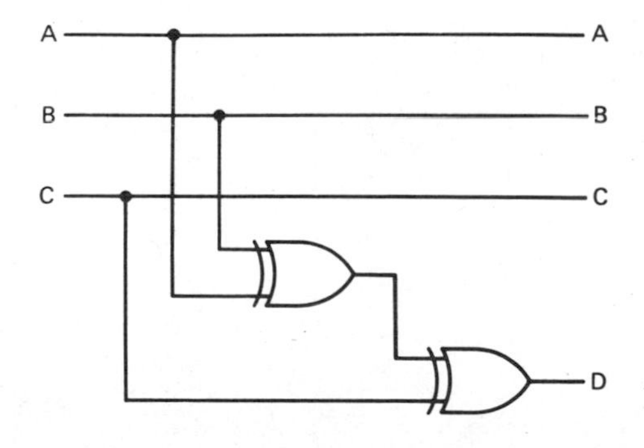

Figure 9-28. See Problem 9-21.

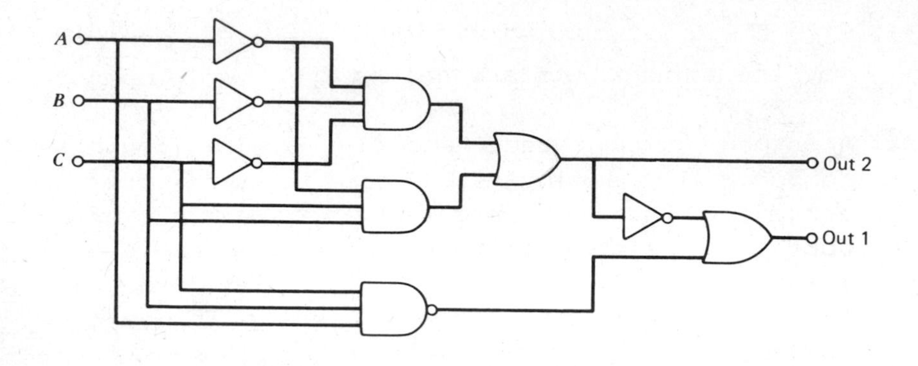

Figure 9-29. See Problem 9-24.

9-24. Construct the truth table corresponding to the logic diagram in Figure 9-29.

X

FLIP-FLOPS AND COUNTERS

Up to this point we have considered only those logic systems in which there is an instantaneous correlation between input and output. Any change in the states of input lines is immediately reflected in a corresponding change in the output. However, there are many situations in which it is necessary to retain or remember the state of logic present at some previous moment. To accomplish this, logical elements with *memory* are required.

The concept of memory is quite distinct from that of time delay, which may account for an output holding its state for a short time after a change of input signals. True logical memory devices must be able to retain information for long periods. Indeed, some kinds hold their state even when the power is turned off.

Digital computers require very extensive memory capability and invariably employ some type of magnetic component for the purpose. Magnetic materials can retain magnetization in either of two possible states corresponding to orientation of the magnetic dipoles in opposite directions. These memory units can be fabricated in many physical forms, including magnetic tape, disks, and cores.

Also required in many applications are various

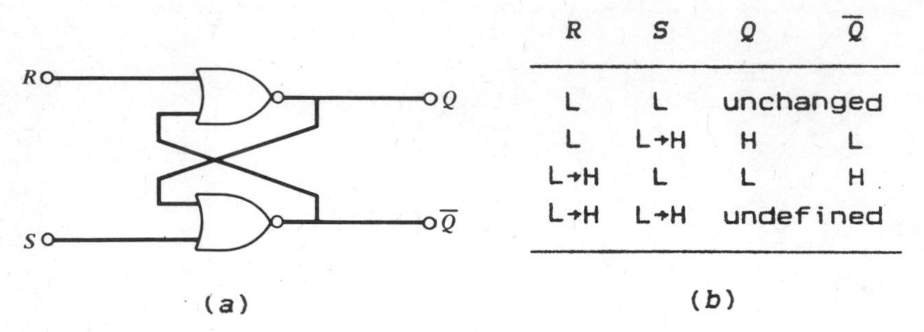

R	S	Q	$\overline{Q}$
L	L	unchanged	
L	L→H	H	L
L→H	L	L	H
L→H	L→H	undefined	

Figure 10-1. (*a*) A **NOR**-based flip-flop, and (*b*) the corresponding state table. The symbol "L→H" indicates a transition from Low to High.

nonmagnetic types of memory, often called "volatile," because loss of power erases their information content. These memory devices are the principal subject of the present chapter.

FLIP-FLOPS

The flip-flop (FF), also called a *bistable multivibrator*, is a device that has been known for a long time, having been invented (as implemented with vacuum-tubes) by Eccles and Jordan in 1919. In its modern

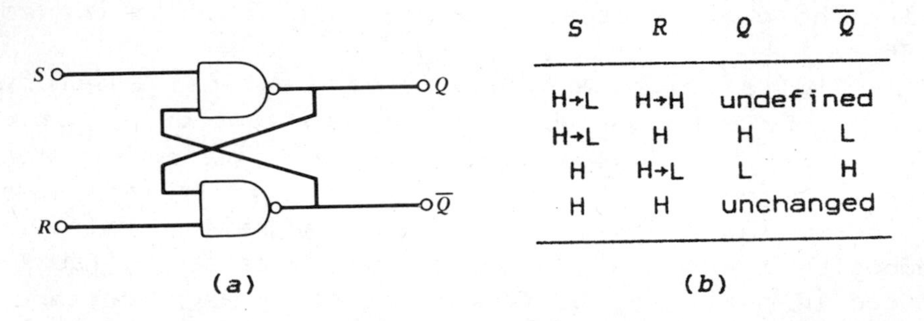

S	R	Q	$\overline{Q}$
H→L	H→H	undefined	
H→L	H	H	L
H	H→L	L	H
H	H	unchanged	

Figure 10-2. (*a*) A **NAND**-based flip-flop. The inputs are normally high. (*b*) The corresponding state table showing the result of momentary grounding of one or both inputs.

form it uses solid-state circuitry, as either discrete transistors or ICs. A number of variations make the flip-flop an extremely versatile circuit element.

The *RS*-Flip-Flop

The basic form can be obtained by cross-connecting two **NOR** gates to form a "**NOR**-FF" or two **NAND** gates to give a "**NAND**-FF." The two types are shown in Figures 10-1 and 10-2. Both varieties are commonly known as *RS flip-flops*. Observe, in Figure 10-1, that the feedback connections of a **NOR** flip-flop restrict the output state in the following way: (1) if both inputs are high, both outputs must be low; (2) if both inputs are low, there are two possible states with equal probability, namely, with output values 0,1 or 1,0; (3) if one input is high and one low, one output is low, and the other high. These facts suggest the possible use of the flip-flop as a memory device, provided the inputs are maintained normally low and are not allowed to go high simultaneously. The results of momentary pulses at one of the two inputs are summarized in Figure 10-1*b*.

The state wherein Q is high is called the *set* condition, whereas if Q is low, the circuit is *reset*. Momentarily pulsing the S input *sets* the circuit to give a Q output of "1," whereas pulsing the R *resets* it, giving $Q = 0$. Whenever the flip-flop is operated under the specified restrictions, the two outputs are complementary; hence the notation Q and $\bar{Q}$.

The **NOR**-*RS* flip-flop can remember whether a pulse to logical "1" has or has not occurred at the S input. Assume that the circuit is initially in its normal state, with both inputs low. The state of the Q output supplies an answer to the question "Has S been pulsed more recently than R?" If $Q = 0$, this may be interpreted as the answer "No, it has not." A pulse at S changes Q to "1," meaning "Yes, it has." The answer is still "Yes" after repeated pulses at S, but a pulse at R resets the circuit to $Q = 0$, so that the answer becomes "No." Alternatively, one can think of the circuit as indicating *which* input has most recently felt a pulse.

This circuit has a major disadvantage. If both inputs are brought to +5 V and are then simultaneously grounded, there is no way to determine which of the two outputs will go high. This situation may well occur in practice, for, as we shall see presently, logic systems are usually controlled by a series of pulses generated for the purpose, so that all logic changes throughout the system occur together. The circuit designer has the responsibility of making sure that

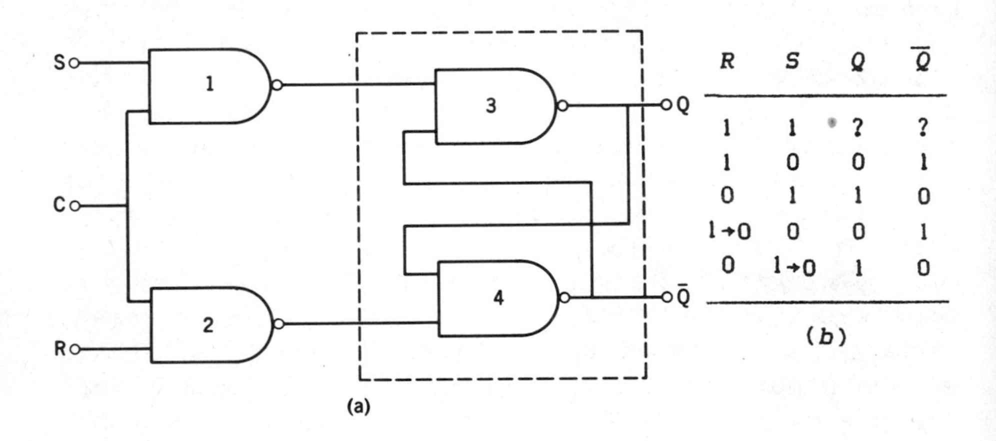

R	S	Q	$\bar{Q}$
1	1	?	?
1	0	0	1
0	1	1	0
1→0	0	0	1
0	1→0	1	0

(*b*)

Figure 10-3. The clocked *RS* flip-flop. (*a*) The schematic; the portion in the dashed box is the basic *RS* unit; input **NAND** gates and a clock input (*C*) have been added. (*b*) The state table. The clock is assumed to be initially in its "1" state. When the clock changes to "0," the preceding state is retained, except for the first case, which is undefined. The 1→0 notation refers to the return to zero following a pulse (the same as designated in Figure 10-1 as "high pulse").

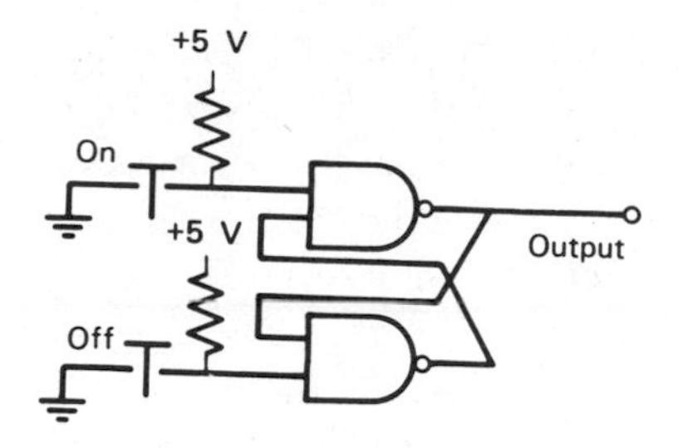

Figure 10-4. An *RS* flip-flop acting as a bounceless switch.

this anomaly cannot occur.

The **NAND**-implemented *RS*-flip-flop (Figure 10-2), shows similar properties, except that here the normal condition calls for both inputs to be high and action to occur on momentary grounding.

Because *RS* flip-flops are so easily implemented from **NOR** or **NAND** gates, they are not commonly manufactured as ICs. If several inputs are available, as when the circuit is constructed from three-input gates, a pulse at any input is effective in triggering the response.

Figure 10-3 shows a slightly more complicated circuit, the *gated*, or *clocked RS flip-flop*, in which two additional gates are inserted at the inputs. A series of synchronizing pulses mentioned previously, called *clock pulses*, are applied to the common input (marked *C*) in order to synchronize or strobe the flip-flop with other parts of the system. When the clock pulse is high, the **NAND** gates act simply as inverters and the overall action is the same as for a simple **NOR** *RS*-flip-flop. When the clock signal drops to zero, the mode of operation changes since both input **NAND** gates are now forced high regardless of the states of *R* and *S*. The output is thus frozen in whatever state it held just before the change in clock level. The two modes of operation, enable and disable, are analogous to the *sample* and *hold* that we encountered in op amp circuitry. If at any point during the sampling period (clock high) *S* was at "1" (while *R* was at "0"), the output *Q* will be "1" and will remain so after the clock goes to zero. If *R* was "1" and *S* was "0," however, *Q* will be "0." If both inputs were at "0" during the sampling period, the state of the outputs would remain the same as before the clocking cycle.

One use of *RS* flip-flops is in interfacing mechanical and electronic logic devices. An example of such use for a switch is shown in Figure 10-4. A mechanical switch is liable to suffer from "bounce," a series of fast on-off transitions that result from vibrations occurring as the switch is closed. This can adversely affect logic systems, as each momentary contact can be interpreted as a separate logic change. If a switch is connected to the input of a flip-flop, this effect is not felt. Since both *R* and *S*, once

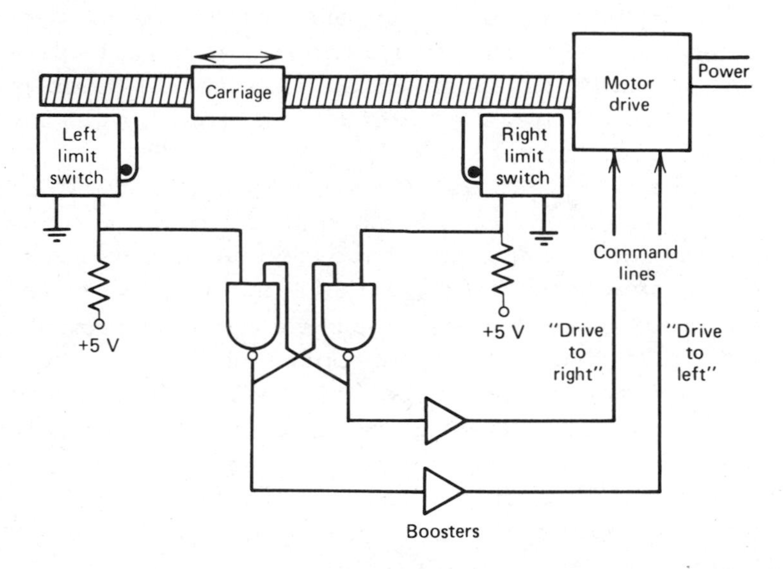

Figure 10-5. A limit control system for an automatic scanning mechansim. The carriage, when driven to the left, eventually closes the "left limit switch," causing the motor to reverse driving the carriage to the right, when it will close the "right limit switch," reversing again, and so on.

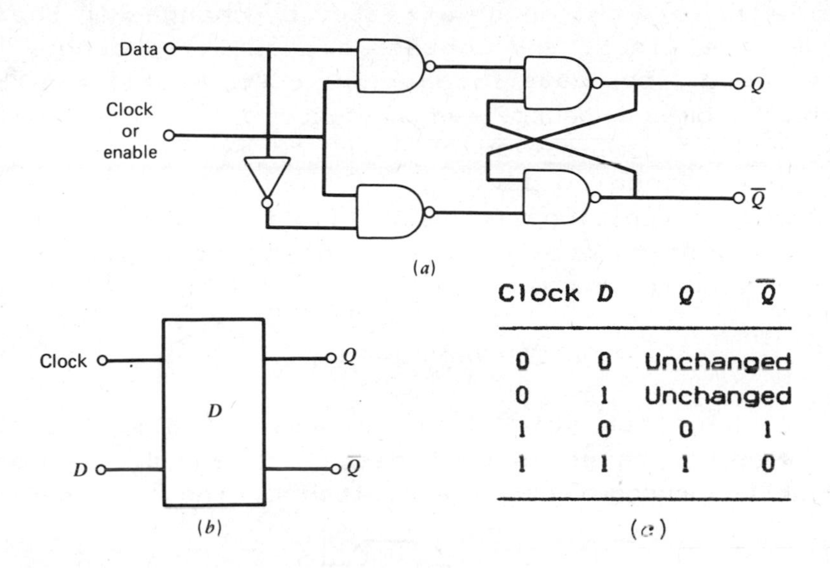

Clock	D	Q	$\overline{Q}$
0	0	Unchanged	
0	1	Unchanged	
1	0	0	1
1	1	1	0

Figure 10-6. (*a*) The *D*-type flip-flop; (*b*) a block diagram; (*c*) the state table.

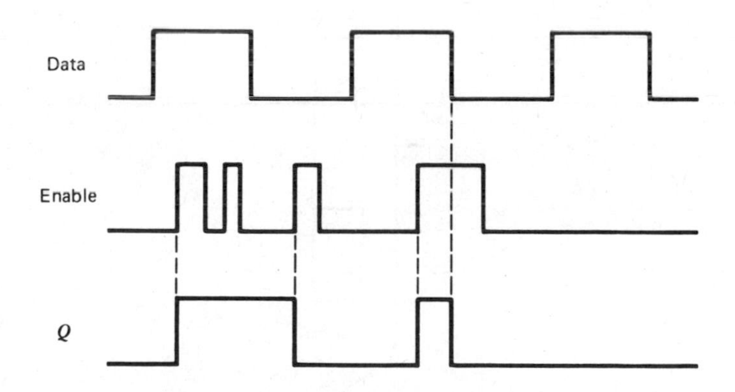

Figure 10-7. Behavior of a *D* flip-flop. Successive "enable" pulses, entering at input *C*, are shown as irregular in order to illustrate various situations that may arise.

activated, are no longer affected by changes in their respective lines, any contact bounce is ignored. In other words, the gate *latches* on to its state, and the device becomes a *bounceless switch.*

Another application of latching mechanisms is in automatic scanning devices. The heart of such an instrument is functionally an *RS* flip-flop (Figure 10-5), which serves to prevent the scanning mechanism from overrunning its range.

The D-Flip-Flop

A characteristic of the clocked *RS* flip-flop is the need for three input lines, *R*, *S*, and *C*. A more versatile arrangement is exhibited by the *D* flip-flop

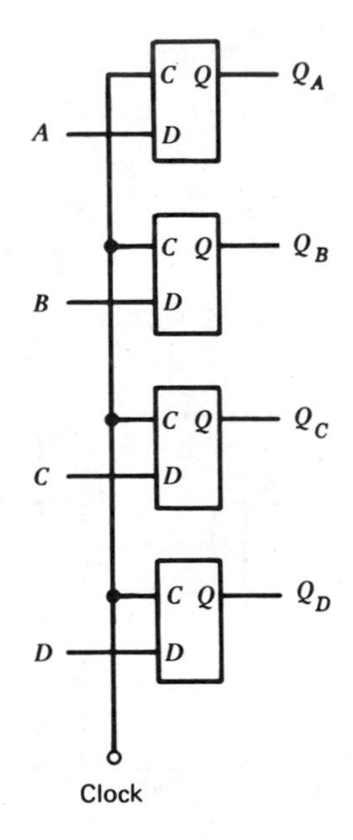

Figure 10-8. A 4-bit register composed of *D* flip-flops. Each of the input lines *ABCD* can be considered to be carrying 1 bit of a four-place binary number. The register will accept this number, provided the clock is high, and retain it until the next clock pulse.

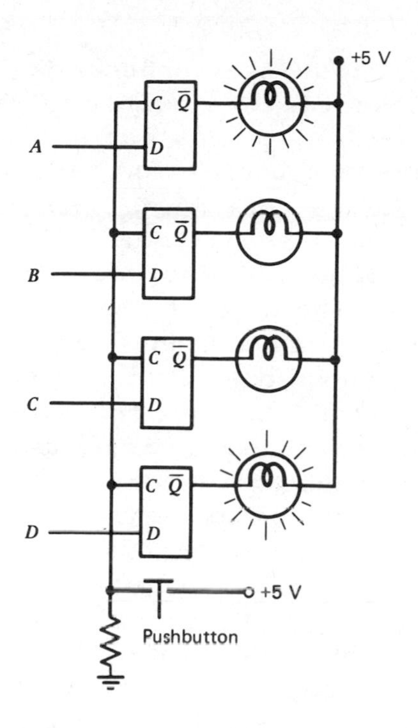

Figure 10-9. A register used to "freeze" four-digit variables by a pushbutton command. In practice it is more power-efficient to connect the lamps as shown between $\bar{Q}$ and +5 V than between Q and ground. The logical result is the same either way.

(Figure 10-6). This requires two lines, but only one of them determines the logic state. The other, the clock line, is used to determine the exact moment when the input state is to be loaded into the memory.

The *D* input (*D* for data) replaces the former *S* input. The state table corresponds to that in Figure 10-3, except that only the second and third entries are possible, and these are active only if the clock is high. When the clock is low, the output is unaffected by changes in the *D* input. In short, the *D* flip-flop accepts and memorizes the value of *D* when the clock is high and retains it in memory when the clock drops to logic "0." The relationship between the various states can be visualized from Figure 10-7.

A group of *D* units connected together can be used to retain the states of a number of lines at one time. Such a combination is known as a ***register*** (Figure 10-8). Registers are useful for retaining various logical data for future reference. For example, one might want to store the instantaneous values of various logic lines that change too fast for direct observation. For this purpose the circuit shown in Figure 10-9 would be appropriate. Its action is an example of the HOLD operation, used when one wants to examine a fast-moving process at leisure. Assume, for example, in Figure 10-9 that at the moment one presses the button, variables *A*, *B*, *C*, and *D* have the values 1, 0, 0, and 1, respectively. Regardless of any subsequent change in levels, lamps 1 and 4 will remain lit.

<u>Edge-Triggered *D* Flip-Flops</u>: As can be seen in Figure 10-6, the output of a *D* flip-flop follows the input continuously as long as the clock signal remains high. In other words, it acts as a *track-and-hold* mechanism. Of more utility in computer circuitry is the *sample-*

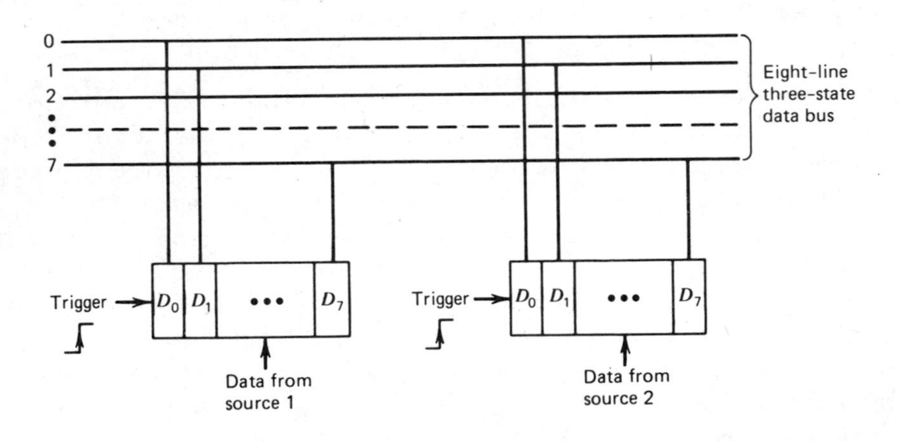

Figure 10-10. Two registers composed of edge-triggered flip-flops sending data onto a bus. A separate bus system (not shown) selects the proper trigger.

and-hold operation, where the loading of information is done only during an extremely short period. Such a device is the edge-triggered *D* flip-flop. An example is the TTL model 74174 which has six such units in one IC package. The sampling operation occurs, as the name suggests, on the 0 to 1 transition, the *edge* of the clock pulse.

Edge-triggered *D*-type flip-flops provided with three-state outputs are widely used for loading data into a computer bus (Figure 10-10). Each group of eight flip-flops forms an 8-bit register. The trigger lines are specifically actuated so that only a single register is activated at a given time. There is no theoretical limit on how many bits a register can have, nor on how many registers can be connected to a given bus.

The JK Flip-Flop

An even more versatile component is the *JK master-slave flip-flop* (Figure 10-11), which actually contains two flip-flops in one unit. In its fully extended form it has no less than seven connections to external circuitry, not counting the power supply. If all inputs, *R*, *S*, *J*, and *K*, are at logic "1," the flip-flop will interchange states whenever a clock pulse moves from "1" to "0." This is called *toggling* or *complementing*. The state does not change on the clock pulse returning to its high state. Toggling can be inhibited by a logic "0" at either *J* or *K*. The circuit is set to $Q = 1$ by grounding the *S* input or reset by grounding *R*. These commands override any signals present at other inputs, including the clock.

The presence of the clock input in the *JK* flip-flop is essential in ensuring lack of interference between an incoming signal and the information previously stored in the flip-flop. This is a consequence of the fact that the unit contains two cascaded elementary flip-flops, the master (composed of **NAND** gates 3 and 4 in Figure 10-11*a*) and the slave (gates 7 and 8). Information is fed into the master only when the clock signal goes high; in this state there is no communication between master and slave. The information is passed on to the slave as the clock goes low. The result of this sequence is that data from both the present and previous cycles are stored simultaneously. Part *c* of Figure 10-11 shows schematically that changes in the output occur only on the downward transitions of the clock signal.

Monostable and Astable Flip-Flops

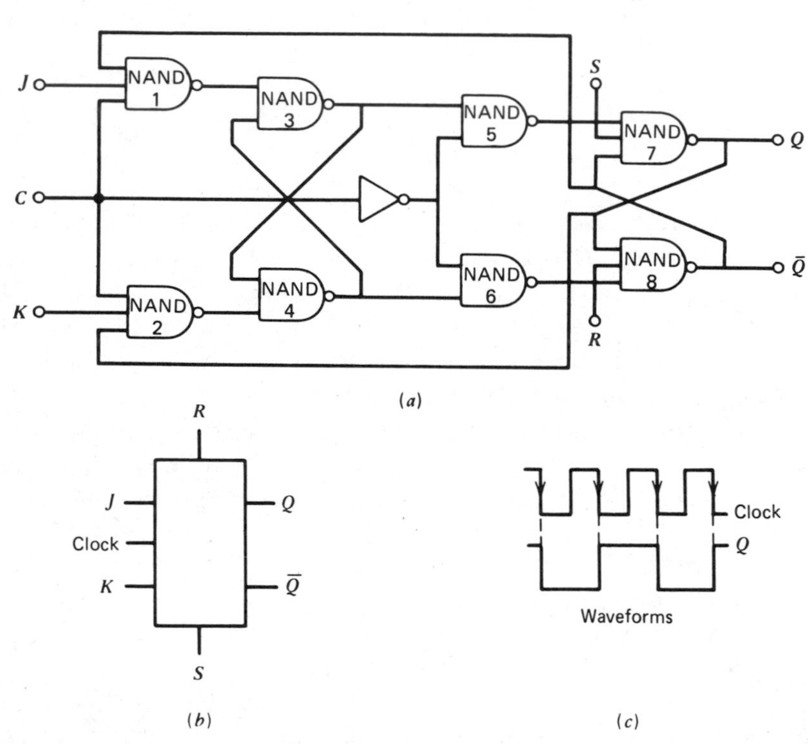

Figure 10-11. An example of a *JK* flip-flop: (*a*) schematic; (*b*) circuit representation; (*c*) waveforms at *Q* when *J*, *K*, *R*, and *S* are all at logic "1." Other models of *JK* flip-flops vary in details of state table and waveforms.

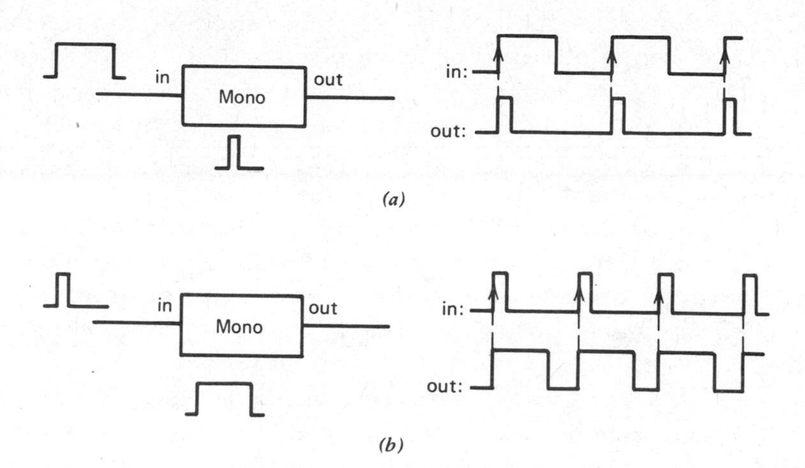

Figure 10-12. Examples of (*a*) pulse shortening, and (*b*) pulse lengthening by means of a monostable that responds to a positive-going logic transition.

The flip-flops described above are inherently bistable devices; they will remain in either of two states indefinitely. Similar circuits with other modes of operation are also possible. For example, some types have internal feedback connections that cause a change in state to be followed automatically by a return to the prior condition. This device, an *astable* flip-flop, or *multivibrator*, requires no input, since it is a type of oscillator producing a continuous train of square waves. On the other hand, if the feedback action is only one-sided, a *monostable* circuit results, in which a change of input generates a single output pulse, and then resets to the original state. The monostable is also called a *one shot* or a *univibrator*.

Any of these flip-flops can be assembled from discrete components, such as in the circuit shown in Figure 7-15, but are also available as ICs at considerably lower cost. A widely used monostable is the TTL 74121, which can give pulses varying from 40 ns to 40 s or more following an input transition. The adjustment of the pulse width τ, is established by an external resistor and capacitor, according to the formula $\tau = 0.7RC$. (The type 555 IC, mentioned in Chapter VII, can serve as either an astable or a monostable pulse generator, although it is not strictly a flip-flop; its operation is described in the next section.)

A monostable can be made to alter the width of pulses, as shown in Figure 10-12. This makes it possible to change the duration of the operation controlled by each pulse without changing the repetition rate.

It was mentioned earlier that an *RS* flip-flop can be used to debounce on-off switch operations. An analogous situation for momentary contact switches can be resolved by application of the circuit shown in Figure 10-13. This prevents multiple toggling of a *JK* flip-flop when controlled by a pushbutton switch, provided only that the pulse width of the monostable is longer than the bounce time.

For applications that require the timing of a series of events with a precision of 1% or so, a chain of monostables as shown in Figure 10-14 presents a convenient and inexpensive solution. In this system, each monostable after the first is triggered by the trailing edge ("1" to "0" transition) of the pulse from the preceding unit. Applications to automation are limitless, especially as one can form branched chains of any desired complexity. Higher timing precision, however, can be obtained by means of shift registers

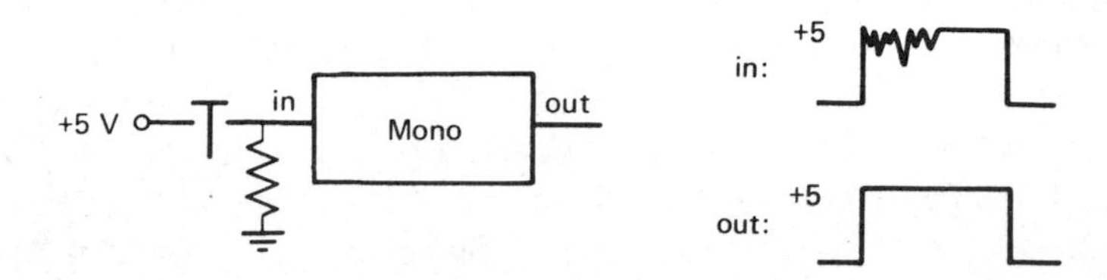

Figure 10-13. A momentary contact pushbutton switch rendered bounceless with a monostable.

and counters, to be considered in the following paragraphs.

THE TYPE 555 TIMER

A very useful and versatile device that can be employed as a monostable or an astable oscillator is the 555, shown in Figure 10-15. The heart of this IC consists of a tandem pair of comparators, *A* and *B*, which actuate a flip-flop. When connected as an oscillator (as shown in the figure), both comparators sense the voltage present on the capacitor *C*. The two comparators use as reference voltages, respectively, one-third and two-thirds of the power-supply voltage. Initially the capacitor charges through R_A and R_B until it reaches V_{max}. At this point comparator *A* sets the flip-flop, which in turn causes transistor *T* to saturate. The capacitor then discharges through R_B and *T*

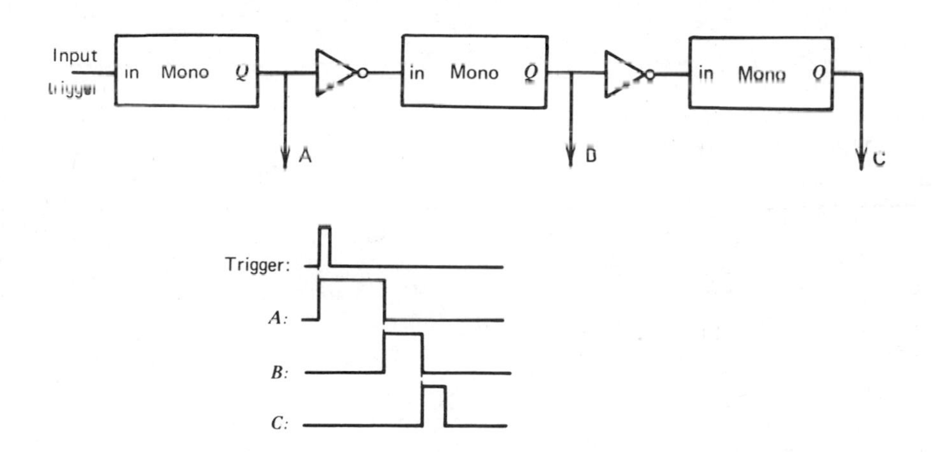

Figure 10-14. A sequential controller using a string of monostables. The action can be initiated by a trigger signal from a manual switch or by a logic signal. Since many monostables offer complementary outputs, Q and $\bar{Q}$, the inverters may not be necessary. Some monostables are triggered by $0 \rightarrow 1$ and others by $1 \rightarrow 0$ transitions.

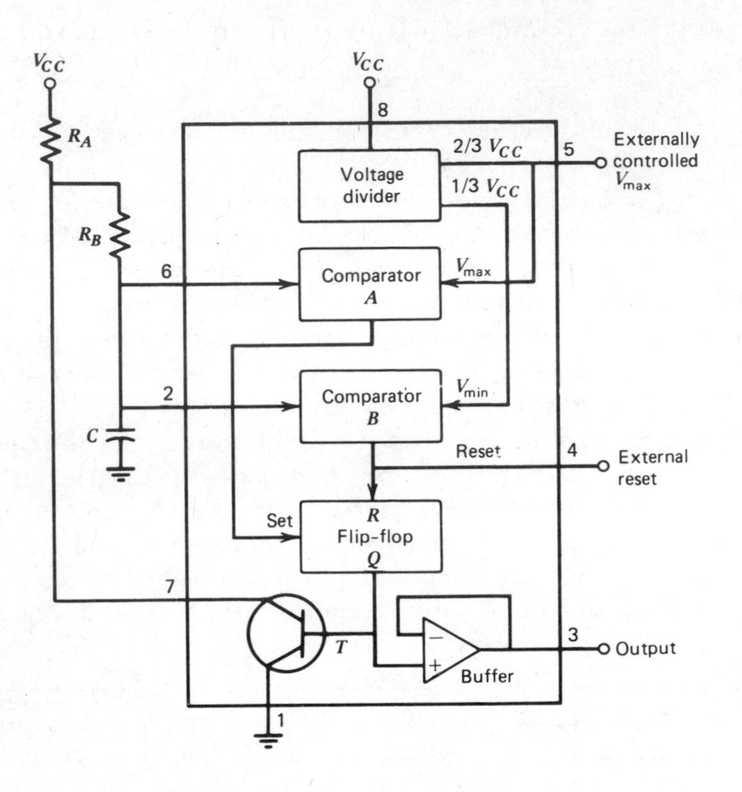

Figure 10-15. Simplified internal operation of a 555 IC, with external components to implement a square-wave generator.

until it reaches V_{min}, at which point it triggers comparator *B*, resetting the flip-flop and completing the cycle. The capacitor then begins to recharge, and the process repeats itself *ad infinitum*, with the capacitor oscillating between one-third and two-thirds of V_{CC}.

The 555 also has provisions for overriding the reset command by a pulse applied to pin 4 and for changing the reference potential for comparator *A* to some value other than two-thirds of V_{CC}, by a voltage applied to pin 5. The output is taken from the flip-flop, through a buffer amplifier, so that considerable current can be drawn to an external load. The output square wave can be made compatible with TTL logic. The duty cycle (i.e., the relative time durations spent in the high- and low-output states) can be varied by changing the relative values of R_A and R_B. The frequency of the square wave is given (in hertz) by $f = 1.44/[(R_A + 2R_B)C]$. The duty cycle is $R_B/(R_A + 2R_B)$(see also Figure 5-7).

For operation of a 555 as a monostable, a signal is applied to pin 2. A momentary "1" to "0" transition at pin 2 triggers the monostable operation. The

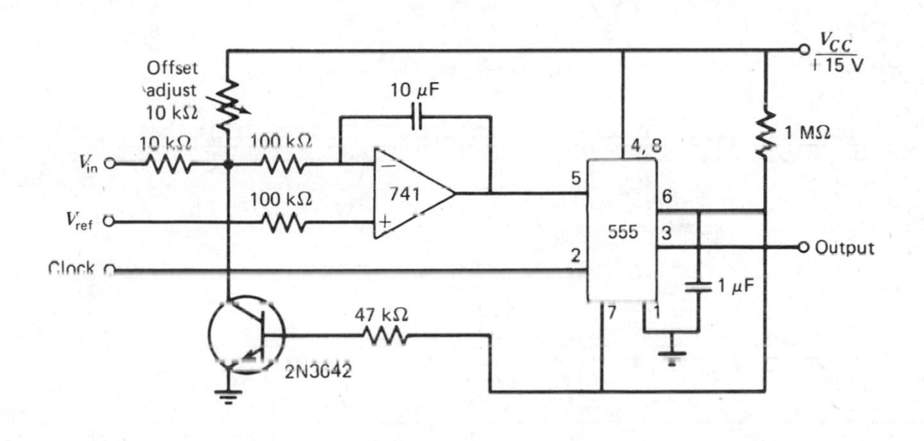

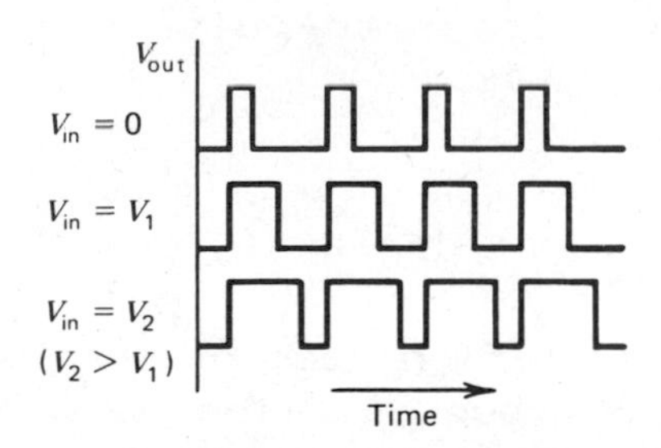

Figure 10-16. A circuit for the production of a train of pulses of duration proportional to the input voltage. (From J. Wyland and E. Hnatek, *Electronic Design*, June 7, 1973, p. 88, by permission.)

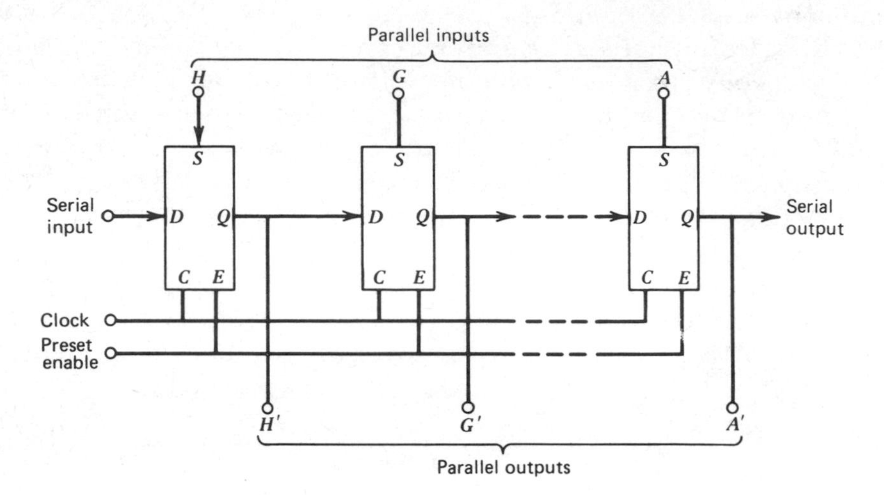

Figure 10-17. A shift register. The logical state at the input is shifted through the system on consecutive zero-to-one transitions of the clock. Data can also be entered in parallel through the *S* inputs.

duration of the pulse can be made very large (many minutes) if R_A and C are both large.

The 555 can be used in a variety of other applications. For example, if pin 5 is fed with a variable voltage, the unit will act as a square-wave frequency modulator, or (for DC inputs) as a voltage-to-frequency converter. A voltage-to-pulse duration circuit is given in Figure 10-16.

SHIFT REGISTERS

Since one flip-flop records one logical value (one *bit*), it takes a group of flip-flops, forming a register, to memorize a word. Thus a group of eight flip-flops can store an 8-bit word. Many types of registers can provide not only storage, but also logical manipulation of those words. The simplest form of such manipulation is shown in the register illustrated in Figure 10-17, an example of a *shift register*. The eight inputs, **A** through **H**, can accept any given pattern of ones and zeros, provided the "preset enable" is held high. After the binary word is loaded, successive clock transitions cause the information to be shifted to the right. The content of the first flip-flop is transferred to the second, the second to the third, and so on, until the last one, where the information exits. In this way information that is loaded into the register in *parallel* (i.e., simultaneously on all entry lines) exits in a *serial* fashion, one bit after another.

A more detailed description of a shift register using *JK* flip-flops is given in Figure 10-18. The Q and $\overline{Q}$ outputs of each flip-flop are connected directly

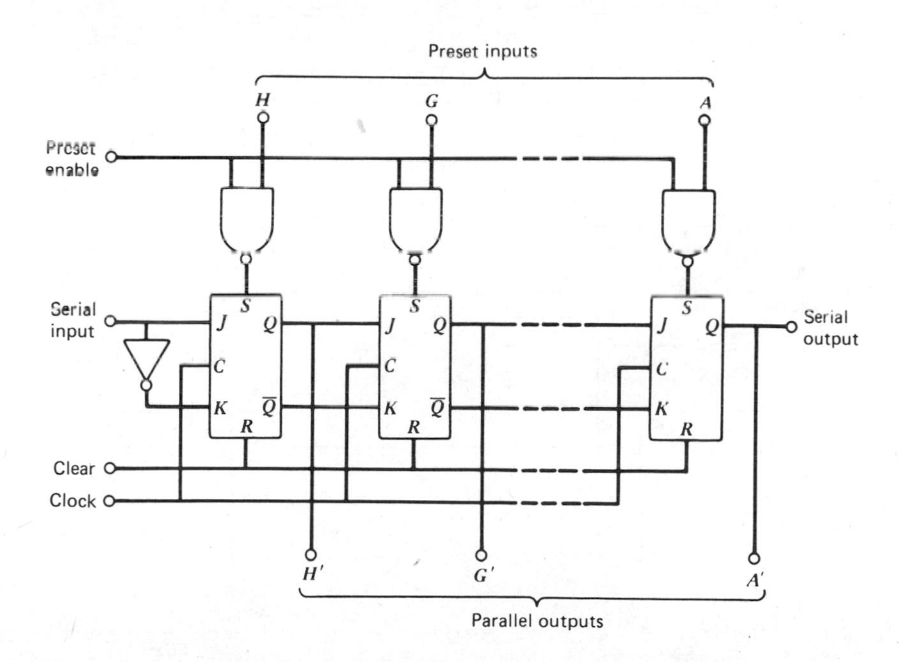

Figure 10-18. A detailed schematic of a shift register. The inverter at the input generates complementary inputs to J and K.

to the J and K inputs of the next stage. Data can be entered in parallel through the *preset* gates or serially through the J and K inputs of the first stage, as shown. Information can be retrieved from the register in either of two ways: (1) taken in a parallel mode (from all the Q outputs simultaneously) or (2) serially from the last Q only. In the latter case, the information is taken one bit at a time in response to successive clock pulses that advance the data synchronously through the register. The line marked "clear" serves to empty the register of information previous to loading new data.

Note that a connection made from the output back to the input avoids loss of information from the last flip-flop as it is fed back into the first. This arrangement, called a *ring counter*, can be used to generate a repeating pattern of ones and zeros. An example of such an application can be seen in Figure 10-19, in which the register is loaded with the binary digits 1000001, which is the equivalent of the letter

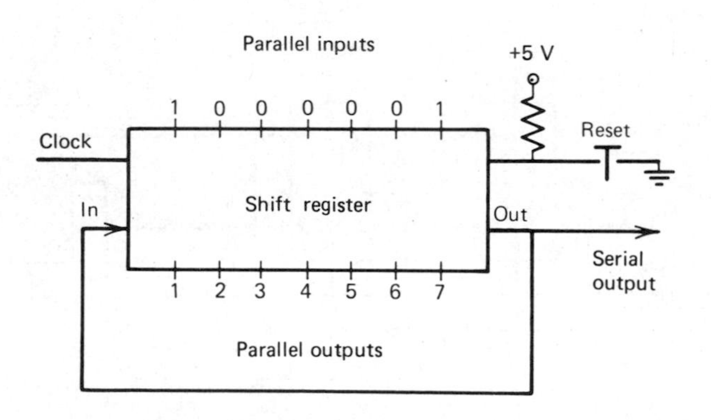

Figure 10-19. A shift register used to generate a repetitive pattern. Once loaded by means of the parallel inputs, the pattern circulates through the register, appearing serially at the output. Note that the content of the register exits in the order right to left.

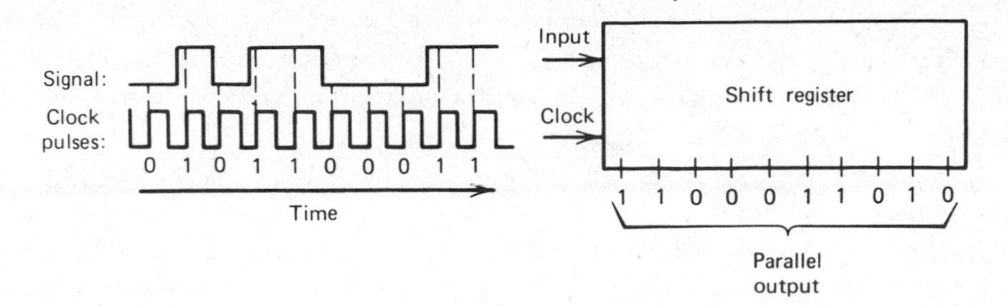

Figure 10-20. Serial entry into a shift register. Each clock pulse triggers the entry of a new value while also shifting previously entered data to the right.

"A" in the ASCII code. These digits are loaded into the register in parallel and then circulated through the register to appear serially at the output, where it can be picked up by an outside circuit (perhaps a computer) whenever an exit gate is enabled.

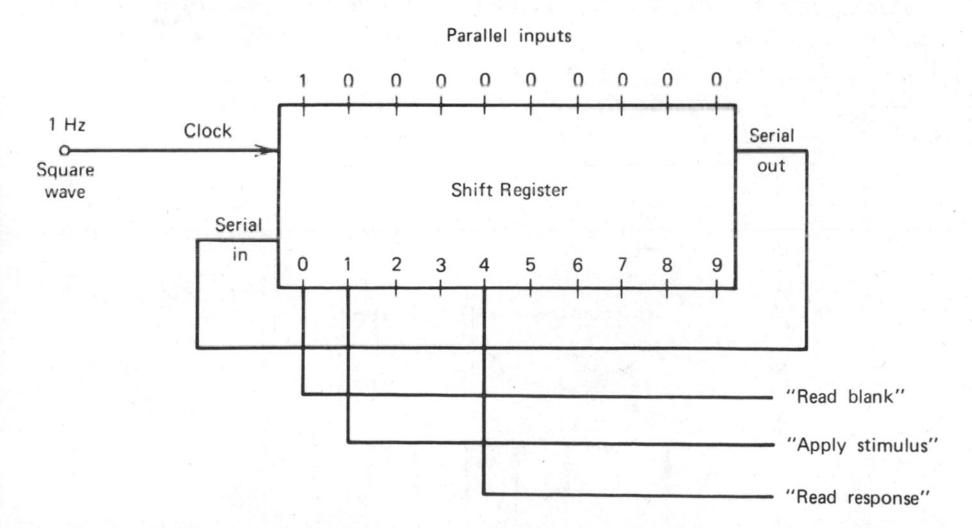

Figure 10-21. A shift register used to control a stimulus-response cycle. The complete cycle is 10 s in duration. The same basic system can be used with most so-called pulse methods, which are of the stimulus-response type. Cycling through the unused outputs provides time for the "relaxation" of the system to its original state.

The inverse operation, *serial-to-parallel conversion*, is also possible. The data are now accepted at the single entry port and read out at the parallel outputs (Figure 10-20).

Shift registers in the ring-counter configuration are useful for controlling a sequence of events. Typical of many experiments, one might want to execute the following operations: (1) take a measurement before the experiment, (2) apply a stimulus, (3) wait for the response, and (4) measure the response. Such a scheme can be implemented in elegant style by a single shift register, as shown in Figure 10-21. The initial pattern, 1 000 000 000 ("read blank"), changes after a 1-s delay to 0 100 000 000 ("stimulus ON"). After an additional 3 s it arrives at 0 000 100 000 ("read response") and after a further 6 s the cycle repeats.

The ring counter has practically no inherent time limitation, so that cycles from microseconds to days are equally possible. Ring counters, however, must be initialized whenever the power is turned on. This can be done by a combination of a "clear" command to elimi-

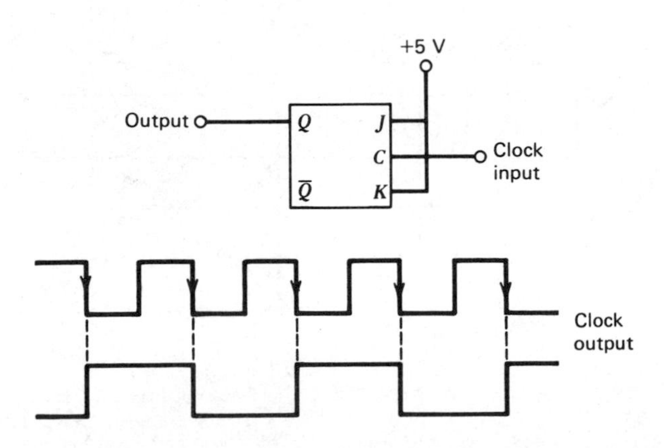

Figure 10-22. A *JK* flip-flop as a divide-by-two unit.

nate all previous highs and a "preset" command to load the desired pattern.

COUNTERS

Consider a *JK* flip-flop operated in the toggling mode, as in Figure 10-22. The output Q will change state for every *falling* edge of the clock transitions. If we compare the input and output, it becomes evident that the frequency of the square wave is divided by 2. Thus a toggling *JK* flip-flop, sometimes referred to as a "T-FF," acts as a divide-by-two. By cascading more T-FF units, division by 4, 8, 16, 32, and so on can be accomplished. Such circuits are called *counters*.

To better understand how a counter functions, refer to Figure 10-23. Here a series of four flip-flops are cascaded to form a divide-by-16. Clearly each successive stage has half as many transitions as its predecessor. The right-to-left mode in the drawing has the advantage that the outputs *A*, *B*, *C*, and *D* have a direct numerical significance. The logical values for a few completed input cycles are:

Cycles	*A*	*B*	*C*	*D*
0	0	0	0	0
1	0	0	0	1
2	0	0	1	0
3	0	0	1	1
• • •	• • •	• • •	• • •	• • •
15	1	1	1	1
16	0	0	0	0

Evidently the outputs represent at any moment the binary number of input cycles. Thus the circuit has two functions: (1) it divides the frequency by powers

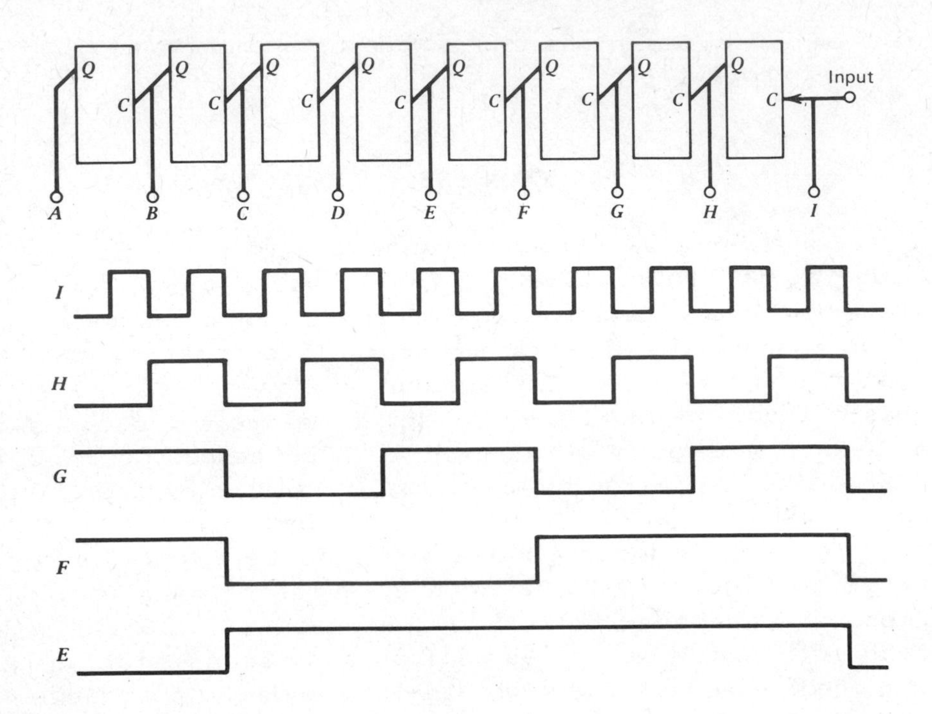

Figure 10-23. A divide-by-sixteen register. Only the first eight input cycles are shown. The right-to-left mode of depicting the register emphasizes the relationship to binary counting.

of two, and (2) it counts the number of input pulses.

It is often desirable to obtain an output reading directly in the familiar decimal system. The underlying difficulty in achieving this is that 10 is not an integral power of 2. Three flip-flops are not enough to count to 10, and four are too many; therefore, some means must be devised for interrupting the counting process at 10. One way in which this can be done is to arrange logic circuits to detect when binary 1010 has been reached, since this corresponds to decimal 10, and to reset all flip-flops at this point (Figure 10-24). For every power of 10, four flip-flops and one resetting circuit are required.

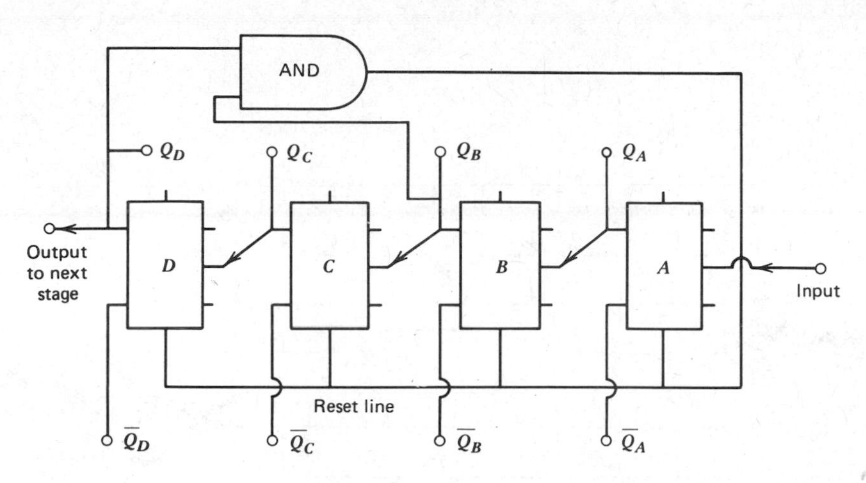

Figure 10-24. One stage of decimal counting, based on resetting all flip-flops by an **AND** gate at the tenth pulse. The output sequence is in BCD code.

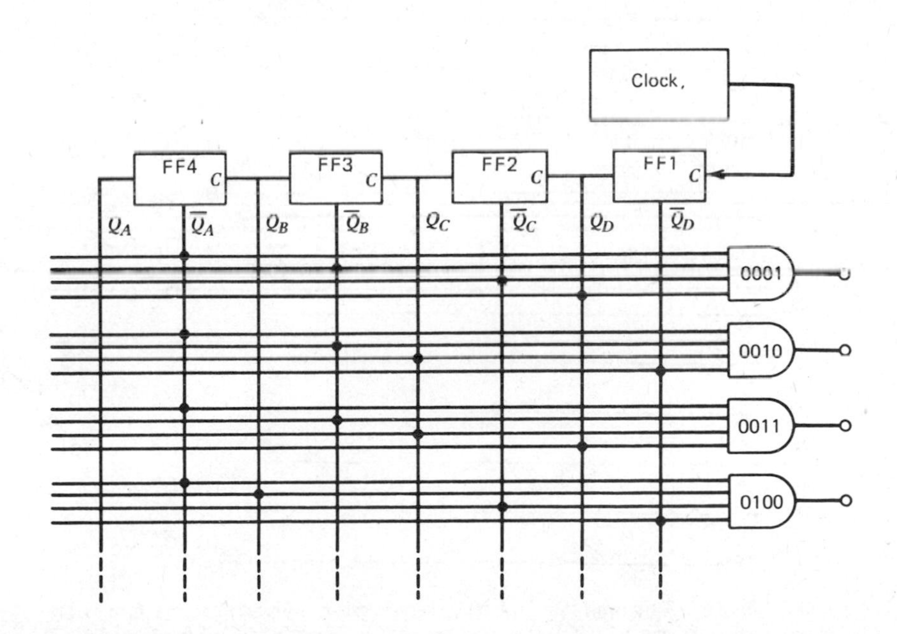

Figure 10-25. Timing circuitry, showing the output gates corresponding to the first four combinations given in Figure 10-26.

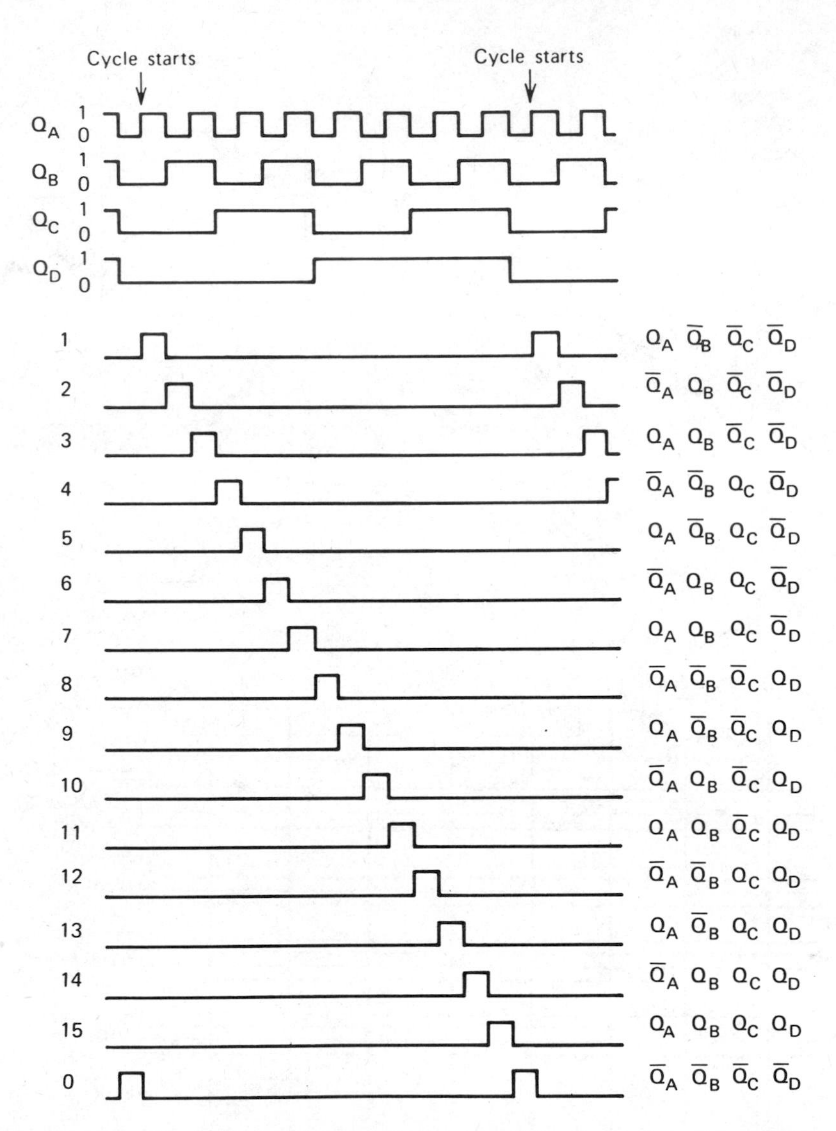

Figure 10-26. Timing sequence obtainable from the 4-bit flip-flop register shown in Figure 10-25, with a 1-Hz clock.

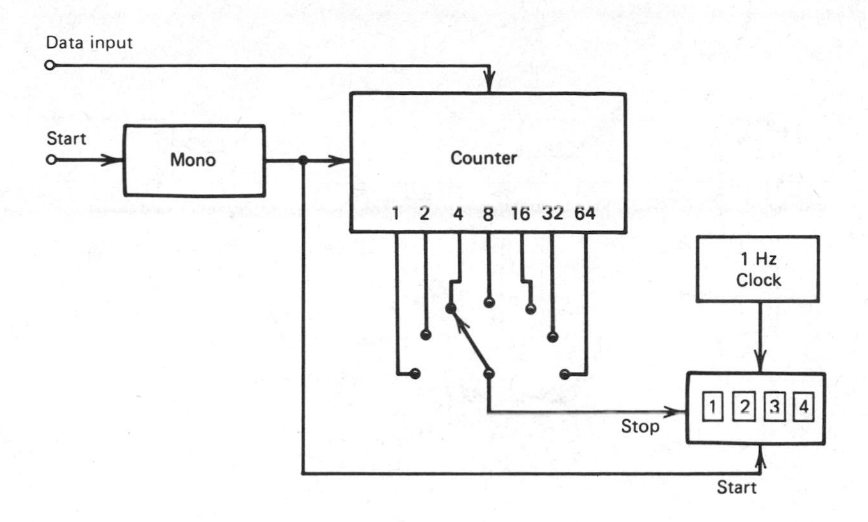

Figure 10-27. An example of a preset counter. Following the number of counts indicated by the switch position, the output of the counter will go high, thus actuating the control that stops the timer.

Sequential Timing

A counting register is useful in quite another way, in that it provides a flexible and convenient manner of programming events with greater precision than afforded by a string of monostables. The use of a ring counter for this purpose has already been mentioned, but open-ended binary counters are more flexible. Every binary number can be uniquely identified by means of an **AND** gate with as many inputs as there are digits in the number. For example, the number 3 (binary 0011) can be selected by a gate **AND**ing $\bar{Q}_A$, $\bar{Q}_B$, Q_C, and Q_D. Figure 10-25 shows circuitry for the implementation of such a decoding procedure. Sixteen **AND** gates would be required to exhaust all the possibilities. The actual timing diagram thus obtained is depicted in Figure 10-26.

Let us suppose that the clock gives a square wave with a 1-s period. The output of flip-flop A is then a square wave with a period of 2 s. Each state (Q_A =

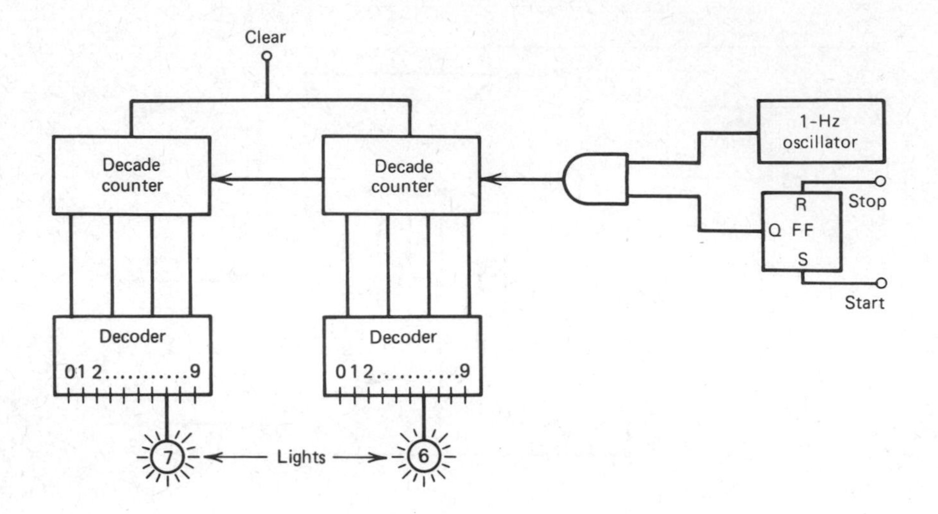

Figure 10-28. A timing circuit indicating "76 s." Only one lamp from each decoder can be ON at a given moment, representing times from 0 to 99 s.

1 and $\overline{Q}_A$ = 0) lasts for one second, which is the duration of the pulses at each of the output **AND** gates. The experiment to be sequenced is made to start at the same instant that the counter starts counting clock pulses. Then, if event *X* is to occur exactly 3 s later and last for 5 s, it can be actuated by the signal from the **AND** gate marked 0011 and turned off by 0101. Unless inhibited by logic, *X* will repeat every succeeding 16 s.

Counters can also be used in a "preset" mode, meaning that they will indicate when a preselected number of input pulses have arrived. The nuclear counter described earlier could benefit by such a preset counter, as shown in Figure 10-27. Here the timer measures the interval needed to reach a certain number of counts. The timer itself can be implemented by the use of counters as shown in Figure 10-28. The 1-Hz clock is gated on and off by an *RS* flip-flop that is manually set to start the counting. The system continues to count the 1-s pulses until turned off by a

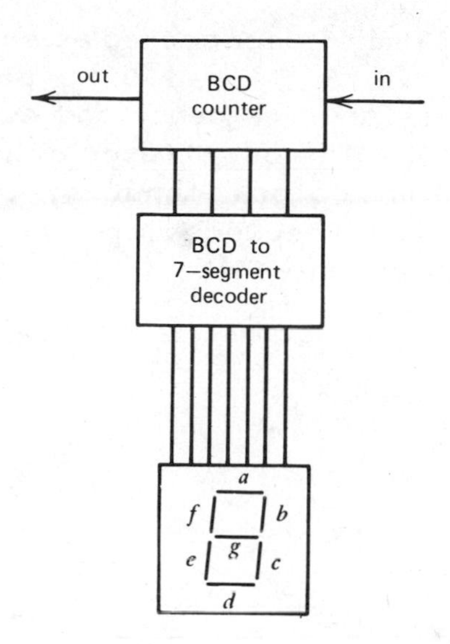

Figure 10-29. An example of a timer or counter driving a seven-segment numerical display. Various combinations of illuminated segments will give any digit from 0 to 9. The decoder is rather complex; for example, it must energize segment *g* for the digits 2, 3, 4, 5, 6, 8, and 9, whereas segment *d* is needed for 2, 3, 5, 6, and 8, and so on.

signal at *R*. The indicating lamps will continue to read the final count until the register is cleared.

Most timers and counters use direct decimal readout through *seven-segment display* indicators, as shown in Figure 10-29.

PROBLEMS

10-1. By comparing Figures 10-15 and 10-16, explain how the 555 operates as a monostable.

10-2. (*a*) A 555 timer is connected as an astable oscillator with $R_A = R_B = 150$ kΩ and $C = 0.01$ μF. What will be the frequency and duty cycle of the output? (*b*) The oscillator of part (*a*) is connected to the clock input of a *JK* flip-flop. Calculate the frequency and duty cycle of the output of the flip-flop.

XI
INTERDOMAIN CONVERSION

In this chapter we discuss methods of information transfer between the analog and digital domains. Such conversion is essential in two ways: as part of the input to a computer for signals that are derived from analog transducers, and in computer outputs that operate continuous control systems.

ENCODING CONSIDERATIONS

The definition of resolution differs in the analog and digital domains. In digital measurements, an inherent limitation lies in the essential discontinuity of the process, whereas there is no similar limitation in the realm of analog signals. Another factor to be considered is that the frequencies present in the analog signal must be compatible with the digital clock rate since this is the highest frequency that can be present in digital systems.

In order to understand the type of problems to be attacked in adapting analog measurements to digital instruments, let us consider the measurement of an optical spectrum. The experiment consists of scanning the wavelength over a range of, say, 500 nm and measuring the corresponding light intensity. The wavelength mechanism can conveniently be varied by means of a stepping motor that advances the scan by successive increments while sending out one pulse per step. The determination of the wavelength then reduces to counting the number of pulses from the start of the scan. It would appear that a good choice would be to let 500 pulses correspond to the desired 500 nm. In the case of data handling by computer, however, a much more important criterion than the availability of round numbers is the requirement that the step is smaller than the expected error in the analog instrument. This will ensure that the stepwise nature of the count does not add to the overall instrumental error. Suppose that statistical considerations require the wavelength step to be no larger than 0.7 nm. In this case the drive mechanism should be designed to give at least 500/0.7 = 714 steps.

The situation is different in respect to the measurement of light intensity. In the first place, we do not know in advance the range of values to be expected. We do know, however, the maximum output of the instrument at full scale. Once the desired resolution is established, say, 0.5% of full scale, it is possible to determine the requirement of the digital system. In

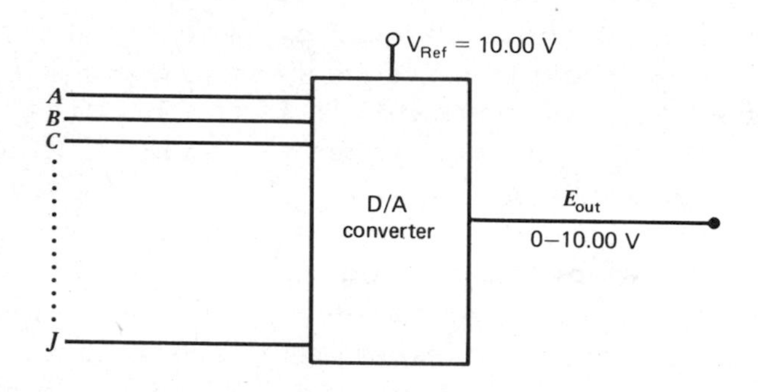

Figure 11-1. The D/A converter. The input has 10 lines, with 2^{10} = 1024 possible combinations. The device can be adjusted to give other output ranges.

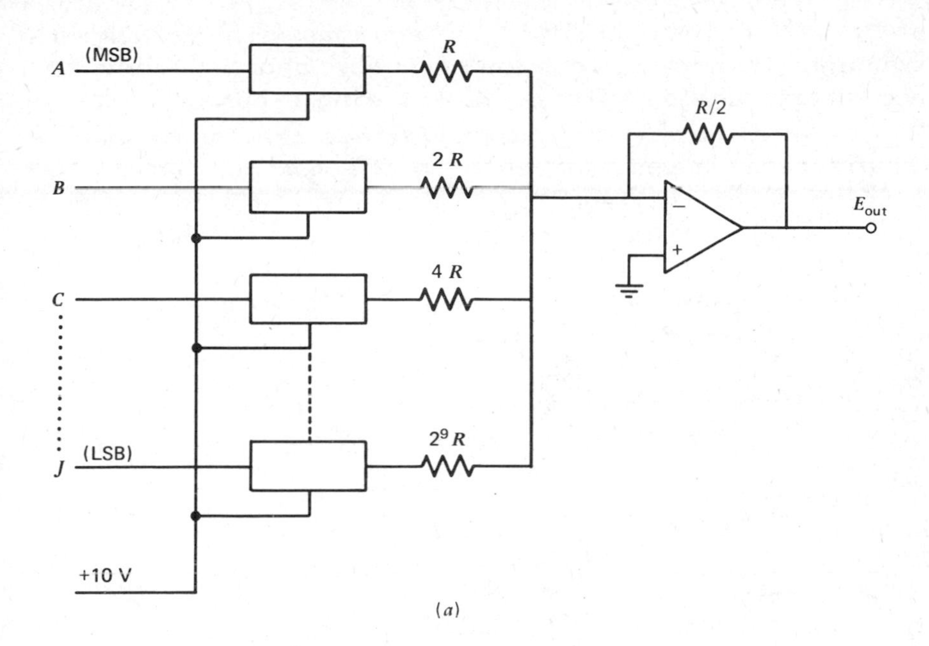

Figure 11-2(*a*). Block diagram of a 10-bit DAC. If a variable signal is substituted for the 10-V reference, the device functions as a programmable amplifier with digital controls.

order to obtain this resolution, the scale must be divided into 200 steps (or more). In practice, the output can be described by a single 8-bit word since this corresponds to the range of values of zero to $(2^8 - 1)$, or 255 steps.

We now describe techniques of interdomain conversion. It is advantageous to treat digital-to-analog conversion before considering the reverse process.

DIGITAL-TO-ANALOG CONVERSION

A component used to perform this transformation is called a *digital-to-analog converter*, a *D/A converter*, or simply a *DAC*. The block representation of a DAC is shown in Figure 11-1. It accepts binary numbers through its multiple input lines, and produces the equivalent analog voltage at its single output. If it has n input lines, the output voltage can assume any of 2^n discrete values from zero to the maximum determined by a reference voltage E_{ref}. Thus for $n = 10$ and E_{ref}

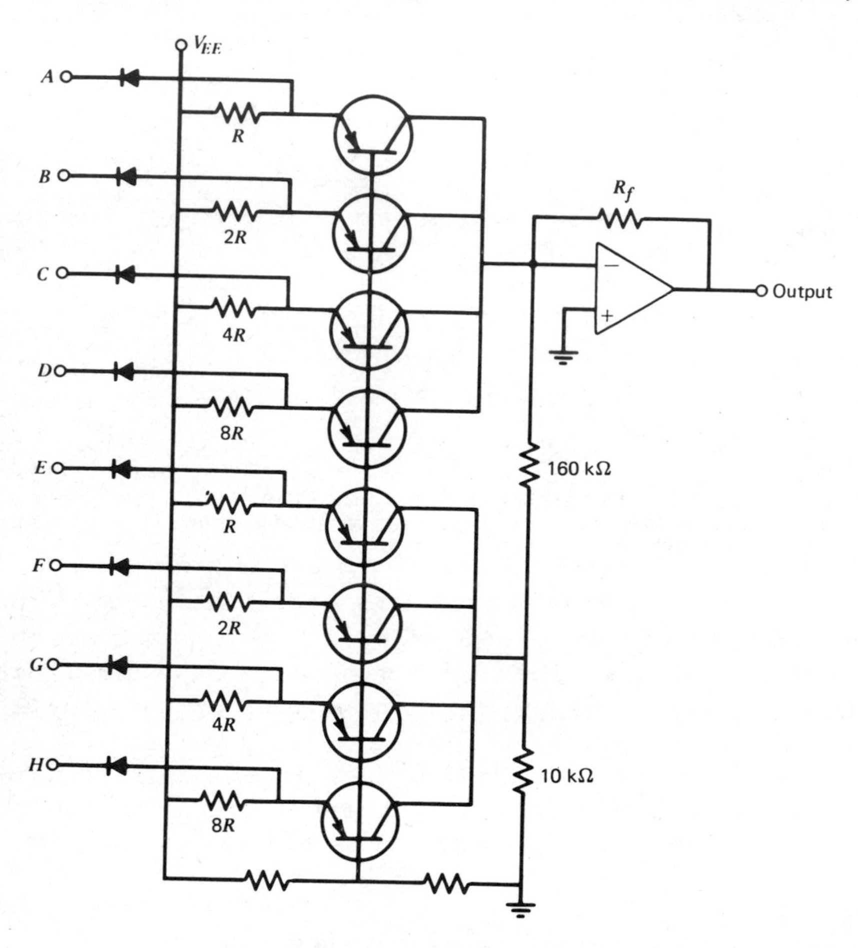

Figure 11-2(*b*). An 8-bit DAC using *pnp* transistor switches, organized as two 4-bit units.

= 10 V, the output can have any of 1024 evenly spaced voltages, including the limits of 0 and +10.00 V, giving a resolution of slightly better than 0.1%. Any voltage between 0 and 10 can be specified to within 9.8 mV. This gives better precision than most analog panel meters, which are seldom rated better than ±2%. If still better resolution is needed, it can be obtained either by selection of a DAC with more input ports or reduction of the reference voltage. For instance, if E_{ref} = 5 V, the smallest output increment will be 4.9 mV, doubling the precision but halving the range.

One type of DAC takes the form of a programmable amplifier, such as that described in Chapter VII. An example is shown in Figure 11-2*a*. A series of *n* transmission gates are controlled by the *n* digital input lines, being turned ON when the signal is high and OFF when it is low. The output voltage can readily be calculated by simple algebra, using the rules for parallel combinations of resistances to determine R_{in} for each input code. For example, if the input is 111000000, so that the three upper channels in the figure are opened, the input resistance becomes

$$R_{in} = \frac{1}{1/R_1 + 1/R_2 + 1/R_3} \tag{11-1}$$

and the output voltage is

$$E_{out} = -E_{in} \cdot \frac{R}{2}\left(\frac{1}{R} + \frac{1}{2R} + \frac{1}{4R}\right)$$

$$= -10 \cdot \frac{1}{2}\left(1 + \frac{1}{2} + \frac{1}{4}\right) = 8.75 \text{ V} \tag{11-2}$$

Note that the reference voltage must be negative if positive outputs are desired.

It may be impractical to use this simple circuit for more than about 4-bit resolution because of the need of very high value resistors, which lead to instability and slow speed. Where greater resolution is needed, it is customary to arrange the gates in groups of four, connected through a current-division network that gives the equivalent results without these difficulties. A circuit for an 8-bit DAC using this design is shown in Figure 11-2*b*. An input is LOW if grounded and HIGH if it is equal to or larger than V_{EE}. If V_{EE} = +5 V, standard TTL levels can be used.

Another type of DAC is shown in Figure 11-3, differing from the previous one primarily in the configuration of the assemblage of resistors, called a "ladder network." This has the practical advantage from the design standpoint of requiring only two resistor values, some of *R*, and others 2*R*. Note that the input closest to the voltage reference has the greatest effect on the result, and so is the MSB (the most significant bit), whereas the LSB (the least significant bit) is the furthest.

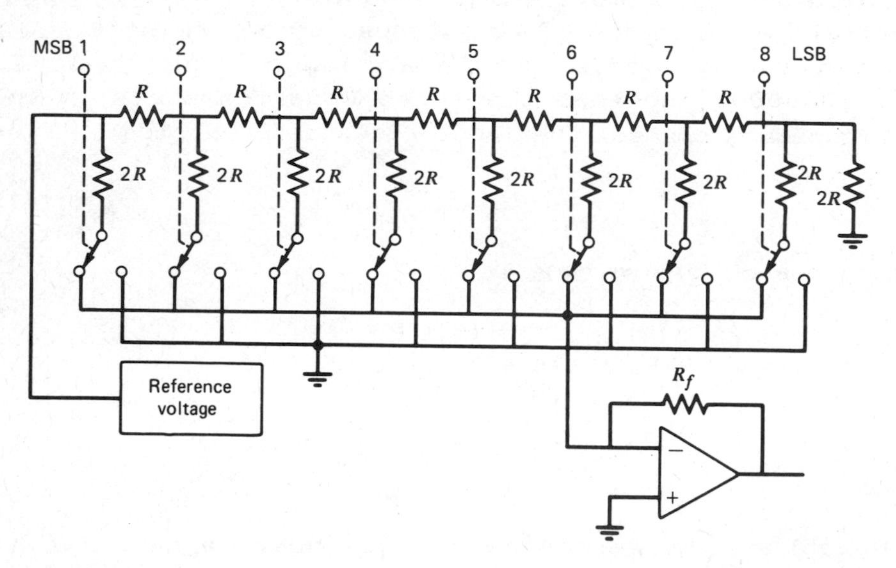

Figure 11-3. An 8-bit DAC with an *R*/2*R* ladder network.

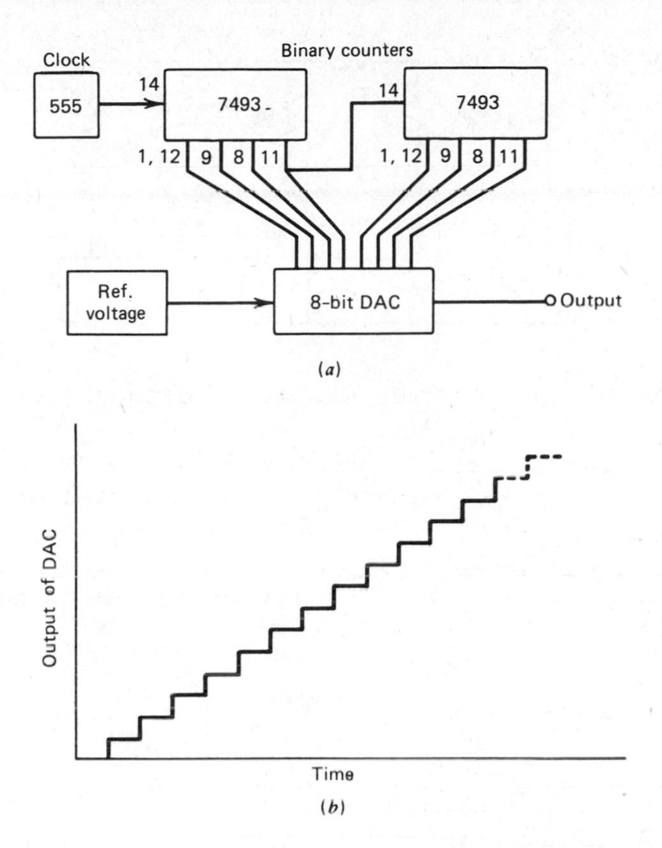

Figure 11-4. A staircase generator using an 8-bit DAC. This will give 256 steps; the voltage interval per step depends on the reference voltage.

Both types of DACs can be operated with a fixed reference voltage, as shown, or with an analog signal applied to the same connection. In the latter case, the output consists of the analog input E_{in} multiplied by the digital number n:

$$E_{out} = n \cdot E_{in} \tag{11-3}$$

So used, the unit constitutes a *multiplying DAC*.

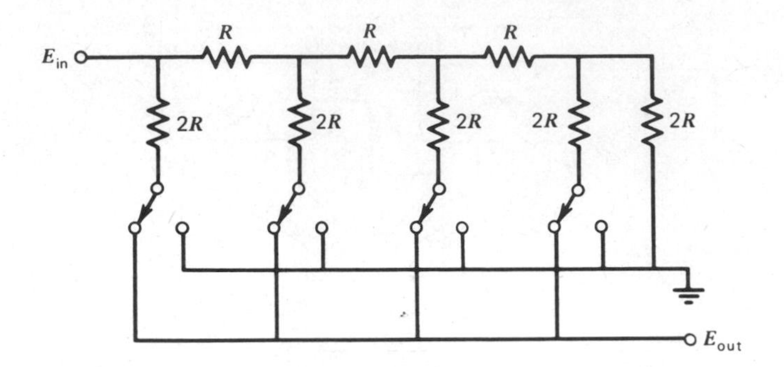

Figure 11-5. An *R*/2*R* DAC as a voltage divider.

DACs can be used in various applications besides converting signals to a computer-compatible format. For example, a DAC driven by a binary counter will produce a "staircase" (Figure 11-4), useful in a variety of experiments. A multiplying DAC can be connected

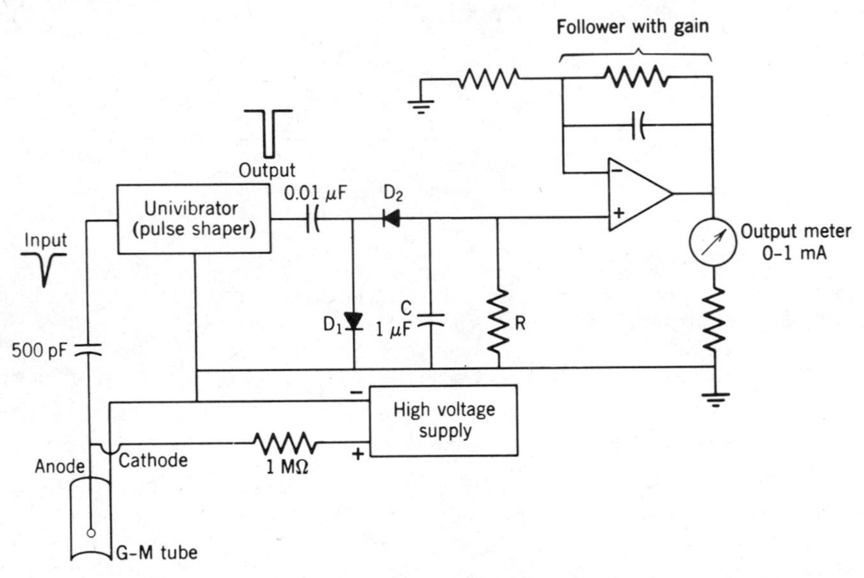

Figure 11-6. A ratemeter as used with a Geiger counter. (From A. Weissberger, Ed., *Techniques of Chemistry*, Vol. I, Part 1B, Wiley-Interscience, New York, 1971, by permission.)

as shown in Figure 11-5 to provide a digitally controlled voltage divider; the model AD7525 (Analog Devices) is designed for this service. It can give an output anywhere between 0 and 1.999 times E_{in}, with a resolution of 0.001, as controlled by a BCD-coded input.

An example of an instrument that has a built-in special-purpose D/A converter is the ratemeter used in nuclear and x-ray work (Figure 11-6). Nuclear particles or photons produce individual pulses through the action of a transducer such as a Geiger-Mueller tube. Each pulse contributes an increment of charge to the capacitor marked C, and the resulting voltage is indicated on the meter. On the other hand, the resistor R discharges the capacitor at a rate proportional to its voltage, so that a steady state is attained in proportion to the number of pulses per unit time. The capacitor and associated diodes constitute a special-purpose DAC, converting frequency to an analog voltage.

Glitches

The signals forming a binary sequence pass through what are called "major" and "minor" transitions, between successive counts. The most damaging of these is the transition between 01111111 and 10000000, where every digit changes. It may well happen that the switches have slightly different turn-on and turn-off times, in which case the output of a DAC may go momentarily to zero (or full scale) before assuming the new

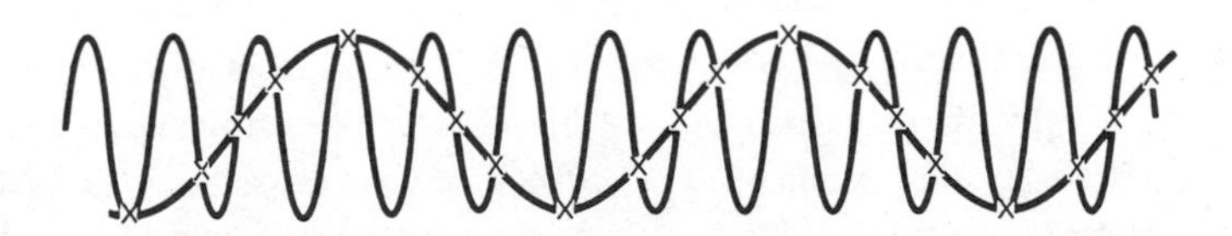

Figure 11-7. Aliasing. A sine wave (1), sampled at a rate slightly less than its own frequency, gives false evidence of a much slower sine wave.

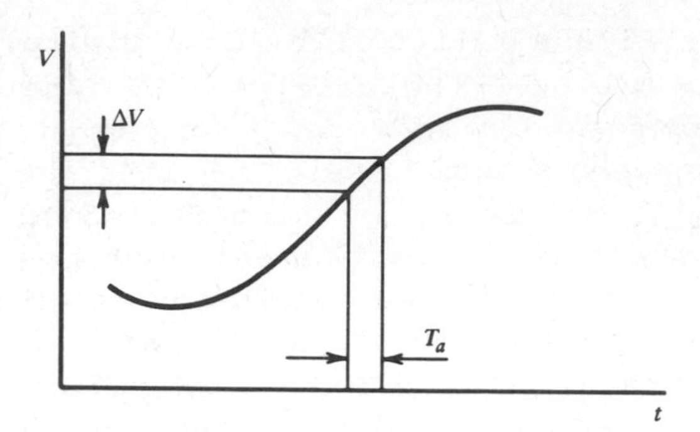

Figure 11-8. The relationship between the aperture time T_a and the rate of change of the input signal.

value. This spike, called a *glitch*, which can be either negative or positive, is likely to cause havoc with following circuitry. A common *deglitching* circuit uses a sample-and-hold amplifier to retain the former output for a few microseconds until the switches have settled down in their new configuration.

ANALOG-TO-DIGITAL CONVERSION

In interdomain conversion, the interaction between the various frequency components of the analog signal and the number of digital readings per second (the clock rate) must be carefully considered. Let us assume that a pure sine wave of frequency f is to be converted into its digital equivalent. The first limitation is that sampling must be clocked at a frequency at least as great as $2f$, to avoid a fault called *aliasing*. If the analog signal were sampled only at the frequency f, namely, once per cycle, a little thought would show that the samplings would all be identical and not demonstrate that the signal is varying at all. Sampling less frequent than this could be interpreted as indicating a sine wave of a submultiple of f, such as half its frequency. Figure 11-7 will clarify these relationships.

Another limitation is concerned with the finite time required for a DAC to accomplish its conversion. We can call this the *aperture time*, T_a. The analog signal will vary during this time by the amount ΔV, as seen in Figure 11-8, depending on the slope of the signal curve at that point:

$$\Delta V = \frac{dV}{dt} T_a \qquad (11\text{-}4)$$

We want to establish the maximum allowable aperture time that will still give the required resolution. Let us consider the signal to be a Fourier series with its greatest frequency ω_{max} rad/s. The largest permissible value of T_a is related to the period T_s associated with the frequency ω_{max} by the relationship

$$T_a = T_s \frac{\Delta V}{V} \qquad (11\text{-}5)$$

where $\Delta V/V$ is the relative resolution required to reproduce the signal adequately. This means that the aperture time must not be greater than the period of the greatest frequency component of the signal, multiplied by the resolution.

Suppose, for example, that we need to find the appropriate aperture time for sampling a 400-Hz sine wave (or a nonsinusoid in which the highest frequency is 400 Hz) with a resolution of 1%. Equation 11-5 gives

$$T_a = T_s \frac{\Delta V}{V} = 2\pi \left(\frac{\Delta V/V}{\omega_{max}} \right) = \frac{0.01}{400} = 25\ \mu s \qquad (11\text{-}6)$$

If the DAC is not fast enough to make a conversion in 25 μs, we must resort to a sample-and-hold (S/H) amplifier. The S/H amplifier is programmed to sample the signal for 25-μs intervals with the proper repetition rate, holding its value for the time required by the DAC to make its conversion.

Comparators and Schmitt Triggers

The comparator, introduced in Chapter IV, can be considered to be a single-bit digitizer (Figure 11-9). When the output of a fast comparator is bounded by a pair of diodes, as shown, it gives TTL-compatible digital levels indicating whether the input is higher or lower than a reference potential. Special comparators, such as the high-speed LM160, give TTL levels directly.

Digital signals generated by mechanical switches or op amps may not have suitable waveforms for direct connection to digital devices. Specifically, the rise time of conventional IC logic is 10 ns or less, and this can hardly be matched by an op amp. This waveform discrepancy can be avoided by the use of a *Schmitt trigger* such as the 7413. Its effect is shown in Figure 11-10*a*. The chief feature is *hysteresis*; as the input is increased from zero, the output stays at logical "1" until voltage V_2 is reached, when it suddenly changes to "0"; but when the applied potential is decreased again, no change takes place until the lower threshold voltage V_1 is reached. Figure 11-10*b* shows how this effect can be used to convert a sine wave into a square wave of the same frequency. This hysteresis effect can be important, tending to make a system more stable, since a considerable input change is necessary to affect the output; it also minimizes the effect of noise. The Schmitt trigger function can be added to a comparator by providing positive feedback resistors (Figure 11-10*c*).

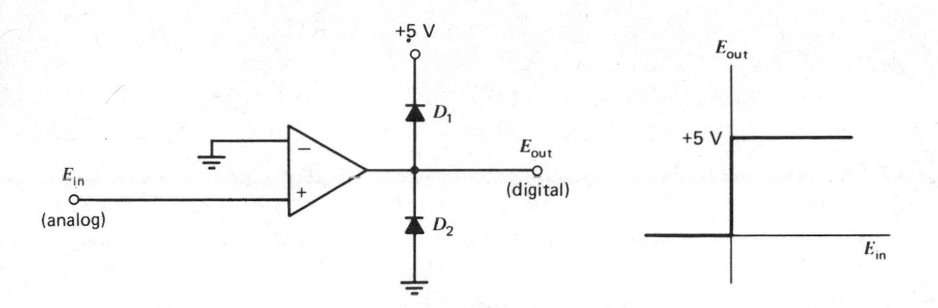

Figure 11-9. A comparator used as a digitizer. The diodes limit the output signals to the two extremes of 0 and +5 V. Diode D_1 can be omitted if D_2 is made a 5-V Zener.

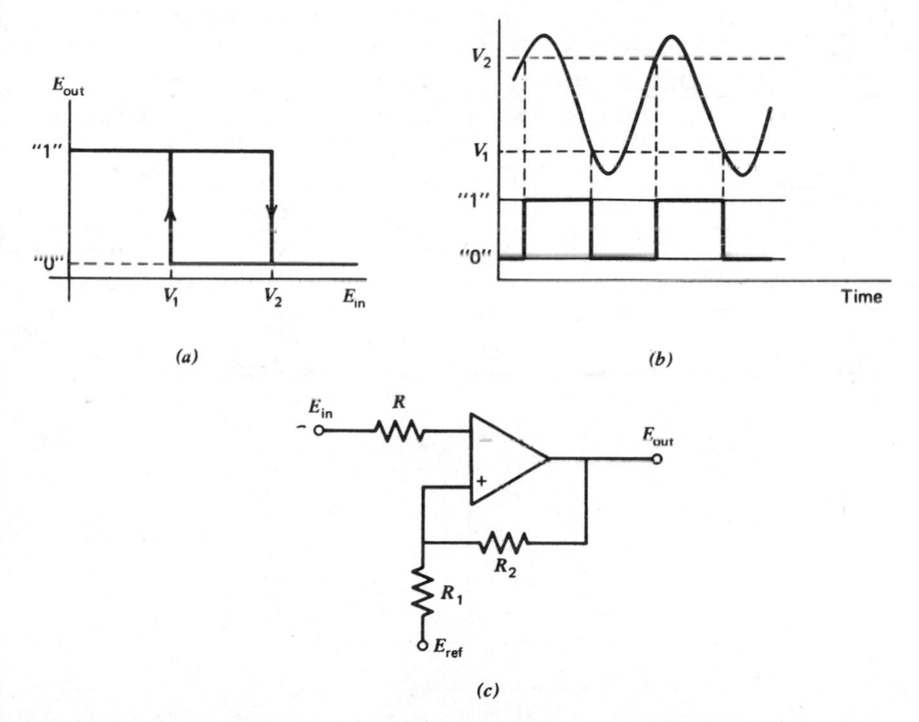

Figure 11-10. (*a*) The wave-shaping function of a Schmitt trigger. (*b*) Formation of a square wave from a sine wave. (*c*) An op amp connected as a Schmitt trigger; for minimum offset error, R should equal the parallel resistance of R_1 and R_2.

ANALOG-TO-DIGITAL CONVERTERS

The counterpart to the DAC is the *analog-to-digital converter*, also called *A/D converter*, *ADC*, or sometimes *digitizer*. There are several types, varying considerably in their method of operation. The simplest can be assembled from a binary counter and a DAC, as in Figure 11-11. The analog input is compared with the signal from the DAC and the output of the comparator enables or disables the **AND** gate transmitting the clock pulses. As long as V_{DAC} is less than E_{in}, the signal sent to the **AND** gate is a logical "1" and the clock pulses are allowed to pass to the counter, so that V_{DAC} continues to increase. When the two become equal, the **AND** gate is disabled. At the same time a signal called an "end-of-conversion flag," (EOC) is produced at an auxiliary output indicating that the conversion is complete. The counter must be reset to zero before starting another measurement.

Figure 11-12 shows a modification that permits continuous counting. An up-down counter is substituted for the unidirectional counter in Figure 11-11.

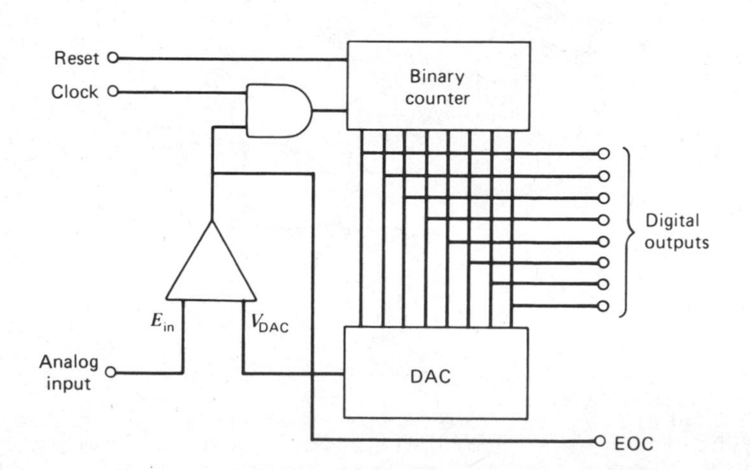

Figure 11-11. A simple form of ADC.

Now the reset command is not needed, for one conversion directly follows another. No matter whether the new measurement calls for a higher or lower count, the system can respond. The comparator is connected directly to the up-down control point of the counter so that if the input increases, the count will go up, or if it decreases, the count will go down until it is equal to the analog input. Since conversion is continuous, there is no "conversion complete" flag, and a data latch may be needed on the output to freeze the digital signals periodically.

Successive Approximation Converters

A third method of A/D conversion uses the principle of successive approximations. This is illustrated in Figure 11-13, where it is supposed that an input voltage of 7.18 V is to be digitized. The operation begins with the register set to zero. When a conversion is started, an initial clock pulse is given to the MSB position causing the register to assume a value of 10000000 = 5 V, corresponding to half-scale, as a first approximation to the input. If, as it happens, the

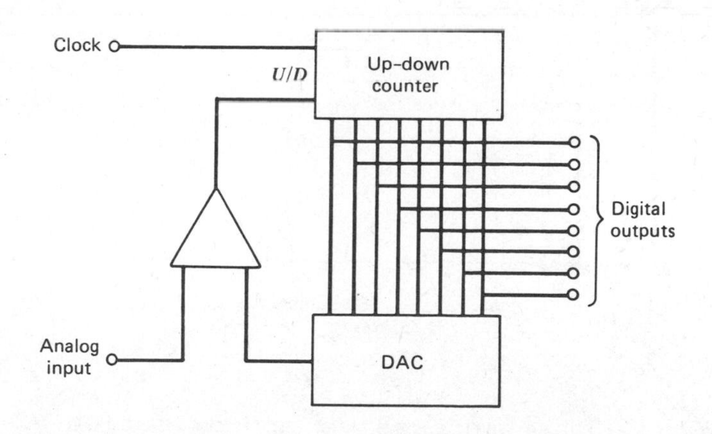

Figure 11-12. An ADC using a bidirectional counter.

input is greater than half-scale, the "1" is allowed to remain as the MSB, and a second approximation is made by giving a tentative "1" to the next bit. This time the contents of the register (11000000 = 7.5 V) is larger than the analog input, and the trial "1" is replaced by a "0." This trial-and-error procedure is

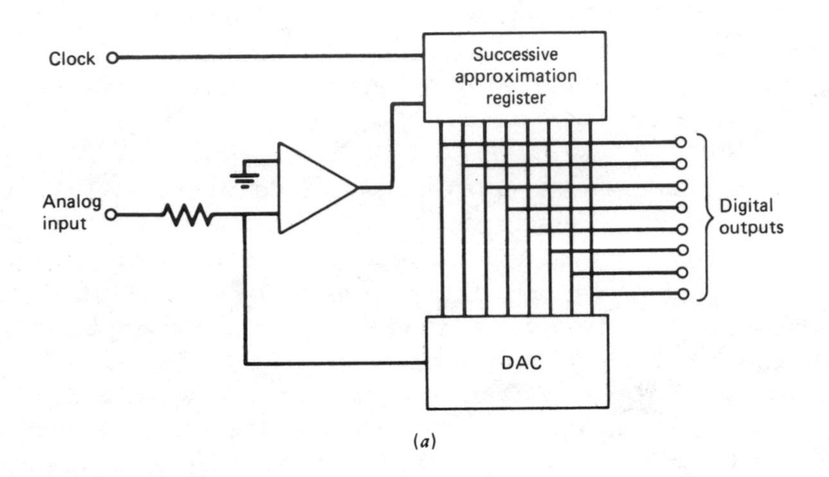

(a)

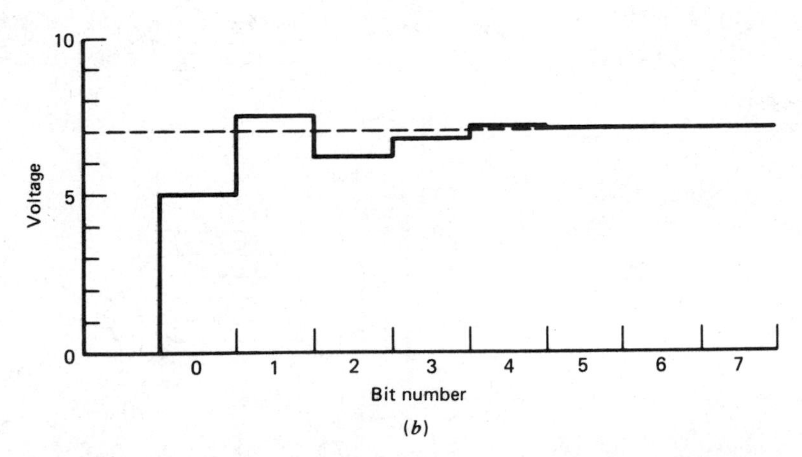

(b)

Figure 11-13. A successive approximation ADC: (a) schematic; (b) the output voltage for successive bits; the resulting binary number is written above the graph.

repeated until the register is filled. Each stage in the register must be latched to retain the binary result until conversion is completed.

Dual-Slope ADCs

The dual-slope A/D converter uses still another approach, as shown in Figure 11-14. In this circuit a comparator is fed by the output of an analog integrator. Initially the analog switch steers the incoming signal to the integrator for a predetermined period t_1. The charge on the capacitor at this time bears a direct proportion to E_{in}. At this point the switch is thrown so as to connect the (negative) reference voltage to the integrator. This starts a downward ramp of predetermined slope, which continues until the comparator senses zero volts at time t_2. The duration of the downward integration ($t_2 - t_1$) is a measure of the capacitor charge and thus of the input voltage. The counter is initially preset in such a way as to reach zero at exactly time t_1; hence at the end of the conversion it displays the desired elapsed time directly.

The dual-slope ADC is highly accurate and inexpensive, but fairly slow. It is often used in digital panel meters and other applications not requiring high speed.

Voltage-to-Frequency (V/F) Converters

The V/F converter is a special type of ADC, which, instead of generating a digital word corresponding to the magnitude of the incoming analog voltage, produces a square wave or string of pulses such that the frequency is proportional to this voltage. Figure 11-15 shows a circuit for one commercial model, the type LM131. The principle of operation is based on a forced equality between the voltage on capacitor C_B and

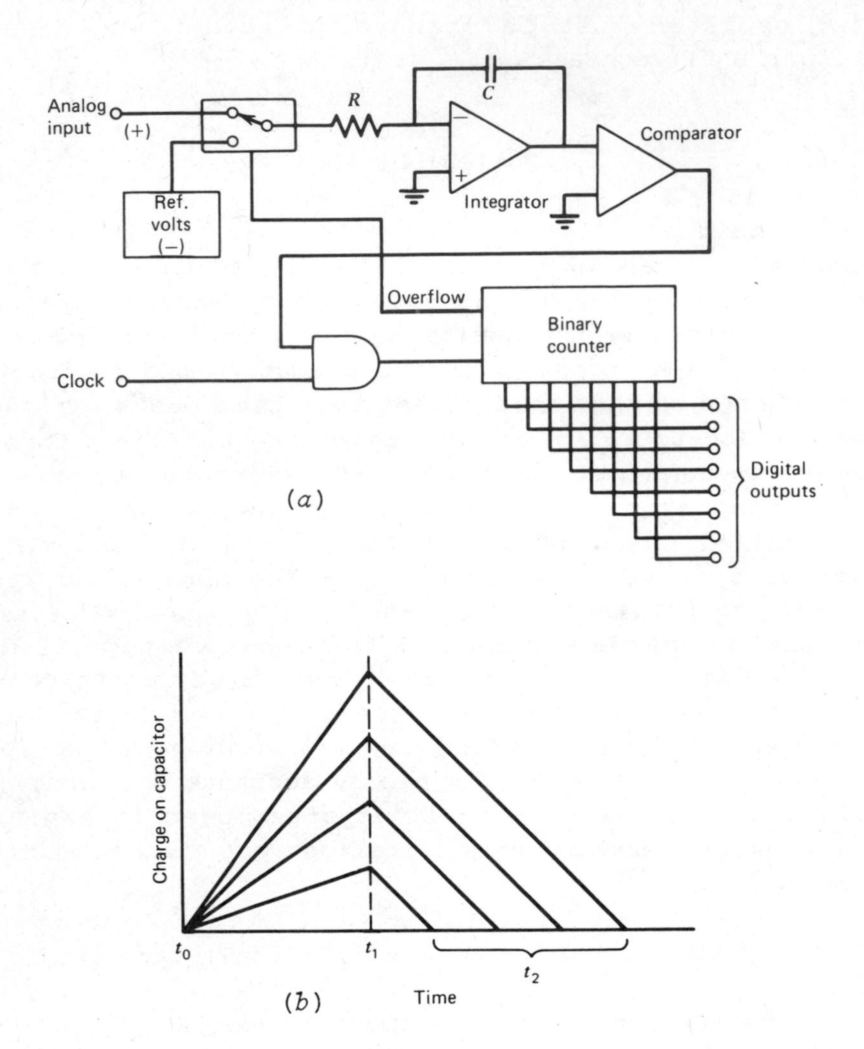

Figure 11-14. A dual-slope ADC: (*a*) schematic (note that no DAC is required); the overflow pulse is produced on crossing zero; (*b*) timing diagram for several input values. Observe that the first period is characterized by a fixed duration and variable slope, whereas the second has a fixed slope and variable duration.

the input voltage E_{in}. Because of this equality, the capacitor discharges through R_B at a rate proportional to E_{in}. Feedback through the system compensates for this discharge current by providing a train of constant current pulses. This, in turn, requires the frequency of the pulse train to be proportional to E_{in}, which is the desired effect.

When starting the operation, the comparator compares the input voltage (E_{in}) against the voltage V_B appearing on the capacitor C_B. If E_{in} is higher, it will cause the monostable to fire. This closes the analog switch, turning on the current I. The pulse from the monostable stays on for time t, during which an amount of charge $Q = It$ will be injected onto the capacitor. This charge will increase the voltage V_B by a small amount. If V_B is still less than E_{in}, the monostable will be fired again and another increment of charge will be impressed on C_B. This process repeats until V_B is just slightly greater than E_{in}. At this point the system will have achieved equilibrium and the monostable will continue sending pulses of charge to

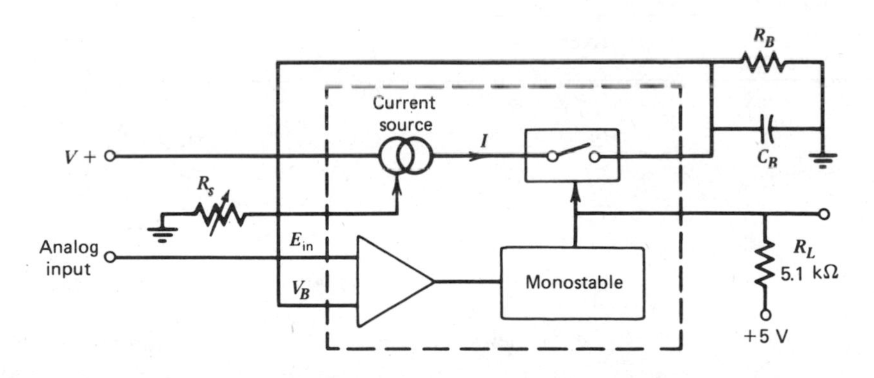

Figure 11-15. Simplified schematic of a voltage-to-frequency converter (LM-131). The range is 0 to 10 kHz, corresponding to 0 to 10-V input. The variable R_s permits adjustment of the full-scale range. The "pull-up" resistor R_L serves to maintain the HIGH output state close to +5 V.

the capacitor at exactly the rate required to compensate for the charge that is disappearing to ground. Hence the frequency of the pulses is proportional to the input analog voltage.

The frequency is proportional to the input voltage within about 1%, and the time required to respond to a large change of input is of the order of 100 ms. The circuit can be linearized (to 0.05%) and the response time decreased, with the aid of an op amp that forces the charge and discharge of the capacitor to be linear.

The 555 timer, the 8038 signal generator, and many varieties of PPL can also be used to convert a voltage to a corresponding frequency. The advantage of the LM131 and similar units lies in their being optimized for this service.

PROBLEMS

11-1. In the generator shown in Figure 11-4, what clock frequency should be selected to produce a staircase with 100 steps per second?

11-2. What would be the E_{out}/E_{in} ratio for the circuit shown in Figure 11-5 if the first and fourth switches are closed (i.e., as shown) while the second and third are grounded?

11-3. The clock for a certain dual-slope ADC (Figure 11-14) runs at 100 kHz, the reference voltage is 10.000 V, and time t_1 corresponds to 10^6 counts. If the time $(t_2 - t_1) = 15.0$ ms, what is the analog input voltage?

XII MICROCOMPUTERS

The term "microcomputer" describes a digital computer that is implemented with a small number of integrated circuits (ICs). The distinction between computers labeled "micro" and "mini" is arbitrary, determined largely by the whims of the manufacturers. The difference is mainly the price; the former tends to become more of a consumer product than the latter. Nevertheless, in terms of computing power and sophistication, microcomputers are advancing at such a fast pace that many of them have now become equal to or even superior to the average minicomputers of one or two decades ago.

A microcomputer contains at its core a module called a *microprocessor*. The microcomputer originated only as recently as 1970, when the Intel Corporation produced the first microprocessor, the model 8008, a unit of rather low power by modern standards. With the appearance in 1974 of the next generation of microprocessors such as the Intel 8080 and Motorola 6800, the era of the microcomputer finally began.

THE ARCHITECTURE OF THE MICROCOMPUTER

There are three fundamental parts essential to a

microcomputer: input/output ports, internal memory, and the central processing unit. Microcomputers may also be equipped with video monitors, magnetic memory, and printers.

The *input/output ports* (I/O ports) constitute the means by which the computer communicates with the outside world. They include facilities for the *input* of information and instructions and for the *output* of processed results.

The *internal memory* is the segment of the computer that receives data from the input port. The data can be in numerical form or in the form of text containing both letters and numbers (alphanumerics). In addition, the memory must store the operating instructions (the program) and the results of data processing.

The *central processing unit* (the CPU) performs the actual manipulation of data in accordance with the programmed instructions.

THE INTERNAL MEMORY

The most important characteristic of the memory is its capacity. This is expressed in terms of the number of words that can be stored. The size of a word varies from one computer to another. The most common is the 8-bit word, consisting of 1 *byte*. Some microcomputers operate with 16-bit words (2 bytes), and larger computers may use 32- or even 64-bit words. We will concentrate on 8-bit computers.

Eight-bit words are stored in memory as binary numbers in the range 0000 0000 to 1111 1111, the equivalent in decimal notation of 0 to 255. The computer itself does not assign any meaning to such numbers, but the user, by means of the program, can make them represent either a number, a string of letters, or a command, as desired. A section of the available memory can be assigned to data and another section to the program commands. It follows that a computer with a large memory capacity can store both large programs and bodies of data. The extent of memory is expressed in "K" units, where each K unit represents 2^{10} = 1024 bytes. Thus a 4K memory can contain up to a maximum of 4 x 1024 = 4096 bytes.

Memory Addresses

Evidently some means must be found to organize the memory so that words can be readily accessible. The process of storing a word of information into memory is called *writing*, and the process of recovering a previously written word is termed *reading*. Each location in memory is given a unique *address* that must be used both in writing and in reading. Each address consists of a four-digit hexadecimal* number from 0000 to 0FFF, the equivalent of decimal 0 to 4095.

The memory of a 4K computer can be "mapped" as follows:

0FFFh
0FFEh
0FFDh
....
....
....
0002h
0001h
0000h

The leading zeros simply indicate that possibly more than three digits may be needed.

A memory position can be loaded with a single word, but it is also feasible to use two or more con-

* Hexadecimal numbers (see Appendix V) can be denoted by a subscripted "16" or by an appended lowercase "h." Thus decimal 17 (= 17_{10}) = $2F_{16}$ = 2Fh.

catenated locations. In this case a larger piece of information can be stored as a unit, but only the address of the first word need be specified.

Types of Memories

From the operational point of view, one can distinguish between random access memory (RAM), which is the principal type that has both write and read capability, and read-only memory (ROM), which does not possess the write option. The ROM is not any less randomly accessible than the RAM, but this nomenclature is firmly established. The ROM is useful for the storage of operating instructions or computer languages that need not be modified by the user.

The reader might ask how it is possible to accommodate in memory the great variety of information encountered in actual use. An 8-bit memory accepts only numbers between 0 and 255. How can one store and retrieve such items as the number "-4.555×10^{-10}," the word "NOT," or the command "add *A* and *B*"?

Codes

The procedure, varying from one microcomputer to another, makes use of two techniques: (1) the *clustering* of a group of words to form one unit large enough to accommodate the information and (2) *coding*, the use of binary values to represent other types of information. For example, the string of characters making the word "NOT" can be stored in 3 bytes, one for each letter. Since there are 256 possible values for each byte and only about 90 alphanumeric characters, there is more than enough space to accommodate one character per computer word, provided a *code* is established. The most commonly used code is ASCII (Appendix VI). For the string "NOT," the memory positions will contain the ASCII codes 4Eh, 4Fh, and 54h, respectively. (Certain computers make use of a different code, called EBCDIC, in place of ASCII.)

The instructions that make up computer programs require as a rule more than 1 byte per instruction. Consider the command "add *A* and *B*." The microcomputer cannot interpret this order until it has been codified by conversion to a series of binary words. These words must contain, in the proper order, information that will identify the operation to be performed and the memory locations (addresses) that contain the data on which the operation is to be executed. In addition, other intermediate steps must usually be performed, such as reading the data and sending it to the CPU, and then storing the results of the calculation in a specified memory location.

A possible sequence is given below, starting at memory location 0100h:

Location	Instruction
0100h	Address follows
0101h	Address of *A* (first bit)
0102h	Address of *A* (second bit)
0103h	Read *A* into CPU
0104h	Address follows
0105h	Address of *B* (first bit)
0106h	Address of *B* (second bit)
0107h	Read *B* into CPU
0108h	Add value of *B* to value of *A*
0109h	Store result at prior location of *B*

The computer will execute this sequence, starting at 0100h and terminating at 0109h, and then proceed with the next instruction (if any) at position 010Ah.

To summarize, the computer memory consists of a collection of bytes that can be loaded with binary numbers. Viewed in this light, the computer content is a seemingly endless ocean of ones and zeros. The great miracle of computation is how this sea organizes itself and operates intelligently by specifying only an initial address. At this address is located the first instruction, and with this starting point, each memory position in succession is given a meaning, and the computer continues its activity until the end of the program.

THE CONSTRUCTION OF MEMORIES

For many years computer memories were principally implemented with a collection of tiny magnetic rings called *cores*. The cores could be magnetized in either of two directions, which corresponded to high and low states. This type of memory is very reliable but is too bulky and expensive for microcomputer use. The lower priced, more compact, types of memory demanded by microcomputers have been achieved with ICs.

The ROM is a simple form of semiconductor memory, characterized by its permanence (nonvolatility), and its low cost. Industrial production, however, is handicapped by the fact that the information to be encoded in the ROM must be present in the production tools themselves, which makes it difficult and expensive to modify or correct a ROM design.

Some ROMs are constructed so as to permit the user to introduce the desired program. Such units, called *programmable ROMs* (PROMs), can be loaded with information only once. More advantageous are *erasable PROMs* (EPROMs), from which the program can be removed by flooding with ultraviolet (UV) radiation. Another type is the EAROM, that can be written on and erased repeatedly by electrical means. Here the distinction from RAM memory begins to fade, but an EAROM can seldom replace a RAM since it is thousands of times slower to

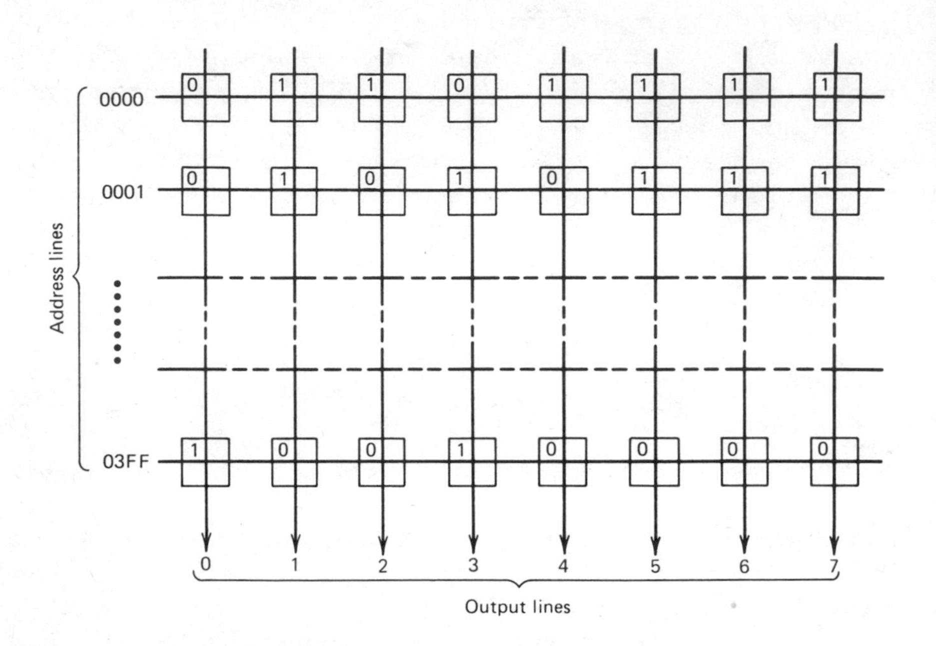

Figure 12-1. Organization of 1024 words of memory, each eight bits in length.

write to. Special equipment is needed for reprogramming EPROMS and EAROMS, and this operation cannot be carried out during the normal use of the computer.

The internal design of a ROM includes a two-dimensional matrix (Figure 12-1) in which the READ command is entered on one side and the output emerges at the bottom. Each cell contains either a "1" or a "0" permanently inscribed. The command to read is entered in a single horizontal line, and the content of the corresponding word is sent down through the vertical lines to the output.

Consider a simplistic implementation of the matrix given in Figure 12-1 made by soldering only those intersections corresponding to logical "1." It might seem that this would give the desired behavior. If, for example, +5 V is connected to line 0001 and the other address lines are left open, the number 01010111 would be expected to appear at the output via the soldered junctions. In reality, a closer examination reveals that the HIGH is propagated in unwanted directions, so that the output is 11111111.

The situation can be saved by preventing interaction between address lines. One way to do this is to replace each soldered junction by a diode, as illustrated in Figure 12-2. The set of diodes connected to any given output line performs the logical operation **OR**, in that any one of them can bring the line to +5 V, even though all the other diodes of the set are reverse biased.

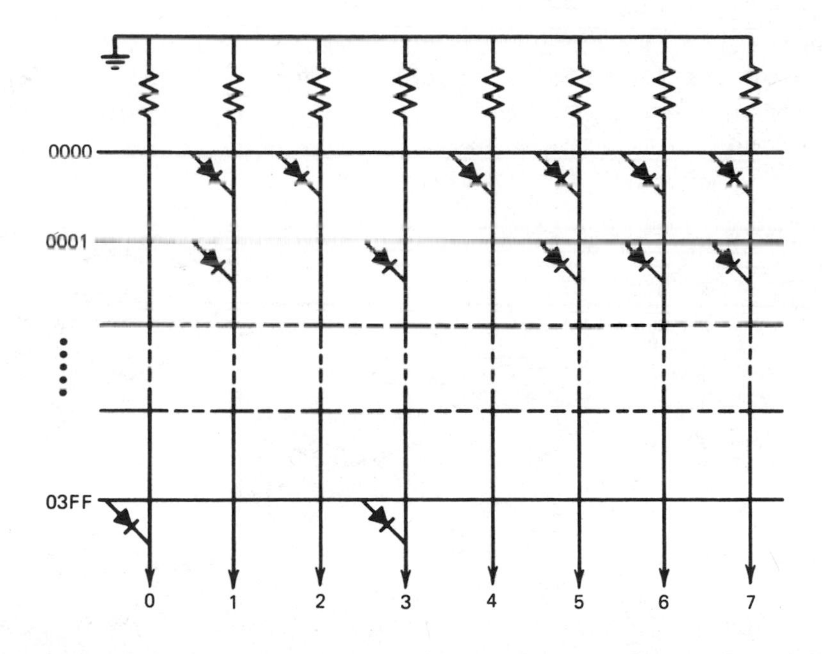

Figure 12-2. An implementation of the ROM diagrammed in Figure 12-1, utilizing diodes. Each output line is normally maintained at ground by the pull-down resistors.

RANDOM-ACCESS MEMORIES

Semiconductor RAMs, in contrast to ROMs, suffer from volatility, meaning that all information content is lost when the power is shut off. There are two major types of semiconductor RAMs, dynamic and static. In *dynamic RAMs* the information is only held for a short time, perhaps a millisecond; thus a dedicated circuit must be provided to "refresh" the memory on a continuous basis. More desirable are *static RAMs*, which will retain information indefinitely. Less expensive, but slower, are *charge-coupled devices* (CCD) consisting of shift registers. Most semiconductor memories are based on MOSFET designs.

Random-access memories require more associated circuitry than do ROMs, in order to implement the addressing, writing, and reading operations. An example of the basic constituent element or "cell" is shown in Figure 12-3. For the information to be written into the cell, three separate lines must be activated: (1) the appropriate data-in line; (2) the write-enable

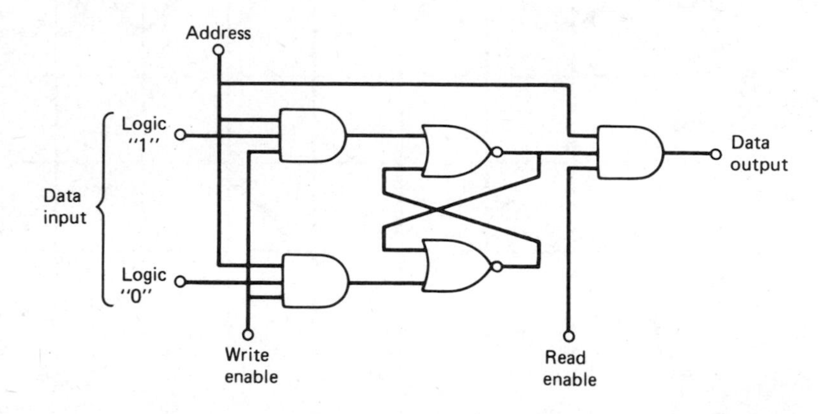

Figure 12-3. A simplified semiconductor 1-bit memory unit.

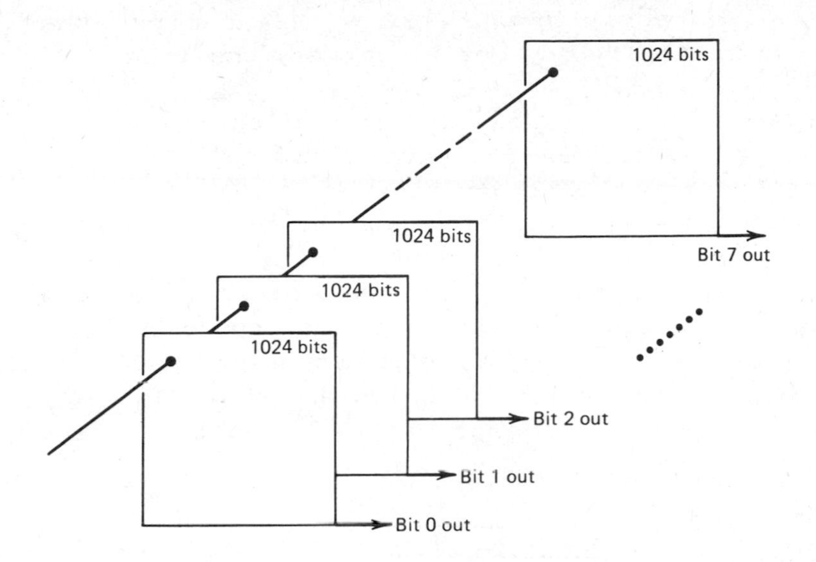

Figure 12-4. An example of RAM memory organization.

line, signifying that the command to write has been received; and (3) the individual address line. Reading the data requires both the address and read-enable lines to be high. A large number of these 1-bit cells can be adjoined in a matrix to form a practical RAM.

Various configurations of memory cells are possible, but it is advantageous to assign one chip per *bit*, rather than to write the words contiguously in the memory. The process is illustrated in Figure 12-4. The first bit for the entire memory is stored in one IC, the second bit on a second IC, and so on. The set of eight ICs is addressed simultaneously, and the output from each is the value of the bit at the particular address. The complete word is reconstructed by combining the outputs.

Each of the 1024 locations can be addressed individually by encoding the 1024 possible signals into 10 lines, as shown in Figure 12-5. The 10 lines are fed into two decoders that generate 32 lines each, thus giving an array of 32 × 32 = 1024 choices that can serve to specify any desired memory position.

Charge-Coupled Devices

Charge-coupled devices (CCDs) warrant special mention because of their high inherent density, meaning that an unusually large number of active elements can be fabricated on a silicon chip of reasonable dimensions. The CCD consists of a long conductive channel, as in a MOSFET, with a large number of insulated gates (perhaps 1000), a few of which are shown in Figure 12-6. An electric charge may or may not be stored be-

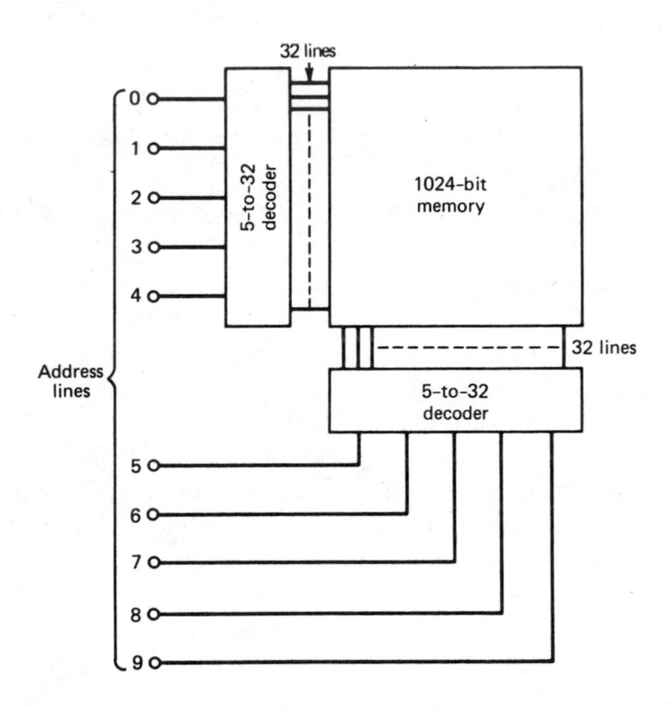

Figure 12-5. A simplified example of 1024-bit memory addressing, using 10 control lines. The whole system of addressing corresponds to the single line marked "word address" in the further simplified schematic in Figure 12-4.

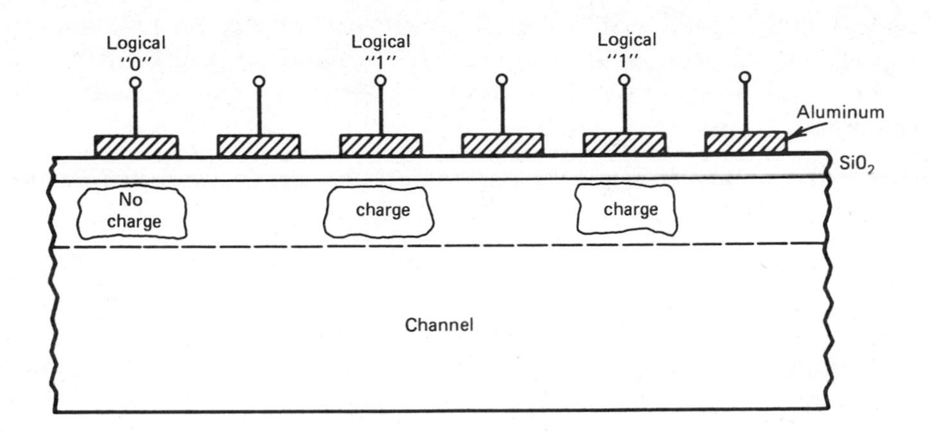

Figure 12-6. Schematic representation of a charge-coupled device.

neath each gate, generating either of two logical states. Since these charges are stable, a memory function can be implemented. The charges remain unchanged until a proper sequence of pulses is applied to the device, which shifts each charge by one position, the operating mode of a shift register. For this reason the data retrieval time is long, possibly as much as 10 ms. More advanced units, arranged in matrices of short shift registers, can reduce the access time to perhaps 0.1 ms. Very large memories (e.g., 64K) can be implemented on a single chip.

THE CENTRAL PROCESSING UNIT (CPU)

The principal function of the CPU is to perform operations on the data present in memory. Great complexity of structure is needed to permit the performance of all the necessary functions, yet the entire CPU is ordinarily implemented as a single IC, a microprocessor.

As a rule the CPU contains the following components: (1) several registers, (2) an arithmetic logic unit (ALU), and (3) a timing and control unit. A small amount of additional memory called a "scratch pad" is often included.

Registers

These are small, fast, easily accessible memories of 8 or 16 bits. They differ from regular memory in that operations can be performed on them. Often one or more registers, called *accumulators*, are optimized to represent the principal connection between the various units within the computer. All I/O operations are handled by the accumulators, as shown in Figure 12-7. The larger the number of registers and accumulators, the more powerful the computer.

In addition to general-purpose registers, there are a few units that serve special functions. Thus the *program counter* (PC) is responsible for running the program in the desired manner. It allows the storage of the instructions in any consecutive set of memory locations with the provision that the beginning address must be stored in the PC. The PC is said to *point to* the first instruction.

The operation of the computer is thus contingent on the repeated interaction of the loop PC-MEMORY-CPU-PC, as illustrated in Figure 12-8, where the program is shown to have reached the third instruction. A special instruction, not shown, indicates the end of the program, stopping the process.

Another useful register is the *memory pointer* (MP), sometimes called the *data counter*. This is quite similar to the PC, except that it indicates a position in memory where *data* rather than instructions, are stored.

Let us suppose that it is desired to move the contents of memory location *M* to location *A* (which might be the accumulator). The command for this pro-

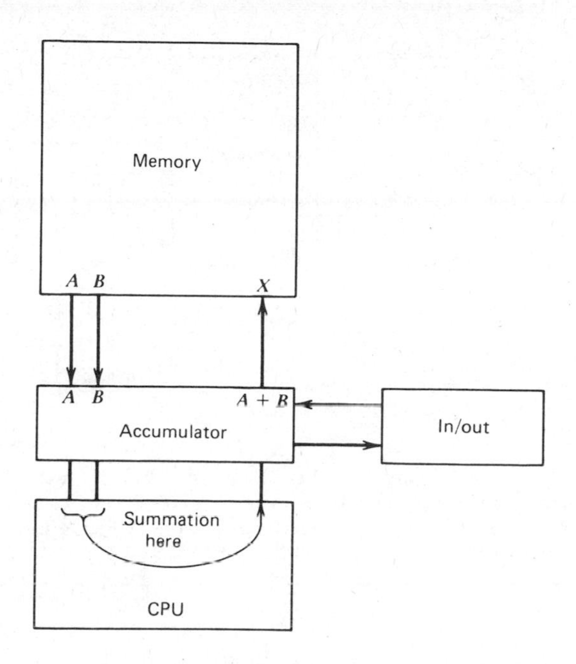

Figure 12-7. The use of an accumulater to interact with both memory and I/O ports. The arrows represent the sequence of operations in taking the values of *A* and *B* from memory, adding them together, and placing the result in memory.

cess can be written in the form "MOV M,A" where "MOV" is an example of a three letter mnemonic. We will assume that this command is represented in memory as 77h. Figure 12-9 shows in simplified form the cycle of operations that executes this instruction. The memory pointer contains the address of *M*, namely, 0101h. The data represented by the number 63h resides at that location. The program counter contains the address 1002h, so that the instruction at that address is executed.

The Arithmetic Logic Unit

This is a collection of gates and other logic elements that can operate on the words in the accumula-

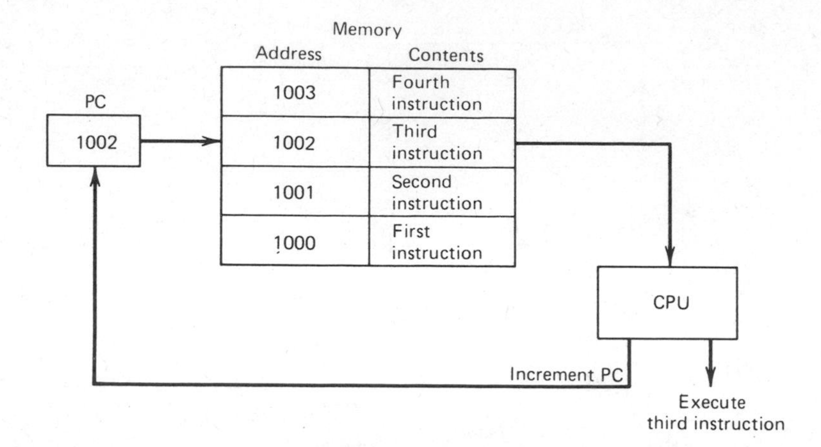

Figure 12-8. A simplified computing cycle. Since the PC contains the address of the third instruction, that is the one that is sent to the CPU for execution. The CPU simultaneously increments the PC to point to location 1003.

tor. The basic functions are relatively simple and become more complicated only by repetition. They are the following:

1. Addition.
2. Right and left shifting, amounting to division and multiplication by 2.
3. Boolean operations such as **AND**, **OR**, and **NOR**.
4. Complementing,* needed in order to implement subtraction.

The interaction between the various components of the ALU is shown in Figure 12-10.

* This is the process by which "1" and "0" are interchanged (one's complement). The two's complement function is obtained from it by adding 1. For example, the one's complement of 000011 is 111100, and the two's complement is 111101.

The Control Unit

The orderly operation of a microcomputer is determined by a controlling unit based on a radiofrequency square-wave crystal oscillator acting as a *clock*. As an example, let us assume a clock frequency of 10 MHz, thus defining periods of 100 ns, as shown in Figure 12-11. A two-part basic sequence is commonly employed, including the *fetch* portion, during which the information is brought in from memory, and the *execute* portion when the instruction is executed. There are two communication systems involved, a 16-bit address network

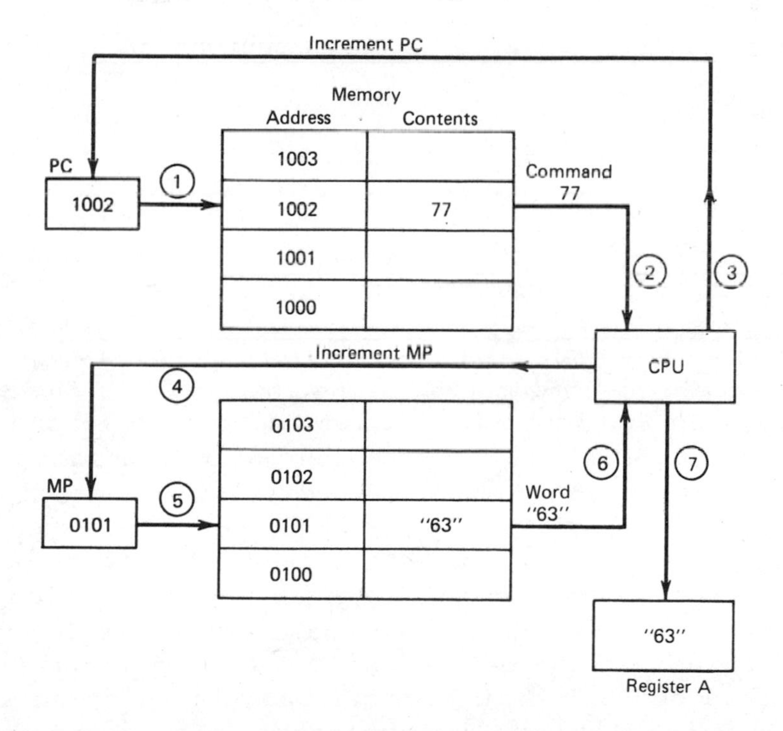

Figure 12-9. The execution of the command "MOV M,A." The encircled numbers show the sequence of operations.

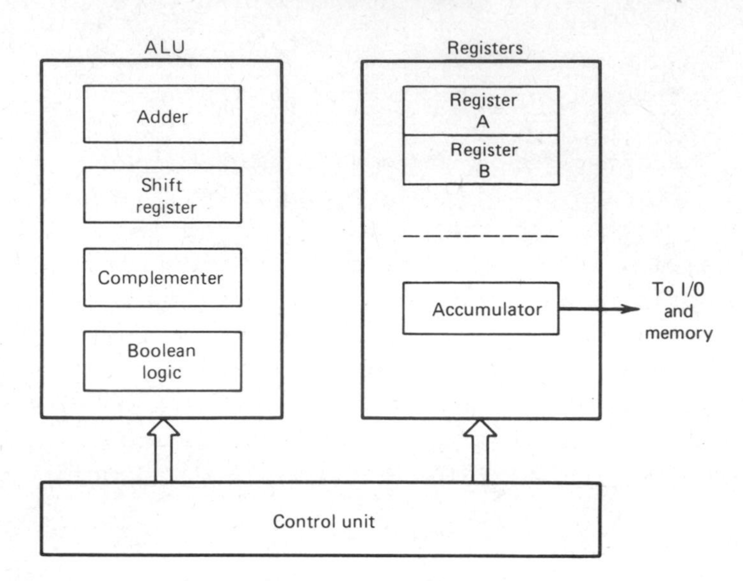

Figure 12-10. The internal architecture of the CPU.

permitting a range from 0 to 65535 and a data system of 8 bits. Each is implemented by a bus, connecting all similar devices. Thus the data bus must have eight conductors and the address bus 16. In addition a number of control lines must be included with each bus to take care of the logistics of data transfer.

The control unit is responsible for the correct interpretation of which devices are sending and which are receiving information along the bus. In the example shown in Figure 12-11, the address is loaded onto the corresponding bus, accompanied by the command to *read* as a pulse synchronous with the clock. On the next 1-to-0 transition the value found at the specified memory address is loaded into the proper register. (This information content might be the instruction "MOV M,A" in Figure 12-9). An analogous cycle occurs in searching for the data specified by the instruction. This second set of binary numbers is interpreted as data and sent to the accumulator. Not shown in the figure is the process of incrementing the program counter by one so that the next instruction can be executed.

INPUT/OUTPUT DEVICES

The raison d'être for the existence of microcomputers is to interact with information from the outside world. The basic elements involved in this process are the I/O ports, of which there are usually several. Each port has its own address, which may be part of the memory, in which a section has been set aside for the purpose. This is called a *memory-mapped* system. In other microprocessors, a separate set of addresses is required. Data from the accumulator can thus be transferred to these ports by means of the data bus whenever the program asks for it.

Also possible, and very effective in instrumentation, is the addition of an *interrupt* feature. In this case the external device is connected to a wire that initiates an *interrupt request*. When this line

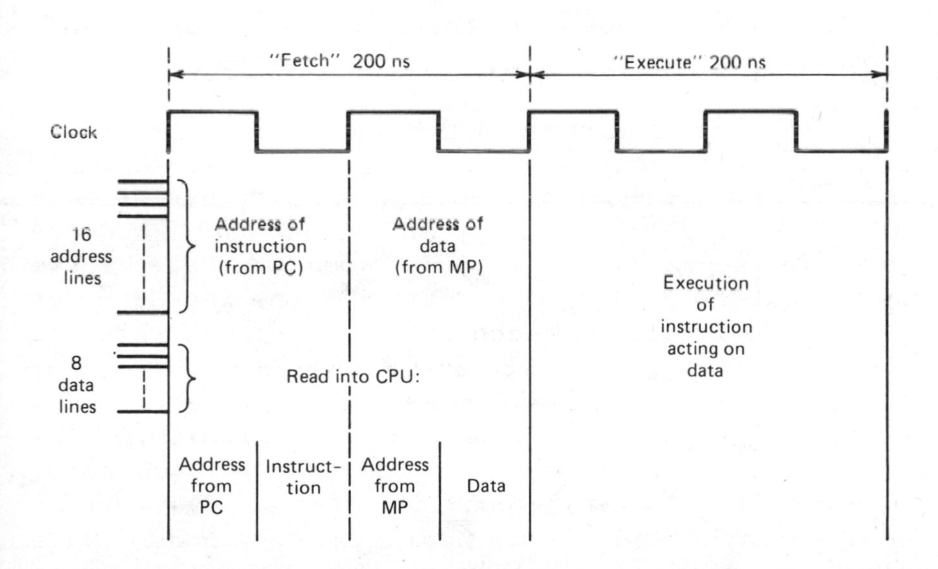

Figure 12-11. An example of computer cycles, in which the "Fetch" and "Execute" cycles each occupy two clock periods.

goes high, the CPU is alerted that data are coming. A signal called "interrupt acknowledge" is sent indicating that the computer is ready to receive input. This causes the data to be transmitted and received at the next 1-to-0 clock transition. The interrupt protocol suspends the current program momentarily and causes a new set of instructions to be executed. This involves: (1) storing the address of the instruction that would have been executed next, (2) storing additional status information and register contents, (3) changing the PC to another address, (4) executing the instructions related to the interrupt, and (5) returning to the main program.

The situation is more complicated if several interrupt lines are present. In this case, a hierarchy of interrupts must be established. Often a special emergency interrupt line is activated when power is switched off. Since DC voltages drop rather slowly, perhaps in 0.1 s, there is time for an emergency set of instructions to shut down the computer in an orderly manner.

Serial I/O Data

Data can be transmitted on a single pair of conductors if the various bits making up each word are loaded in sequence. The only information that the two communicating devices must share is the frequency at which the bits succeed each other. If this were not specified, it would be impossible in such a sequence as 01110 to estimate how many ones there are between the two zeros, since they form a continuous HIGH. The speed of bit transmission is called the BAUD rate, expressed in bits per second. A special character is required to delineate the beginning of words. Once this is received, the entire string of ones and zeros can be correctly cut into 8-bit slices representing consecutive words.

XIII TRANSDUCERS

In its broadest sense, a transducer is a device that can convert energy from one form to another. Our present interest, however, is more restricted. We are concerned only with the transfer of information between energy domains. Input transducers, as encountered in laboratory instruments, convert information from the physical system under examination into easily measured electrical signals. They may be classified according to the kind of information desired, as pressure transducers, temperature transducers, and so on. We follow these major categories, with emphasis on the physical and electronic principles involved.

TEMPERATURE TRANSDUCERS

Nearly every physical quantity X has a temperature coefficient that can be expressed as a power series expanded about some point (X_0, T_0):

* Portions of the material in this chapter are adapted from D. H. Sheingold, Ed., *Transducer Interfacing Handbook*, Analog Devices, Inc., 1981, and from G. W. Ewing, "Transducers," in *Treatise on Analytical Chemistry*, I. M. Kolthoff and P. J. Elving, Eds., Part 1, Vol. 4, Chapter 4, Wiley-Interscience, 1984.

$$X(T) = X_0[1 + \alpha(T - T_0) + \beta(T - T_0)^2 + \cdots] \qquad (13\text{-}1)$$

where X_0 is the value of the variable at temperature T_0 and $X(T)$ that at temperature T. In many cases the square and higher terms can be neglected, and the temperature coefficient can be expressed as

$$\alpha = \frac{1}{X_0} \cdot \frac{dX}{dT} \cong \frac{1}{X_0} \cdot \frac{\Delta X}{\Delta T} \qquad (13\text{-}2)$$

where Δ represents a small finite increment. Over small temperature excursions, in most systems, α is nearly constant.

The most widely used temperature transducers are thermocouples, resistance thermometers, and semiconductor devices. We will discuss each of these in turn.

Thermocouples

The number of free electrons in a conductor varies with temperature, and the effect is different in different metals. Hence any two unlike metals in contact will display a difference of potential that is a function of temperature. A pair of metals connected so as to permit measurement of this potential difference is a *thermocouple*. The response of a thermocouple is somewhat nonlinear, deviating by up to 0.2 K over the range 10 to 40°C. A typical temperature coefficient, that for iron-constantan, is $\alpha = 0.6\ (\mu V/V) \cdot K^{-1}$, often expressed at 0.6 ppm/K.

A few representative thermocouples are listed in Table 13-1, together with their temperature coefficients and useful ranges.

TABLE 13-1

Some Common Thermocouples

Type[a]	Junction Metal	Useful Range (°C)	Voltage[b] Swing
B	Pt,6% Rh-Pt,30% Rh	38 to 1800	13
C	W,5% Re-W,26% Re	0 to 2300	37
E	Chromel-constantan	0 to 980	75
J	Iron-constantan	-184 to 760	50
K	Chromel-Alumel	-184 to 1260	56
R	Pt-Pt,13% Rh	0 to 1600	18
S	Pt-Pt,10% Rh	0 to 1540	16
T	Cu-constantan	-184 to 400	26

a These symbols (except "C") have been assigned by ANSI, The American National Standards Institute.

b Voltages are in millivolts for the indicated ranges.

Any electrical circuit containing two metals necessarily has two junctions between these metals, either directly or with the interposition of a third metal. As a result, thermocouples must be used in pairs, giving differential rather than absolute measurements. Figure 13-1 shows two possible circuits. In (*a*) a microammeter is connected in series with the two thermocouple junctions, whereas in (*b*) it is replaced by an amplifier. In both cases additional junctions are introduced that are potential sources of unwanted thermal effects.

Suppose that, in circuit (*b*), metal *x* is the alloy constantan and *y* is iron. Metal *z* is copper. This means that two additional *x-z* junctions are introduced, and that both are copper-constantan. These will cause no difficulty provided these two *x-z* junctions are held at the same arbitrary temperature. The same situation holds in circuit (*a*) since the internal wiring of the meter, including its terminals, can be assumed to be copper and to be at the same temperature. The meter reading in (*a*) is dependent on the resistance of the wires and must be calibrated for a particular value of resistance. This is not a problem in (*b*), as the measurement is made without drawing current.

Traditionally one of the two junctions is held at

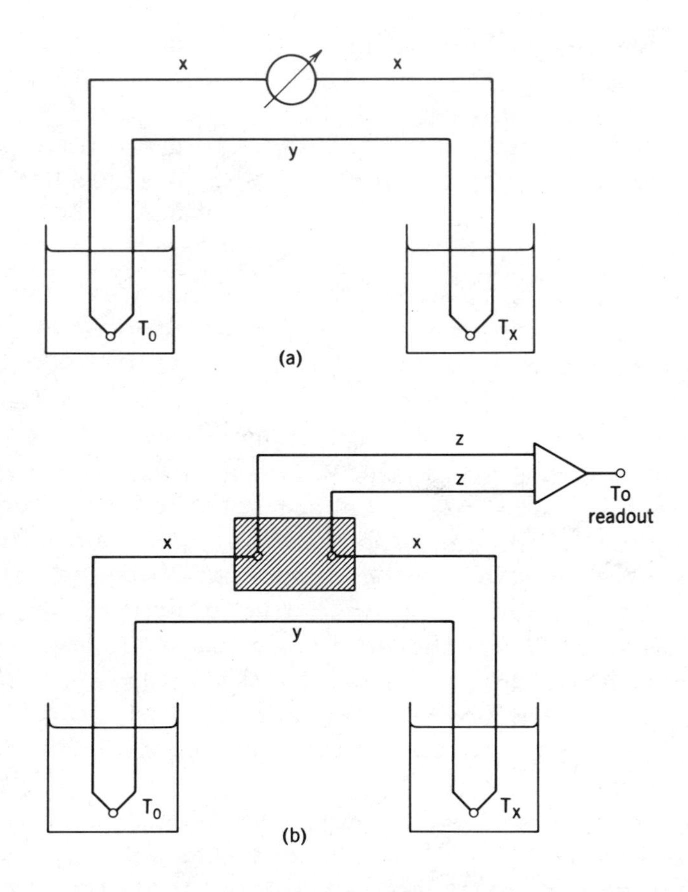

Figure 13-1. A two-junction thermocouple: (*a*) simple arrangement and (*b*) with an intermediate metal. The shaded area must be maintained at a constant temperature. The letters *x*, *y*, and *z* represent different metals.

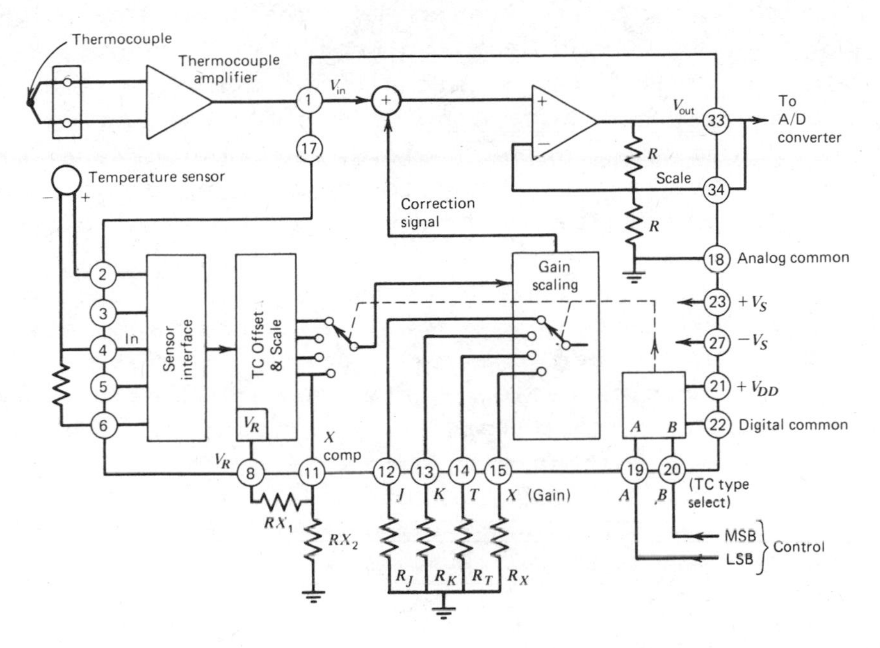

Figure 13-2. Circuitry for thermocouple measurement with cold-junction compensation. (Analog Devices, Inc.)

0°C in an ice bath so that the measured voltage will be directly proportional to the Celsius temperature. This is cumbersome, however, and is usually replaced by an electronic circuit that monitors the temperature of the reference junction and supplies a compensating voltage.

Compensatory circuits of this type are available to match each of the commonly used thermocouples; an example is the model 2B56 of Analog Devices shown in Figure 13-2. The item marked "temp sensor" is a special semiconductor device, the model AD590, to be described later. A two-digit binary input at pins 19 and 20 controls a pair of multiplexing switches that select external gain resistors suitable for four types of thermocouple, labeled *J*, *K*, and *T*, corresponding to the standard designations given in Table 13-1. This adjusts the gain in the sensor circuit and provides a corrective signal to the primary thermocouple. Note that the output can be taken from pins 33 and 34 tied together, or from pin 33 alone if an additional gain of 2 is desirable. Note also that separate analog and digital grounds and power supply connections are provided; this is usual practice in a component such as this, that contains both types of circuitry, to avoid deleterious interactions.

Since *any* two conductors form a thermocouple, care must be exercised in all low-level circuits to eliminate unrecognized thermal sources of potentials. Joints between two solder-coated copper wires may present such a source unless the two segments of copper are in direct contact with each other.

Thermocouples have a wide variety of applications, from rough monitoring of a furnace to very sensitive multijunction "thermopiles" for detecting weak radiation.

RESISTANCE TEMPERATURE DETECTORS (RTDs)

The temperature coefficients of resistance of electrical conductors vary widely, being positive for metals and generally negative for semiconductors. The materials most useful for temperature measurement are platinum resistors and thermistors.

Platinum Resistance Thermometers

The value of α for platinum is about 0.0039 K^{-1}; hence it can provide a more sensitive means of measuring temperature than can a thermocouple. It is used widely in precision thermometry under the designation RTD. Nickel has a coefficient that is about twice as large, but this metal is inferior to platinum in long-term stability and reproducibility.

The resistance of an RTD can be measured conveniently with constant-current excitation. A circuit for direct readout is shown in Figure 13-3. Amplifier 1 provides a low-impedance source of voltage. The adjustable input resistor to op amp 2 converts this to a current of 18 mA, which passes through the platinum resistor in the feedback loop. Single-transistor current boosters are used with both op amps to ensure sufficient current.

Greater precision is obtainable with a Wheatstone bridge circuit (Figure 13-4). The sensitivity can be calculated as follows. Assume that $R_1 = R_2 = R_3 = R_T$ = 100 Ω (at 25°C). Assume also a gain of 100 for the instrumentation amplifier. Resistors R_1 and R_2, acting as a voltage divider, fix the potential of point x at 5 V, whereas that at y is

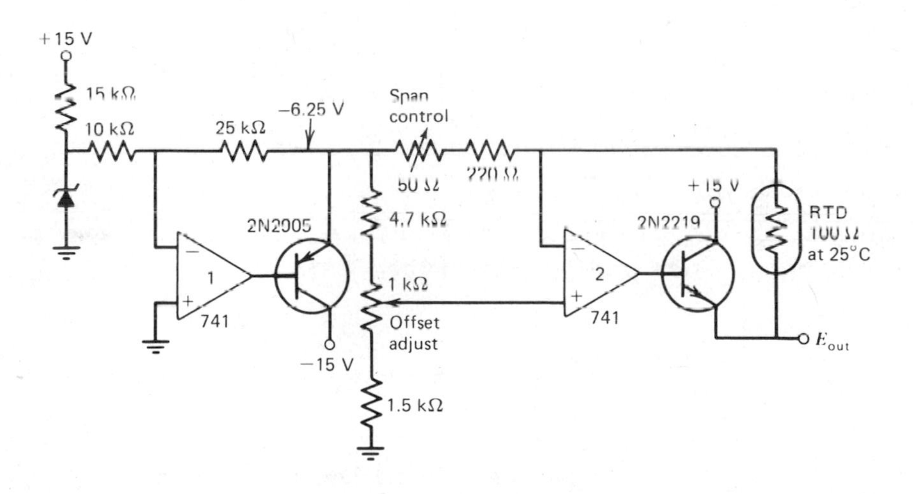

Figure 13-3. A simple measuring circuit for a platinum resistance thermometer. The output is 0 to 1.8 V for temperatures 0 to 266°C. (Analog Devices, Inc.)

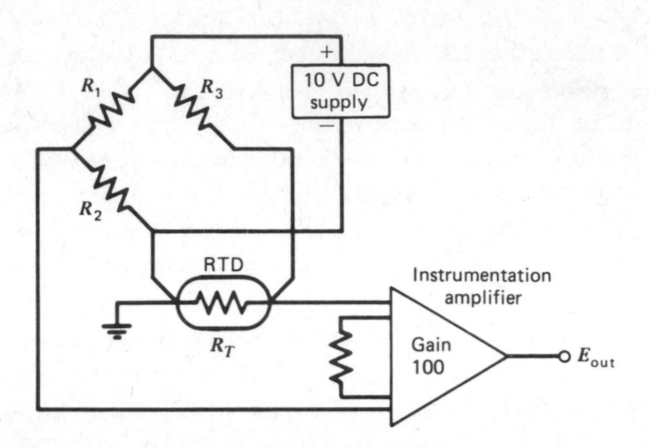

Figure 13-4. A bridge circuit for a platinum resistance thermometer. The junctions marked *A* and *B* are placed as close as possible to the sensor R_T, so that the bridge excitation current will not flow through the cable connecting R_T with the amplifier.

$$E_y = 10\left(\frac{R_T}{R_3 + R_T}\right) = 10\left(\frac{R + \Delta R}{2R + \Delta R}\right) \tag{13-3}$$

where $R = 100\ \Omega$ and ΔR is the increment of R_T corresponding to the smallest temperature change that can be measured. This can be rewritten as

$$E_y - E_x = 10\left(\frac{1 + \Delta R/R}{2 + \Delta R/R}\right) - 5 \tag{13-4}$$

The ratio $\Delta R/R$ can be replaced by its equal derived from Eq. 13-2:

$$E_y - E_x = 10\left(\frac{1 + \alpha\ \Delta T}{2 + \alpha\ \Delta T}\right) - 5 \cong 5\alpha\ \Delta T \tag{13-5}$$

Let us suppose that the output can be read to the nearest 1 mV, so that $E_{out,min}$ = 0.001 V and assume that α = 0.003. Substitution of these values gives a minimum measurable ΔT of 7×10^{-4} K. This is about as small a temperature difference as can be achieved in practice without unusual precautions.

Thermistors

The theory of semiconductors predicts that the resistance will follow an exponential relationship:

$$R \propto \exp \frac{\Delta E}{rkT} \tag{13-6}$$

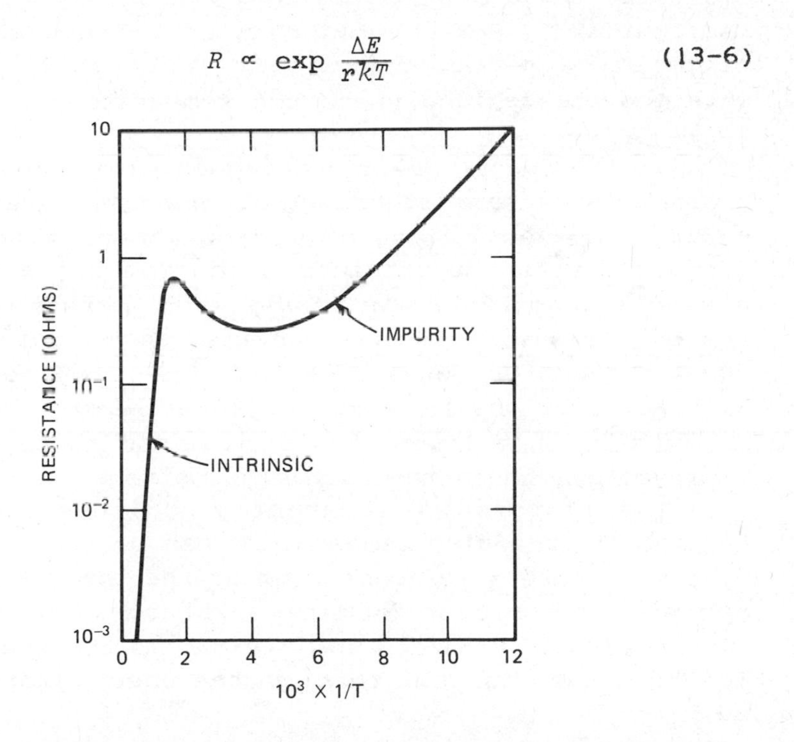

Figure 13-5. The resistance-temperature characteristic of a typical thermistor. The resistance is controlled by the impurity content (doping) at low temperatures, but by the intrinsic conductance at higher temperatures. (Courtesy of Reinhold Publishing Company.)

in which ΔE is one-half of the forbidden energy gap between valence and conduction bands, r is a constant with the value 1 when the number of majority carriers (holes or electrons) is less than the number of impurity sites and with the value 2 when this number is exceeded. The quantity k is the Boltzmann constant, and T is the Kelvin temperature. However, this relationship is obeyed only over restricted ranges. Figure 13-5 shows the curve obtained experimentally for silicon doped with phosphorus;* the curve follows Eq. 13-6 in the two approximately linear regions on the log plot, but these are joined by a region where the slope is negative. For the majority of semiconductors the temperature range of greatest interest lies in the intermediate region, where the temperature coefficient is negative.

Semiconductor devices designed to make use of this property are *thermistors*. They are fabricated by sintering together oxides of various transition metals. Thermistors can be obtained with room temperature resistance from 10 Ω to 10 MΩ and coefficients from −3 to −6% per kelvin. This sensitivity, some hundred times greater than for metal RTDs, is offset for many precision applications by somewhat lower stability and reproducibility, although these characteristics are being improved by modern manufacturing methods.

Thermistors are inherently nonlinear, but their response to changing temperature can be rendered linear by a simple network consisting of the thermistor and an additional resistor forming a voltage divider, as in Figure 13-6. Provided the divider is buffered against loading, the current through the thermistor is given by

* S. A. Friedberg, in "Temperature, its Measurement and Control in Science and Industry," H. C. Wolfe, Ed., Vol. 2, Chapter 20, Reinhold, New York, 1955.

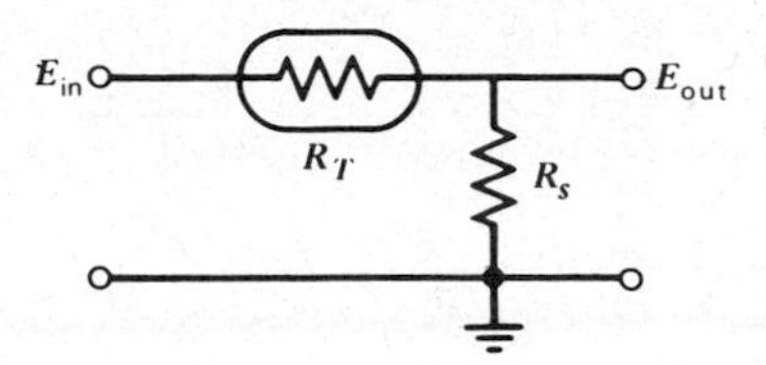

Figure 13-6. A basic circuit for linearizing a thermistor.

$$\frac{E_{out}}{R_S} = \frac{E_{in} - E_{out}}{R_T} = \frac{E_{in} - E_{out}}{A \cdot R_{25}} \qquad (11-7)$$

which can be rewritten as

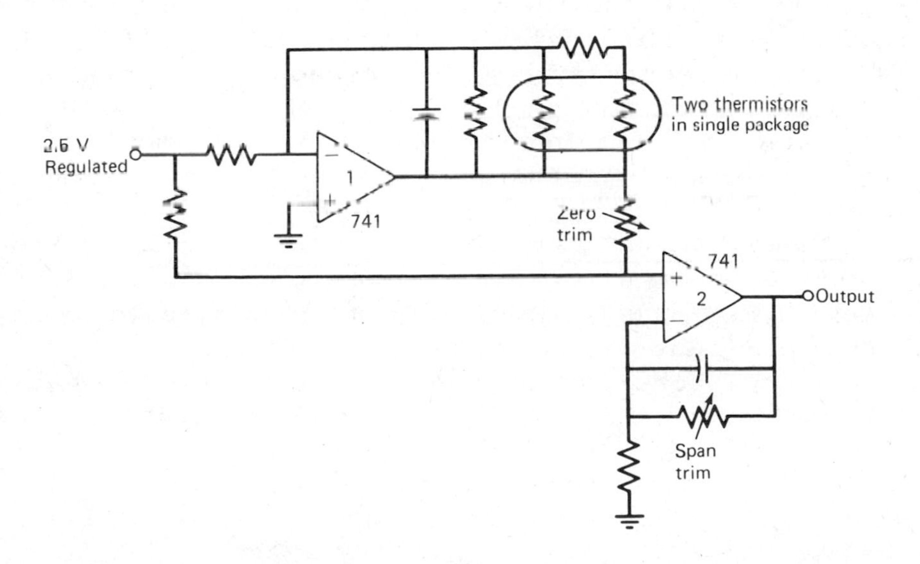

Figure 13-7. A circuit for temperature measurement using a two-thermistor linearizing network. The resistance values in the feedback system must be selected to complement a particular pair of thermistors. (Analog Devices, Inc.)

$$\frac{E_{out}}{E_{in}} = \frac{1}{A(R_{25}/R_S) + 1} \tag{13-8}$$

In these equations A is the ratio of R_T, the thermistor resistance at temperature T, to that at 25°C (R_{25}). Over a limited range, the value of A is given by

$$A \cong p \exp \frac{q}{T} \tag{13-9}$$

where T is the Kelvin temperature.

For evaluation of the constants p and q, A must be determined experimentally by resistance measurements at two temperatures. For a typical thermistor, the coefficients might be $p = 1.49 \times 10^{-6}$ and $q = 4 \times 10^3$ K. Substitution of the resulting expression for A into Eq. 13-8 and differentiation with respect to temperature leads to the conclusion that taking the ratio R_{25}/R_S = 6.1 will give a combination exhibiting a linear resistance-temperature characteristic over a range of 50 K or so.

Linearization* over a longer span can be obtained by combining two thermistors with different properties together with one or more resistors. A circuit using such compensation is given in Figure 13-7.

Thermistors can be used for temperature measurement in circuits similar to those described previously for RTDs. With both RTDs and thermistors it is essential that the current through the measuring element be kept as small as possible, to avoid self-heating.

* The term "linear" in this context indicates obedience to a relation of the type $y = ax + b$, in which b is not necessarily zero.

JUNCTION THERMOMETERS

It was pointed out in Chapter IV that the current-voltage relationships across a *pn* junction follow a logarithmic law. Semiconductor theory shows that the basic relationship is

$$I = I_S(\exp \frac{qV}{kT} - 1) \tag{13-10}$$

where q is the electronic charge; k is the Boltzmann constant; T is the Kelvin temperature; and I_S is the *reverse leakage*, or *saturation current*. If the current I is held constant at a small value, the voltage becomes a linear function of temperature. The coeffi-

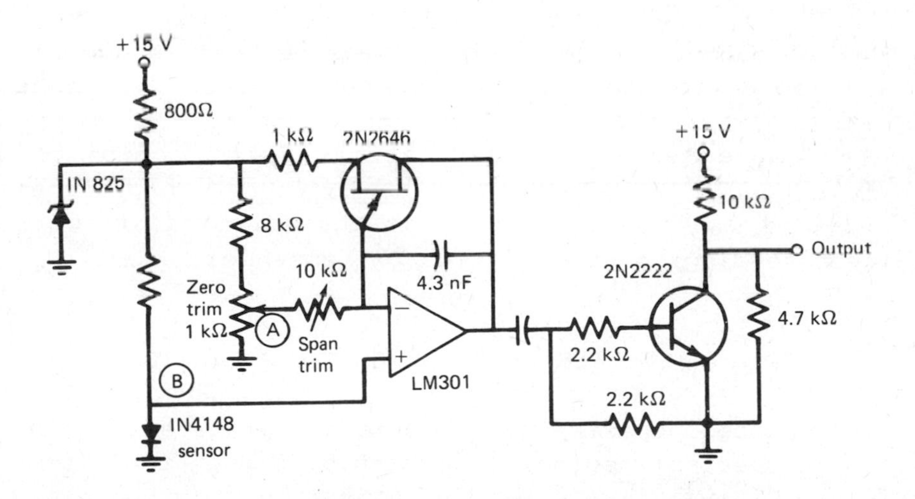

Figure 13-8. A temperature-to-frequency converter using a 1N4148 diode as sensor. This circuit is available as an IC, type AD537. (Analog Devices, Inc.)

cient dV/dT is of the order of -3 mV/K. Note that this is independent of V, so that the general definition of α does not apply.

This relationship holds for any *pn* junction operated at constant current, whether isolated in a diode or forming part of a transistor. Thus a common transistor, such as the popular 2N2222, mounted on a suitable probe will make a very satisfactory thermometer. Figure 13-8 shows a circuit that will give an output frequency proportional to the temperature of a sensing diode. It makes use of a unijunction transistor to discharge the capacitor of an integrator at intervals. The rate of charging of the capacitor, and thus the frequency, is determined by the difference in potential between the points marked A and B, with the latter determined by the forward voltage drop across the diode. A V/F IC could be used in the same way. In either case, the readout requires only a rather simple frequency counter.

The Analog Devices AD590 is a special IC that gives an output current in microamperes that is numerically equal to the Kelvin temperature over the range from -55 to +150°C. Direct connection to a microammeter in series with a 9-V battery forms one of the simplest accurate electronic thermometers. Additional offset circuitry is required if the output is to give direct readings on the Celsius or Fahrenheit scale.

TRANSDUCERS FOR ELECTROMAGNETIC RADIATION

Included here are those devices that detect radiation in spectral regions from X-rays through the ultraviolet, visible, and infrared, converting the contained information into electrical signals.

Electron Emission Devices

According to the classical photoelectric effect, electrons are ejected from a conductive surface by incident photons, provided the energy available is greater than the energy barrier at the surface, the *work function*. The quantum efficiency of this process is usually low (less than 1%) for metallic conductors but much greater (20 to 30%) for semiconductor surfaces.

In a practical detector for visible and UV radiation, the photosensitive surface (the *photocathode*) is enclosed in an evacuated, transparent envelope, together with a collector (the *anode*). The anode is given a positive bias of about 100 V to aid in collecting the electrons, but the number of electrons emitted from the cathode is determined by the radiant flux, not the bias voltage. Hence the phototube is best regarded as a *current* source, and measurement of the radiation

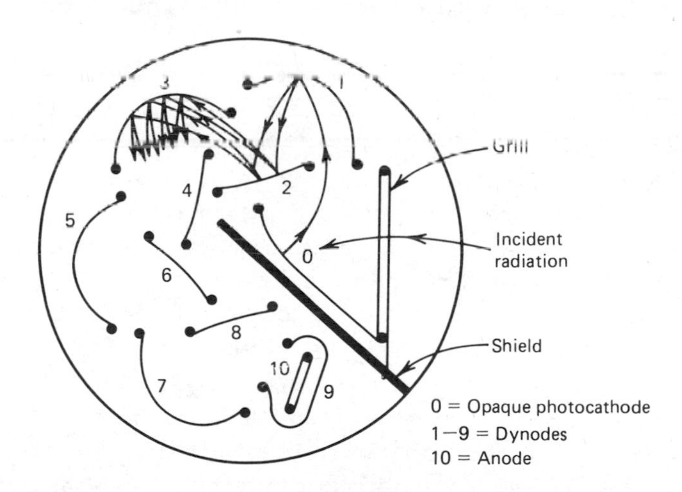

Figure 13-9. Design of a "squirrel-cage" photomultiplier with a side window (on the right). (RCA Corporation.)

is accomplished by observing the current.

The sensitivity of a phototube can be greatly increased by inclusion within the envelope of a series of added electrodes called *dynodes* at successively more positive potentials. The geometry is such that electrons from the photocathode impinge on the first dynode with sufficient kinetic energy to cause secondary emission of three or four electrons per impact. These secondary electrons are constrained to fall on the second dynode, where they produce further emission, and so on through a chain of perhaps 10 or 12 dynodes. This device is known as a *photomultiplier tube* (PMT). There are many different physical designs of PMT that need not concern us here; that shown in Figure 13-9 is typical.

Since there is an amplification of three- or four-fold at each dynode, the overall gain of a PMT can be several million. The gain can be varied by adjusting the voltages applied to the dynodes, but as usually operated at a constant voltage, the current is still determined by the number of photons incident on the cathode.

The current from phototubes and photomultipliers can be measured by means of an op amp acting as a current-to-voltage converter. The basic circuits are shown in Figure 13-10.

The internal gain of the PMT gives an added degree of freedom as compared with the simple phototube, permitting operation at constant anode current (Figure 13-11). For such service, it is connected in the feedback loop of an op amp that is supplied with constant input voltage. Variations in cathode current caused by changes in illumination are then automatically compensated by changed photomultiplier gain, so that the anode current is maintained constant. The amplifier, with booster, must be able to supply about 2000 V. Constant current operation causes the output voltage to be proportional to the logarithm of the radiant flux,

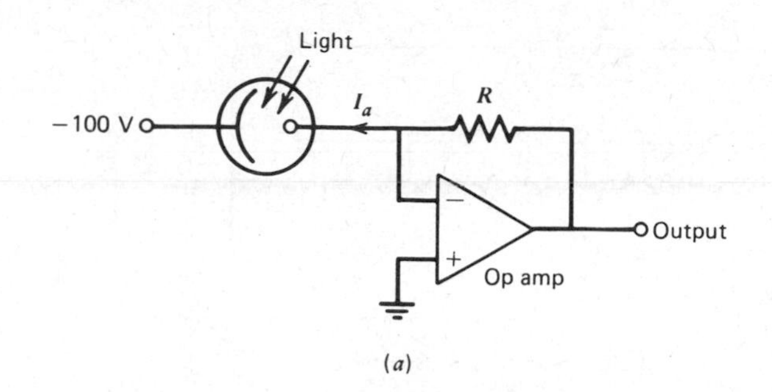

(a)

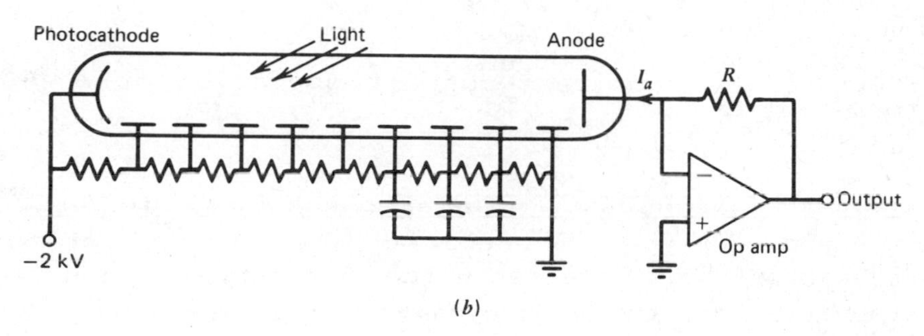

(b)

Figure 13-10. Current-to-voltage converter circuits for (a) a vacuum phototube, and (b) a photomultiplier. The resistor string in (b) provides the required voltages for the dynodes; the capacitors are needed only if the signals have a high-frequency component.

which is a valuable feature for solution spectrophotometry.

Semiconductor Photodetectors

The simplest kind of phototransducer in this group is the *photoconductive cell*. This consists of cadmium sulfide or one of a few similar compounds, in the form of either a single crystal, a pressed powder, or a vacuum-deposited film. The resistance as a function of illumination P is given by

$$R = kP^m \qquad (13\text{-}11)$$

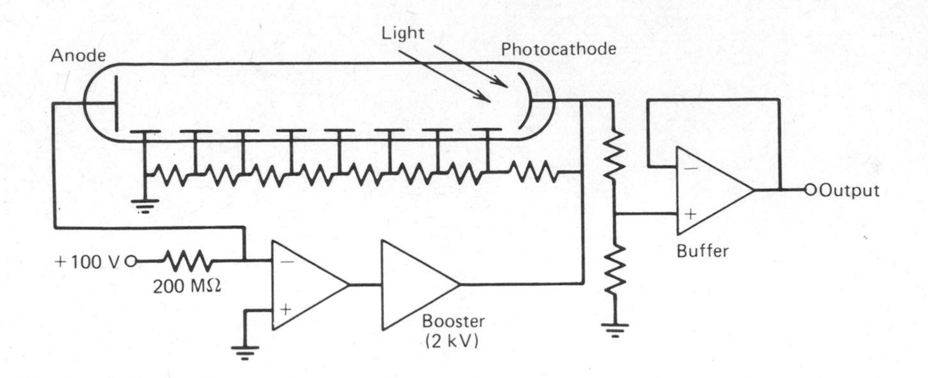

Figure 13-11. Constant-current operation of a photomultiplier tube.

where k is a constant. The exponent m may lie between -0.4 and -3.0 for CdS prepared by different methods. Because of the simple measuring circuitry, the inherent stability, and the wide dynamic range, photoconductors are very useful in such applications as photographic light meters and various switching and control devices.

Another type of semiconductor photocell is based on the properties of a *pn* junction under illumination. The incident radiation can create hole-electron pairs in the semiconductor material, and if this happens close to the junction, the electrons will be drawn into the lower energy *n* material, while the holes migrate in the opposite direction. The net effect is to produce a difference of potential across the junction. The output can be measured either as a voltage (open circuit), in which case the reading is proportional to the logarithm of the illumination, or as a current (short circuit), which gives linear response. Photodiodes are limited to very small active areas, usually less than 1 mm^2, determined by the practical dimensions of suitable junctions.

The *phototransistor*, *photodarlington*, and *photo-FET* combine the small size and fast response of the photodiode with the amplifying ability of the transistor or FET to give a device with a higher gain.

Infrared Detectors

In the infrared (IR) region, photons have too little energy to cause the ejection of electrons from a conductive surface. They may, however, be sufficiently energetic to produce charge separation in a semiconductor. For this application, the device must be cooled with liquid nitrogen (77 K) or even liquid helium (4 K) so that the thermal energy of the silicon will not outweigh the small energy of the photons.

The need for cooling can be obviated if the collective rather than individual effects of photons are utilized. Absorbed radiation causes increased molecular vibration, thus raising the temperature, so that measurement of the radiant flux becomes a measurement of an increase in temperature. For high sensitivity, the heat capacity of the detector must be as small as possible so that the available radiant energy will raise the temperature enough to permit accurate measurement. The final measurement can be made with a thermocouple or resistance thermometer affixed to a tiny flake of blackened gold foil to absorb the radiation.

A more recently introduced IR transducer is the *pyroelectric* detector. This depends on the *rate* of heating of a small crystal of an active material such as lithium tantalate or niobate. The crystal structure is such that the slight deformation caused by a pulse of energy will induce a charge on foil electrodes cemented to the crystal. This detector can respond much faster than the heat-effect detectors mentioned previously, which is a great asset to designers of IR spectrometers and imaging systems. The pyroelectric detector is a high-impedance device and must be mounted immediately adjacent to a FET preamplifier.

X-Ray Detectors

The penetrating X-radiation requires a detector with greater stopping power than those described above. This is provided by *lithium-drifted* silicon or germanium diodes, designated by the symbols Si(Li) and Ge(Li). These are *pn* junction devices, reverse biased so that a depletion region exists between the *p* and *n* material. The radiation must be absorbed within the depletion area, producing many electron-hole pairs. The incorporation of lithium results in a thicker depletion region and hence a greater probability of absorbing the X-rays. The detectors must be cooled to liquid nitrogen temperature to prevent the lithium from diffusing away from the active region. This low temperature must be maintained during the entire life of the detector, regardless of whether it is in operation. These detectors give excellent energy resolution.

X-rays can also be detected by the production of ion pairs in a gas. In the *ionization chamber* a bias

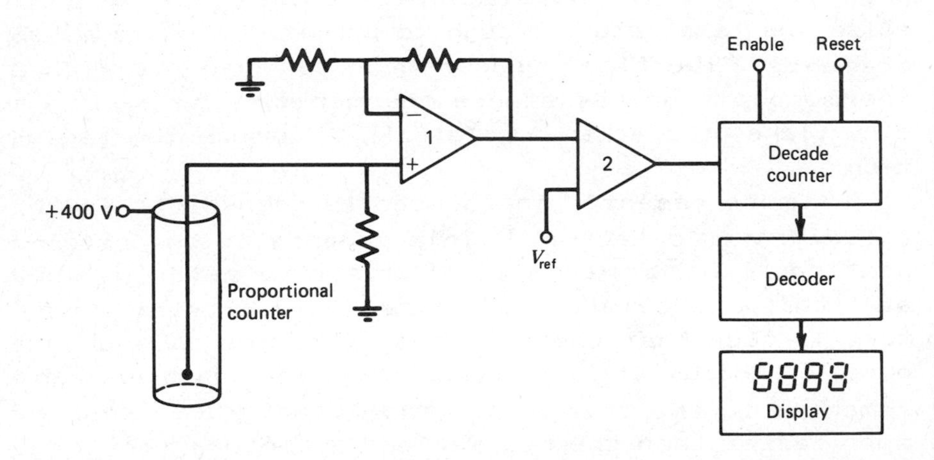

Figure 13-12. The use of a proportional counter. Amplifier 1 and comparator 2 must be optimized for high speed operation. (Note that the word "counter" is used in two different senses.)

voltage is applied across a pair of metallic electrodes. The device acts as a current source controlled by the radiant flux, very much like the vacuum phototube in the optical spectrum. The current is determined solely by the number of ionizing events per second.

At higher voltages the electrons generated by the photon acquire enough kinetic energy to produce additional ions by collision. The charges are removed quickly by the field, so that the arrival of each photon is marked by a pulse of current proportional in magnitude to the energy of the photon. The response speed is great enough that more than 10^5 pulses can be resolved per minute. In this mode the device is called a *proportional counter*. Figure 13-12 shows such a device with its associated electronics.

At still higher voltages (~1000 V) the pulses become larger and a saturation effect appears, so that the energy content of the incident photon no longer controls the size of the pulses, which become uniform. This is the operating region of the *Geiger-Mueller*, or simply *Geiger*, counter. The pulses are so large that external amplification is unnecessary. A measuring circuit is shown in Figure 11-6.

VACUUM TRANSDUCERS

As a rule, *vacuum gauges* must be active rather than passive, depending on the molecular properties of the residual gas. Their response cannot be assumed to be identical for different gases, and in most cases gauges are calibrated in terms of a specified gas, such as dry air.

The *Pirani gauge* operates on the principle of heat conduction through a gas. A thin metal filament is mounted in the vacuum system being measured and connected in a Wheatstone bridge circuit along with a similar

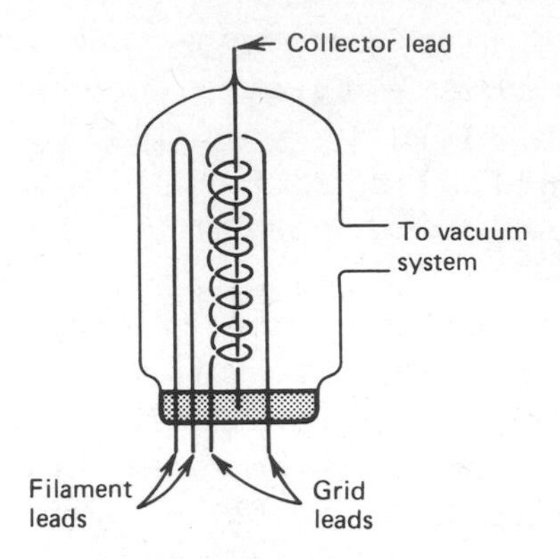

Figure 13-13. The Bayard-Alpert vacuum gauge.

filament in a permanently evacuated tube as reference. The bridge current heats the two filaments equally, but they are cooled differently by conduction of heat through the gas. The filaments can be replaced by small bead thermistors, to take advantage of their greater temperature coefficient. This *thermistor gauge* has been found useful over nearly six decades of pressure, down to 0.1 Pa (10^{-3} torr).

The *ionization gauge* (also known as the *Bayard-Alpert gauge*) consists of a heated filament, a spiral wire called a grid, and a collector wire mounted as shown in Figure 13-13. Electrons emitted from the hot filament are accelerated toward the positive grid, but most pass through the spiral and are decelerated by the reverse field between grid and collector. They tend to oscillate around the wires of the grid several times before finally being caught. This lengthened trajectory gives the electrons opportunity to strike gas molecules and ionize them. The charges so formed are accelerated to impinge on the collector, thus producing a current that can be measured with a sensitive amplifier. The range covered is from about 1 mPa to 10 nPa (10^{-5} to 10^{-10} torr).

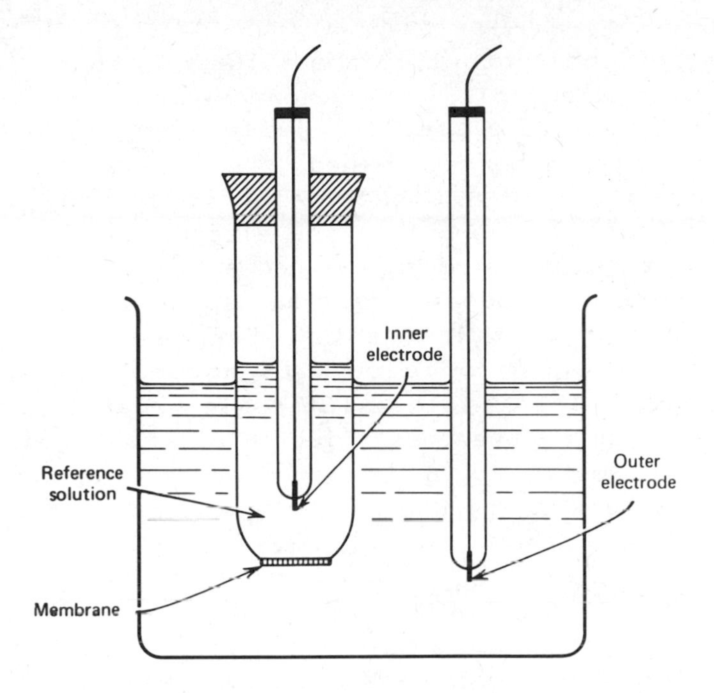

Figure 13-14. A selective ion electrode cell. The container for the reference solution, together with the membrane and inner electrode, is commonly referred to as the "indicator." The outer electrode is the "reference."

ELECTROCHEMICAL TRANSDUCERS

This category includes various electrode systems that produce an electrical signal functionally related to the concentration of some ionic species in solution.*

The concentrations of many ions in solution (more precisely their activities) can be determined by measuring the potential difference between two electrodes. One of these electrodes, the reference, should be inde-

* More details and theoretical background can be found in the present authors' *Electroanalytical Chemistry*, Wiley-Interscience, New York, 1983.

pendent of the content of the solution, to serve as a comparison potential. The second electrode, the indicator, must have a special chemical relation to the ionic species to be determined.

Some types of indicator electrodes consist of a membrane surrounding one of the electrodes. The membrane must be tailored to the properties of the ion to be measured; the details of the chemistry involved are beyond our present interest. Figure 13-14 gives a schematic diagram of a cell using such a membrane electrode. The sample solution is separated by the membrane from a reference solution of known concentration. The potential E developed across the two electrodes is related to the concentration C by the equation

$$E = A + BT \log C \tag{13-12}$$

where A and B are constants and T is the Kelvin temperature.

The voltages produced by any one of these systems seldom exceed 2 V. Their resistances may run as high as hundreds of megohms. A measuring circuit for use with

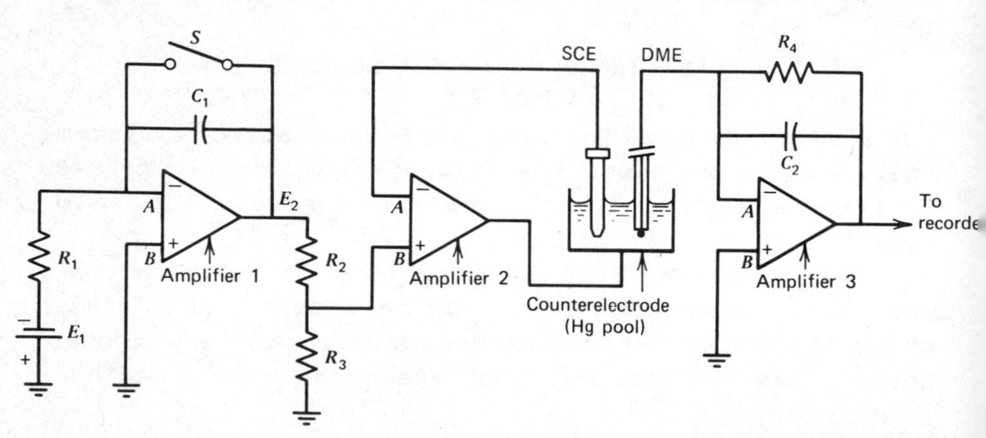

Figure 13-15. An op amp circuit for a polarograph (cf. Figure 5-2).

such high-impedance sources must have an input of not less than 1000 MΩ. Commercial instruments are called *pH meters* or *specific ion meters.*

Other electrochemical systems require the determination of a current-potential characteristic, the shape and magnitude of which is related to the composition of a solution. The prime example of such a method is *polarography.* This technique involves the use of a dropping mercury electrode together with reference and auxiliary electrodes (Figure 13-15).

In operation, op amp 1 serves as an integrator and generates an excitation ramp. This is applied to the electrodes by op amp 2 acting as a potentiostat. The resulting current is converted to a voltage by op amp 3. Study of the output provides information about the composition of the solution.

PROBLEMS

13-1. The temperature coefficient α of resistance of platinum is 0.003 K^{-1}, whereas that of nickel is 0.006 K^{-1}. Which would give the larger change in resistance per degree, a coil of platinum wire with a room-temperature resistance of 10 kΩ, or one of nickel with resistance of 1 kΩ?

13-2. Consider the two possible connections between a vacuum photodiode and an op amp as shown in Figure 13-16. In both a potential of 100 V is impressed upon the phototube, and the current is converted to a voltage for measurement. Comment on the relative advantages and disadvantages of the two circuits.

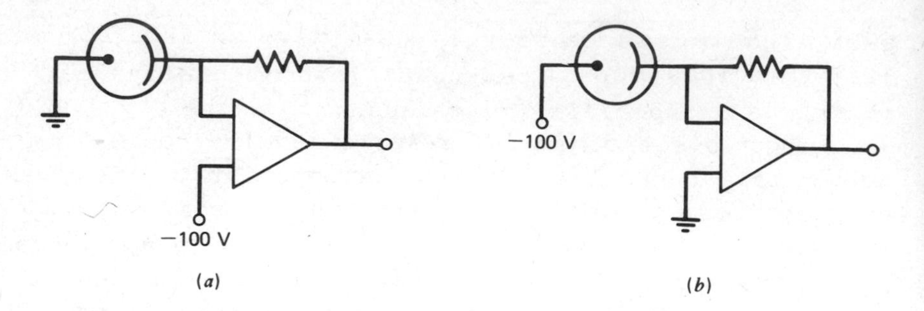

Figure 13-16. See Problem 13-2.

13-3. Figure 13-17 shows an op amp circuit that has been used to measure the electrolytic conductivity of solutions. Show that the conductance of the segment of solution between the central pair of electrodes is given by $G = 1/R = (E_{out}/E_{ref})(1/R_{out})$.

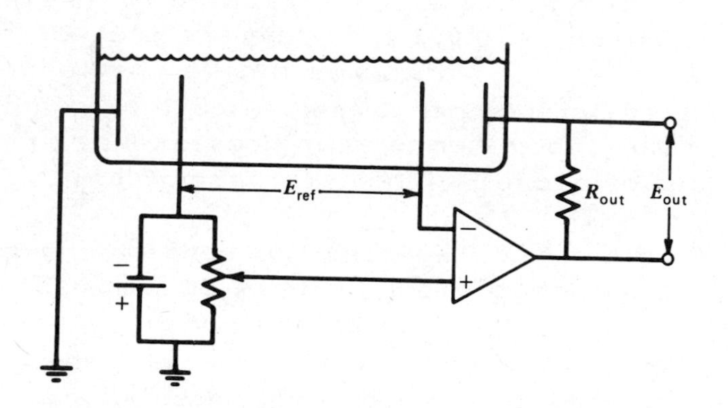

Figure 13-17. See Problem 13-3.

XIV MATHEMATICAL BACKGROUND

Students of electronics per se typically enroll in courses that alternate between mathematical theory and the more practical engineering aspects of the subject. In this book we have perforce covered the complete subject, new for the reader, in a single scan. For this reason it was necessary to eliminate from the main treatment or discuss only briefly several important mathematical tools, which we cover in this chapter and the next.

FOURIER SERIES

Figure 14-1 illustrates the *principle of superposition*, mentioned in Chapter VI. The rectangle represents a linear circuit. Both input and output consist of two different frequencies (but the number of frequencies is not limited to two). A DC potential may also be present. According to the superposition principle, each frequency component of the input will produce its corresponding output, independently of the other components present. The output is the sum of all the contributions.

This situation is enhanced by the fact that all periodic functions of interest to us can be expressed

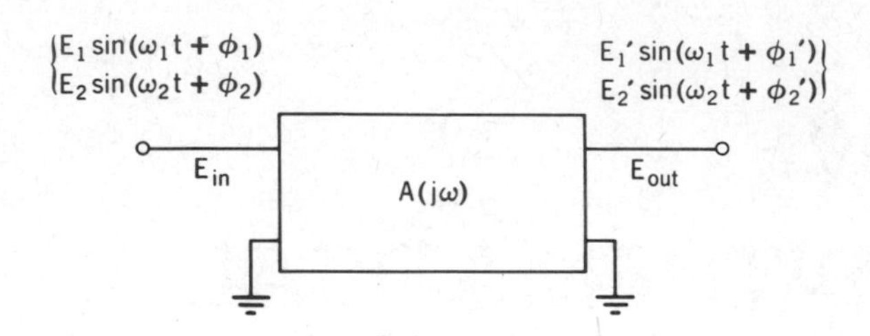

Figure 14-1. Illustration of the principle of superposition. The box is assumed to contain a linear circuit (A is the transfer coefficient).

as a sum of sinusoidal terms treated individually. This is a statement of the *Fourier theorem*, which also indicates that the expansion is unique (i.e., that for each function there corresponds a single expansion, and for each expansion a single function). The theorem applies also to a one-time signal if it is considered to constitute the first period of a fictitious repetitive function.

The Fourier expansion can be carried out in terms of the variable θ, defined as $\theta = \omega t$, where t is time and ω is the repetition frequency in radians per second. The expansion is given by the infinite series

$$F(\theta) = A_0/2 + A_1 \cos\theta + A_2 \cos 2\theta + \cdots$$

$$+ B_1 \sin\theta + B_2 \sin 2\theta + \cdots \qquad (14\text{-}1)$$

where the As and Bs are numerical coefficients. The components of the expansion consist of a DC term and sines and cosines of θ and its multiples (harmonics). Thus a signal with a repetition rate of 100 Hz, regardless of its waveform, can always be expressed as a sum of sines and cosines of 100, 200, 300 Hz, and so on, in addition to a DC term.

The results of Fourier expansion can be verified experimentally. By actual measurment, if the 100-Hz signal is applied to a tunable filter, the output of the filter is found to contain sine waves not only at 100 Hz, but also at 200, 300 Hz and so on, with amplitudes in accord with the theory. The coefficients can be computed if the analytical expression for $F(\theta)$ is known. They are given by the following integrals:

$$A_0 = \frac{1}{\pi} \int_{-\pi}^{\pi} F(\theta)\, d\theta \qquad (14\text{-}2)$$

$$A_n = \frac{1}{\pi} \int_{-\pi}^{\pi} F(\theta) \cos n\theta\, d\theta \qquad (14\text{-}3)$$

$$B_n = \frac{1}{\pi} \int_{-\pi}^{\pi} F(\theta) \sin n\theta\, d\theta \qquad (14\text{-}4)$$

Thus if $F(\theta)$ is equal to $(\theta^2 + 2\theta)$ between the limits of $-\pi$ and $+\pi$, a fictitious function can be assumed that consists of repetitions of $F(\theta)$. The expression $(\theta^2 + \theta)$ is then substituted in each integral and the integrals evaluated. The variable θ disappears when the limits of integration are introduced. A set of coefficients A_0, A_1, B_1, and so on, is obtained, and one can write

$$(\theta^2 + 2\theta) = A_0/2 + A_1 \cos \theta + \cdots$$

$$+ B_1 \sin \theta + \cdots \qquad (14\text{-}5)$$

There are an inifinity of coefficients, a circumstance that may appear to make theoretical computations impossible. In fact, in many cases the numerical values of the coefficients decrease rapidly with increasing n and can be neglected after a relatively small number of terms. Also, sometimes a whole group of terms may be zero, such as all sines or all even-numbered cosines.

As an example of Fourier expansion, consider the triangular wave shown in Figure 14-2. The analytical expression for this function is

$$F(\theta) = M - \frac{2\theta}{\pi}M \qquad 0 < \theta < \pi$$
$$F(\theta) = M + \frac{2\theta}{\pi}M \qquad -\pi < \theta < 0 \tag{14-6}$$

where M is the amplitude. The Fourier coefficients can be computed by splitting the integrals into two. Thus

$$A_0 = \frac{1}{\pi}\int_{-\pi}^{0}\left(M + \frac{2\theta}{\pi}M\right)d\theta + \frac{1}{\pi}\int_{0}^{\pi}\left(M - \frac{2\theta}{\pi}M\right)d\theta \tag{14-7}$$

The integration shows the value of A_0 to be zero, a not unexpected result, since there is no DC component.

The A_n terms are obtained by writing

$$A_n = \frac{1}{\pi}\int_{-\pi}^{0}\left(M + \frac{2\theta}{\pi}M\right)\cos n\theta \; d\theta + \frac{1}{\pi}\int_{0}^{\pi}\left(M - \frac{2\theta}{\pi}M\right)\cos n\theta \; d\theta \tag{14-8}$$

Observe that the two integrals in this equation differ only in the sign of θ and in the integration limits. Actually, since they are equal, it is necessary to evaluate only one of the two and double it to obtain the desired result:

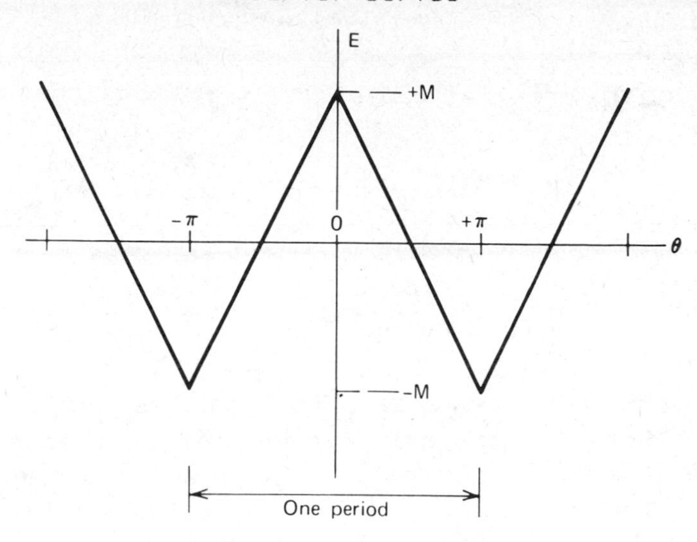

Figure 14-2. A triangular wave. $F(t)$ is shown as a function of $\theta = 2\pi f = 2\pi/T$, where T is the period.

$$A_n = \frac{2}{\pi}\int_0^{\pi}\left(M - \frac{2\theta}{\pi}M\right)\cos n\theta\ d\theta$$

$$= \frac{2M}{\pi}\int_0^{\pi}\cos n\theta\ d\theta - \frac{4M}{\pi^2}\int_0^{\pi}\theta\cos n\theta\ d\theta \qquad (14\text{-}9)$$

$$= \frac{2M}{n\pi}\sin n\theta\Big|_0^{\pi} - \frac{4M}{n^2\pi^2}\int_0^{n\pi} n\theta\cos n\theta\ d(n\theta)$$

The first term has been integrated, and the second rewritten in terms of the new variable $n\theta$. This requires a change in the upper limit since when $\theta = \pi$, the variable $n\theta$ becomes $n\pi$. Substitution of the limits shows the first term to be zero. The second can be estimated from the formula

$$\int x\cos x\ dx = \cos x + x\sin x \qquad (14\text{-}10)$$

as found in integral tables. It follows that

$$A_n = -\frac{4M}{n^2\pi^2}\left[\cos n\theta - n\theta \sin n\theta\right]\Big|_0^{\pi}$$
$$= \frac{4M}{n^2\pi^2}(1 - \cos n\pi) \tag{14-11}$$

The coefficients are zero for even values of n since the factor $(1 - \cos n\pi)$ becomes zero. When n is odd, the same factor becomes 2. Consequently the coefficients are

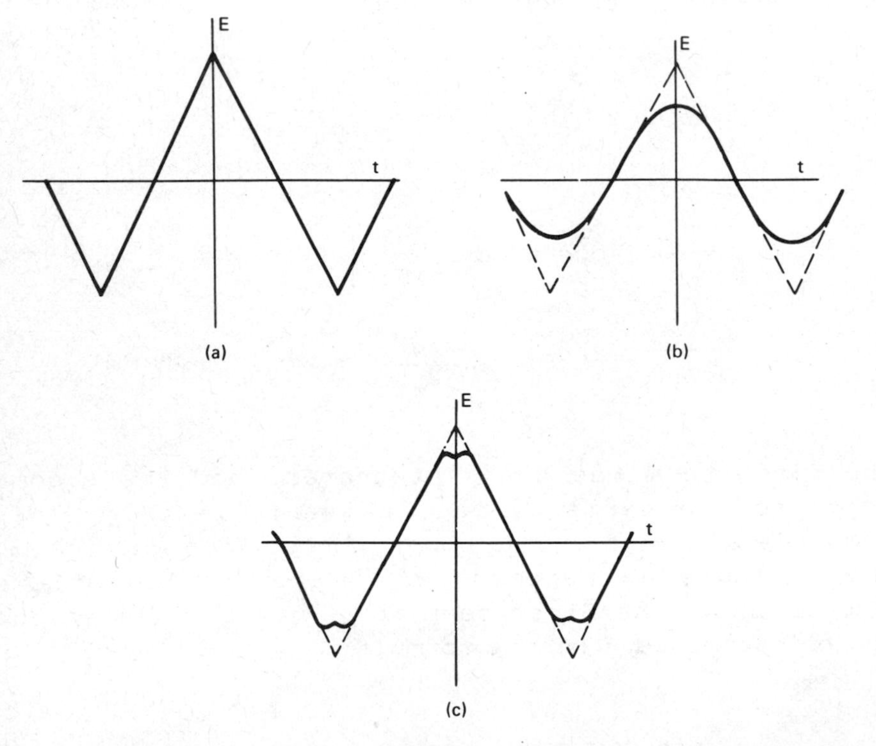

Figure 14-3. (*a*) A portion of a triangular wave; (*b* and *c*) approximations obtained by using only one and two terms, respectively, of the Fourier expansion.

$$A_0 = 0, \quad A_1 = 8M/\pi^2, \quad A_2 = 0, \quad A_3 = 8M/(9\pi^2),$$

$$A_4 = 0, \quad A_5 = 8M/(25\pi^2), \quad \cdots$$

In addition, it can be shown that all B terms drop out. Thus the Fourier series for the triangular wave is given by

$$F(\theta) = \frac{8M}{\pi^2}\left(\frac{1}{1}\cos\theta + \frac{1}{9}\cos 3\theta + \frac{1}{25}\cos 5\theta + \cdots\right) \qquad (14\text{-}12)$$

Note that the terms decrease rapidly. A good approximation can be obtained by retaining only the first two or three terms (Figure 14-3).

Returning to the more general case, it is apparent that functions less similar to a sine wave than the example just considered will require a larger number of terms. To reproduce the original function without distortion, a network or amplifier must fulfill the following conditions:

1. The gain or attenuation must be constant over all harmonic frequencies at which the Fourier terms are significant. This means that the bandwidth must be some 5 or 10 times larger than the basic repetition frequency of the signal.
2. The phase relationships must remain constant. This condition is more difficult to fulfill, as the effect of phase shift is to distort the signal.

TRANSIENT SIGNALS

Both the DC description and the AC phasor representation of circuits have the implicit requirement that the signals involved be stable with respect to time, or in other words, that they be in a *steady state*.

On the other hand, a circuit subject to a sudden change passes through a transition period during which its behavior varies. This can occur, for example, when a switch is closed or when a signal source is connected to a previously grounded input. Both the DC and the AC phasor approaches are useless in the description of transients since they cannot cope with time as a variable.

A very general mathematical method for treating transients is to solve the differential equation of the circuit. This permits theoretical calculations of steady-state signals as well as of transients. We consider here only the case of linear systems consisting of resistors, capacitors, and inductors.

The differential equation of a circuit (more precisely the integrodifferential equation, since it may contain integrals) can be obtained by appropriate combinations of the following relations between voltages and currents:

$$I = C(dE/dt) \tag{14-13}$$

$$E = (1/C) \int I \; dt \tag{14-14}$$

$$I = (1/L) \int E \; dt \tag{14-15}$$

$$E = L(dI/dt) \tag{14-16}$$

$$E = IR \tag{14-17}$$

Such formulas can be combined for a given circuit to obtain the overall differential equation. For example, in the circuit shown in Figure 14-4 the capacitor is initially charged to E_0 volts, but at time zero is suddenly connected to a voltage source E through a resistor R. At steady state no current can flow, and the capacitor will be charged to the new

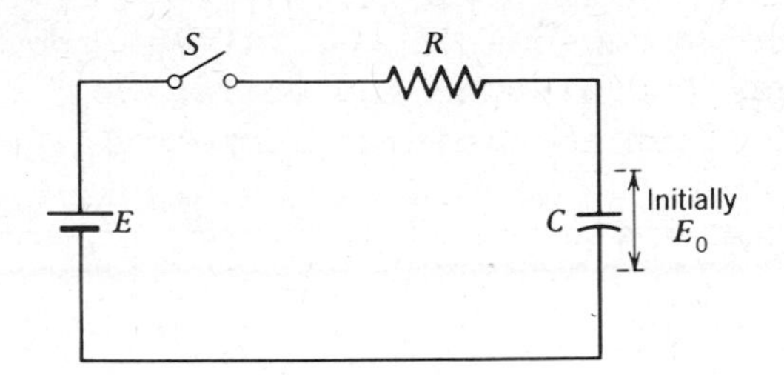

Figure 14-4. An *RC* circuit illustrating transient and steady-state conditions.

voltage. During the transition period, the time dependence of voltage and current will follow the predictions of the differential equation.

The Kirchhoff voltage law, evaluated around the circuit, gives

$$E - IR + \frac{1}{C} \int I \, dt \qquad (14\text{-}18)$$

To solve this equation, we differentiate both sides, obtaining

$$0 = R \frac{dI}{dt} + \frac{I}{C} \qquad (14\text{-}19)$$

or

$$I = -RC \frac{dI}{dt} \qquad (14\text{-}20)$$

The solution of this equation (as can be proved by substitution) is

$$I = A \exp \left(- \frac{t}{RC}\right) \qquad (14\text{-}21)$$

The constant A depends on the initial conditions of the circuit, specifically on the voltage E_0. For t = 0, A becomes equal to the initial current I_0, which is given by $(E - E_0)/R$. Consequently the desired expression for the current is

$$I = \frac{E - E_0}{R} \exp\left(-\frac{t}{RC}\right) \tag{14-22}$$

From this expression we can obtain the voltage on the capacitor as a function of time by using the relationship

$$E_C = \frac{1}{C} \int I \, dt + E_0 \tag{14-23}$$

Substituting the value of I from Eq. 14-22 into Eq. 14-23 gives

$$\begin{aligned} E_C &= \frac{E - E_0}{RC} \int_0^t \exp\left(-\frac{t}{RC}\right) dt + E_0 \\ &= -(E - E_0) \exp\left(-\frac{t}{RC}\right) \Big|_0^t + E_0 \\ &= E\left[1 - \exp\left(-\frac{t}{RC}\right)\right] + E_0 \exp\left(-\frac{t}{RC}\right) \end{aligned} \tag{14-24}$$

Thus as time increases, it appears that the current approaches zero as expected, whereas the voltage across the capacitor approaches E. Similarly, one can show that for time $t = RC$ (after one time constant) the current is reduced to $1/e$ of the initial value, where e is the base of natural logarithms. Figure 14-5 gives graphs of both current and voltage.

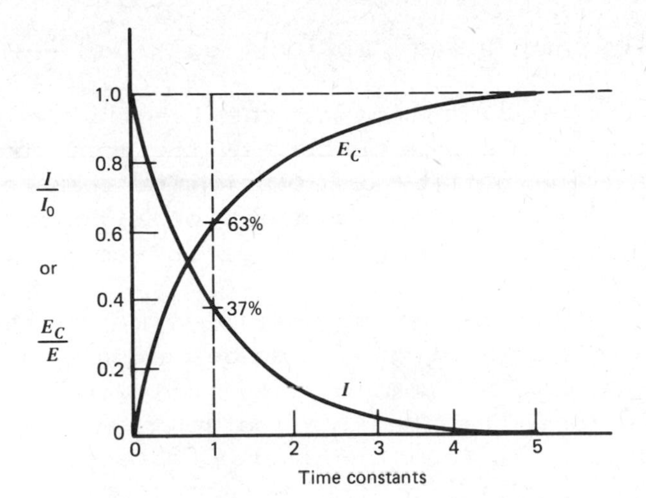

Figure 14-5. The time behavior of the *RC* circuit shown in Figure 14-4.

The solution of the differential equation in this example was relatively simple. Circuits with more components pose considerably more mathematical difficulty.

LAPLACE TRANSFORMS

The solution of more complicated differential equations can be greatly simplified by the mathematical technique known as the *Laplace transformation*. This method is applicable to any linear differential equation, and can be considered analogous to the use of logarithms in multiplication. To clarify the concept of transformation, let us review the operations performed in computing the product xy by logarithms:

1. Transformation of the numbers into their logarithms: $x \longrightarrow \log x$; $y \longrightarrow \log y$.
2. Application of the combinatorial rules: $\log x + \log y = \log (xy)$.

3. Inverse transformation: $\log^{-1}(xy) \rightarrow xy$.

This permits replacement of the lengthy process of multiplication of large numbers by the more convenient addition. The utility of logarithmic transformation is more evident in the process of exponentiation. How, without logarithms (or a calculator) could one evaluate $3^{4.71}$?

Similarly, the rather difficult manipulation of differentials and integrals can be replaced by operations in the Laplace domain, where only simple algebra need be performed. The final solution of the differential equation is found by inverse Laplace transformation. The overall procedure is as follows:

1. Take the transform of each side of the differential equation.
2. Make the necessary algebraic manipulations.
3. Take the inverse transform.

The Laplace transformation is an operation by which any function of time $F(t)$ is converted into a new function $F(s)$. Whereas the variable t is evidently real and positive, the new variable s is complex. The function $F(s)$, called the *transform*, has many advantageous properties, the most important being the close similarity between the function and its derivatives. Such similarities occur only occasionally for time functions.

Laplace transforms obey the following simple relationships, in which £ symbolizes the taking of the transform, and E represents any function:*

* Conventionally a script L is used to designate the taking of the transform; we use £ only because of the limitations of our word-processor.

$$\pounds\{E(t)\} = \bar{E}(s) \quad \text{or} \quad \pounds(E) = \bar{E} \tag{14-25}$$

$$\pounds\{dE/dt\} = s\bar{E} - E(0) \tag{14-26}$$

$$\pounds\{\int E\,dt\} = \bar{E}/s + K/s \tag{14-27}$$

where $E(0)$ is the initial value of E (i.e., at time t = 0) and K is the value of the integral just after time zero. Thus apart from initial conditions, differentiation and integration of a function leave the transform unchanged except for multiplication by s or $1/s$.

The Laplace transform of a numerical coefficient is the coefficient itself: $\pounds(kE) = k\bar{E}$. In contrast, the transform of a constant E_0 representing an initial condition is E_0/s. This is referred to as the *transform of a step function* since it applies to voltages or currents that did not exist before time zero, when they were suddenly switched on.

Let us see how the simple transform relationships between a function and its derivative can be used in solving a differential equation. Consider again Eq. 14-18:

$$E = IR + \frac{1}{C}\int_0^t I\,dt \tag{14-28}$$

Noting that E is switched on at time zero and is thus a step function, we can write, making use of Eq. 14-27, with $K = E_0$:

$$\frac{E}{s} = R\bar{I} + \frac{\bar{I}}{sC} + \frac{E_0}{s} \tag{14-29}$$

This equation contains only $\bar{I}$ and the variable s, plus constants. It can be solved for I with the functional relationship $I(s)$ explicitly shown. An inverse transform furnishes the solution of the differential equation (I as a function of time). Thus Eq. 14-29 can be rewritten as

$$\bar{I} = \left(\frac{E - E_0}{s}\right)\left(\frac{1}{R + 1/sC}\right) \tag{14-30}$$

which can be further modified to the more tractable form

$$\bar{I} = \left(\frac{E - E_0}{R}\right)\left(\frac{1}{s + 1/RC}\right) \tag{14-31}$$

In order to find the inverse transform, a table must be consulted, such as can be found in many handbooks. A few examples are given in Table 14-1 and in Appendix IV, in which the symbol a represents a constant. From the table, it can be seen that the solution of Eq. 14-31 is

$$I = \frac{E - E_0}{R} \exp\left(-\frac{t}{RC}\right) \tag{14-32}$$

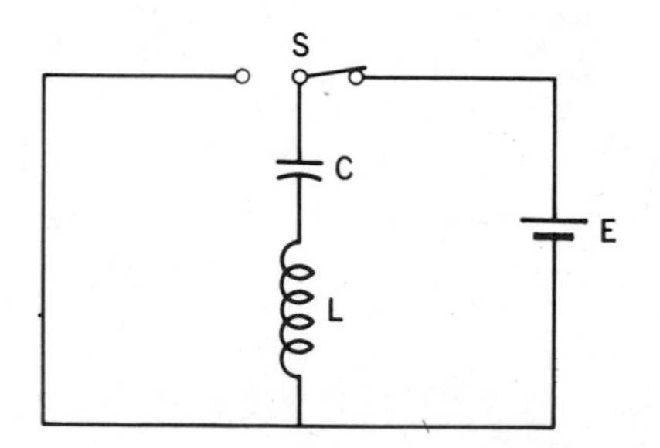

Figure 14-6. An LC circuit in which transients can be induced by means of the switch.

TABLE 14-1

Examples of Laplace Transforms

$F(t)$	$\mathcal{L}\{F(t)\} = \overline{F}(s)$
t	$1/s^2$
$\exp(-at)$	$1/(s + a)$
$(1/a)[1 - \exp(-at)]$	$1/[s(s + a)]$
$t \exp(-at)$	$1/(s + a)^2$
$(1/a) \sin at$	$1/(s^2 + a^2)$
$\cos at$	$s/(s^2 + a^2)$
Step, amplitude a	a/s

which is identical to Eq. 14-22, obtained by conventional means.

As another example, consider Figure 14-6, in which an *LC* circuit is connected initially to a voltage E, so that the capacitor is charged to that voltage. At time zero, the switch is thrown to the left, short-circuiting the *LC* combination. The current is zero prior to switching, and it remains so for an infinitesimal time thereafter because the inductance opposes any instantaneous increase in current; consequently we can state that $I(0) = 0$. In contrast, the voltage, initially E, suddenly vanishes, the equivalent to a step of $-E$ volts. Keeping the initial condition in mind and applying the Kirchhoff voltage law, we can write

$$E \to 0 = \frac{1}{C} \int_0^t I \, dt + L \frac{dI}{dt} \qquad (14\text{-}33)$$

and its Laplace transform

$$-\frac{E}{s} = \frac{\overline{I}}{sC} + Ls\overline{I} \tag{14-34}$$

(Note that it would be quite possible to skip a step and write the Laplace transform directly.) Solving for I gives

$$\overline{I} = \frac{E}{L} \cdot \frac{1}{s^2 + 1/LC} \tag{14-35}$$

The inverse transform is found from Table 14-1, putting $a^2 = 1/LC$:

$$I = -E\sqrt{C/L}\ \sin\frac{t}{\sqrt{LC}} \tag{14-36}$$

This result indicates no transient, but only a stable sine wave. The negative sign is an indication of the phase.

When more involved differential equations are encountered, the process of solution requires some manipulations to bring the various polynomials in s to one of the canonical forms found in the tables. If desired, it is possible to invoke the definition of the Laplace transform and evaluate the integrals

$$\overline{F}(s) = \int_0^\infty F(t)\ \exp\ (-st)\ dt \tag{14-37}$$

$$F(t) = \frac{1}{j2\pi}\int_{a-j\infty}^{a+j\infty} \exp\ (-st)\ \overline{F}(s)\ ds \tag{14-38}$$

The expression of the inverse transform is somewhat forbidding, even though it may not be as difficult as it appears. It consists of evaluating the line integral in the complex plane along the "vertical" parallel to the j axis at a distance a, from $-\infty$ to $+\infty$.

To illustrate, let us evaluate the direct transform of $F = \exp(-at)$:

$$
\begin{aligned}
F &= \int_0^\infty \exp(-at) \exp(-st)\, dt \\
&= \int_0^\infty \exp[-(s + a)t]\, dt \\
&= \frac{1}{a + s} \exp[-(s + a)t] \Big|_0^\infty \\
&= \frac{1}{a + s}
\end{aligned}
\tag{14-39}
$$

which is the transform indicated in the table.

s-Domain Impedance

We have previously defined the notion of complex impedance, which represents the ratio of steady-state AC voltage to current for reactive circuits. Such treatment is completely satisfactory, especially in light of the Fourier theorem. This approach cannot be used, however, in connection with transients.

Transients exist in reactive systems as the result of any sudden change, particularly that occurring when a signal source is connected to the system. For instance, after applying a sine wave to the input of a circuit, the output may be small in amplitude at the beginning, increasing progressively to its steady state. After a sufficient amount of time has elapsed, the current predicted by phasor calculations will be attained.

A very general concept, capable of dealing with steady state conditions as well as transients, is the *s-domain impedance*. This approach is derived from the differential equation for the circuit and as such does not have the limitations inherent in the phasor treatment.

The differential equation describing the circuit is written as before, from considerations of Kirchhoff's voltage law. Thus for a series *RLC* circuit, the equation is

$$E = IR + L\frac{dI}{dt} + \frac{1}{C}\int_0^t I\,dt \qquad (14\text{-}40)$$

The Laplace transform can be written directly, denoting each resistor by R, each inductor by sL, and each capacitor by $1/sC$:

$$\overline{E} = [R + Ls + \frac{1}{sC}]\overline{I} \qquad (14\text{-}41)$$

The quantity in brackets is called the s-domain impedance, or simply the impedance. It is identical in form to the complex impedance $[R + j\omega L + (1/j\omega C)]$, but with a somewhat different meaning. This identity of form ($j\omega$ replaced by s) makes the s-domain impedance an extension of the conventional complex impedance as well as of Laplace transform techniques.

Both types of impedance are complex quantities, but the phasor describes relationships between vectors, whereas the s-impedance describes differential equations. The latter is more widely applicable than the former, in that it includes transients. The *s*-impedance can be set up directly without the need to write out the differential equation, by the use of the following rules:

1. The impedances of single elements are R, sL, and $1/sC$, as the case may be.
2. Series and parallel combinations are given by the well-known rules for combinations of impedances.

For example, the *RC* series circuit will have an impedance given by

$$\overline{Z} = R + \frac{1}{sC} = \frac{sRC + 1}{sC} \tag{14-42}$$

The response to a signal can be obtained by using Ohm's law:

$$\overline{I} = \overline{E}/\overline{Z} \tag{14-43}$$

If the signal is a DC step of amplitude E, then

$$\overline{I} = \frac{E}{s} \cdot \frac{sC}{1 + sRC} \tag{14-44}$$

(This particular transform was encountered earlier, with initial conditions present.) Setting $E_0 = 0$, we can use the solution of Eq. 14-32:

$$I = E_0/R \exp\left(-\frac{t}{RC}\right) \tag{14-45}$$

This equation completely describes the transient. The steady state can be obtained by setting t to infinity, giving $I = 0$, a result to be expected for a capacitive circuit given a DC signal.

To illustrate the use of Laplace transforms with AC systems, let us consider the case of a single capacitor. Table 14-1 tells us that the signal $(1/\omega) \sin \omega t$ has the transform $1/(s^2 + \omega^2)$. An applied signal of $B \sin \omega t$ can be written as $\omega B[(1/\omega) \sin \omega t]$, with the transform $\omega B/(s^2 + \omega^2)$. The transform of the current then becomes

$$\bar{I} = \bar{E}/\bar{Z} = \frac{\omega B/(s^2 + \omega^2)}{1/sC} \tag{14-46}$$

or

$$\bar{I} = \omega BC \cdot \frac{s}{s^2 + \omega^2} \tag{14-47}$$

The inverse transform as given by the table is

$$I = \omega BC \cos \omega t \tag{14-48}$$

which indicates that the current leads the voltage by 90° and has an amplitude of ωBC. The same result obtained from phasor computations is

$$I = \frac{B}{1/j\omega C} = \omega BC \angle 90° \tag{14-49}$$

It is interesting to note that there is no transient in this system, a fact not ascertainable from the phasor treatment.

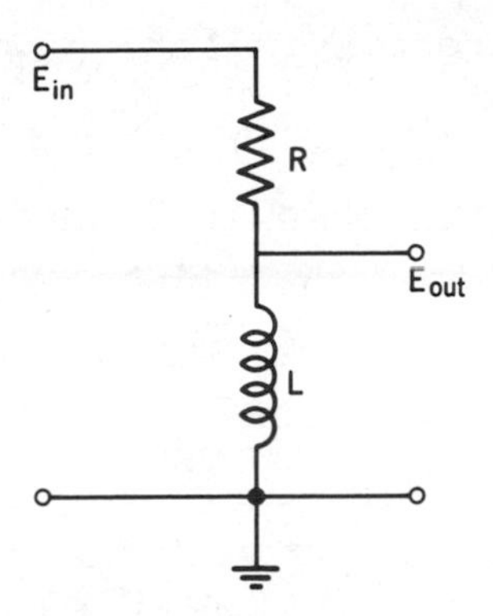

Figure 14-7. An *RL* voltage divider, assumed to be connected to a high-impedance load.

Transfer Functions

The transform method is applicable not only to impedances but also to any other ratios of voltages and currents, such as gains and transimpedances. This represents an extension in the use of the concept of transfer coefficient to a new quantity in the *s*-domain called the *transfer function*, defined as the ratio of output to input transforms.

An example of the use of the transfer function is the *s*-analog of the voltage divider equation. For the circuit shown in Figure 14-7 the attenuation is described by the transfer function *G*:

$$G = \frac{E_{out}}{E_{in}} = \overline{Z}_L/\overline{Z}_T = \frac{sL}{R + sL} \qquad (14\text{-}50)$$

where Z_T is the total *s*-impedance and Z_L that of the inductor.

Such expressions are very useful for describing transient processes. For steady-state AC, the phasor form is usually more convenient, giving the same result with s replaced by $j\omega$.

Consider the effect of a step function E_0/s applied to the circuit shown in Figure 14-7. The transform of the output is

$$\overline{E}_{out} = \frac{E_0}{s} \cdot \frac{sL}{R + sL} = \frac{E_0}{(R/L) + s} \qquad (14\text{-}51)$$

The output is initially equal to the input but decreases exponentially to its steady-state value of zero, as can be seen from Table 14-1, putting $a = R/L$.

The transfer function is particularly useful in connection with operational amplifier circuits. Consider an amplifier configured as an inverter with one feedback and one input impedance. The impedances themselves can be made up of any number of resistors and capacitors. The input current transform is

$$\overline{I}_{in} = \overline{E}_{in}/\overline{Z}_{in} \qquad (14\text{-}52)$$

and the feedback current transform is

$$\overline{I}_f = \overline{E}_{out}/\overline{Z}_f \qquad (14\text{-}53)$$

Writing $I_{in} = -I_f$, we see that

$$\overline{E}_{in}/\overline{Z}_{in} = -\overline{E}_{out}/\overline{Z}_f \qquad (14\text{-}54)$$

or

$$\overline{E}_{out} = -\overline{E}_{in}\overline{Z}_f/\overline{Z}_{in} \qquad (14\text{-}55)$$

This is the transform equivalent of Eq. 4-7. It enables us to compute the response of the circuit to any step input, usually referred to as the response to a DC signal Thus the circuit shown in Figure 14-8a has a transfer function

$$G = -\frac{R + (1/sC)}{R} \qquad (14\text{-}56)$$

Such an op amp circuit, when connected to a DC source of voltage E_0/s, gives an output of

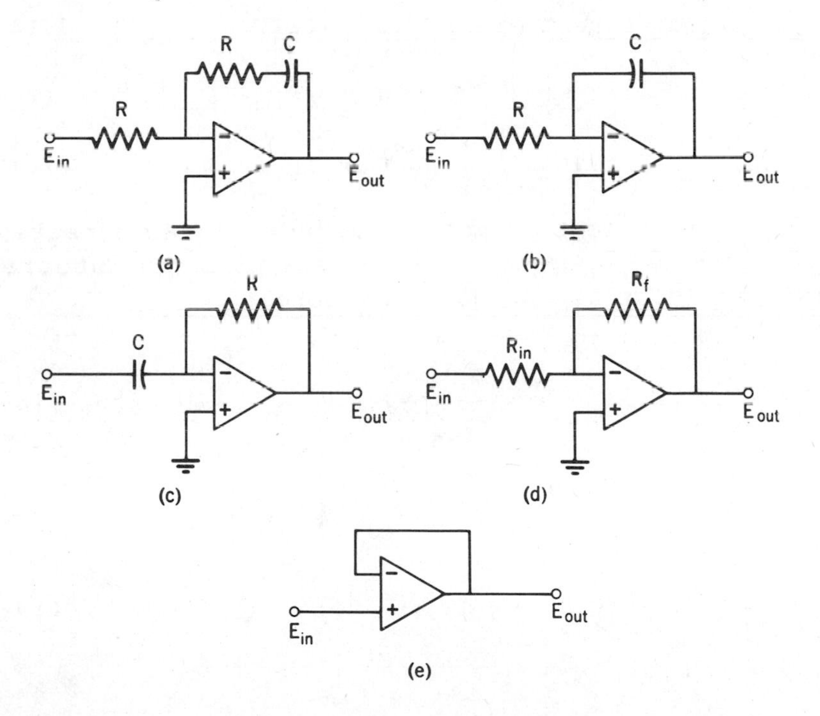

Figure 14-8. Typical op amp circuits illustrating the transfer function.

$$\overline{E}_{out} = G\overline{E}_{in} = -\frac{E_0}{s} \cdot \frac{R + (1/sC)}{R}$$
$$= -\frac{E_0}{s} - \frac{E_0}{s^2RC} \quad (14\text{-}57)$$

By inverse transform, it is seen that the output is a step ($-E_0$) at time zero, followed by a superimposed ramp $-(E_0/RC)t$, generated by the second term.

The transfer functions for the rest of the circuits shown in Figure 14-8 can be shown to be:

(*b*) (integrator) $G = -1/sRC$ (14-58)

(*c*) (differentiator) $G = -sRC$ (14-59)

(*d*) (inverter) $G = -R_{in}/R_f$ (14-60)

(*e*) (follower) $G = 1$ (14-61)

The transient response can be obtained directly by inverse transformation. For example, an integrator given an input E_{in} produces an output

$$\overline{E}_{out} = \left(\frac{E_0}{s}\right)\left(\frac{1}{sRC}\right) = \left(\frac{E_0}{RC}\right)\left(\frac{1}{s^2}\right) \quad (14\text{-}62)$$

which has as its inverse transform

$$E_{out} = \left(\frac{E_0}{RC}\right)t \quad (14\text{-}63)$$

XV

BOOLEAN ALGEBRA AND KARNAUGH MAPS

In our discussion of binary logic, we have referred briefly to the use of Boolean notation. Now we wish to develop this approach further and show its potential utility in the design and optimization of logic systems.

Boolean algebra is a mathematical system useful in dealing with binary numbers since there are only two digits, "0" and "1." In Boolean, as in ordinary algebra, we can use letters to represent numbers. In ordinary algebra, we follow explicit rules to express the common operations of addition, subtraction, multiplication, and division in terms of letter symbols. In Boolean algebra we are concerned instead with the logical operations expressed by the terms **AND**, **OR**, **NAND**, **NOR**, **XOR**, and **NOT**. It is thus appropriate in describing digital logic gates. In fact there are one-to-one correspondances between the voltages in logic gates and the results of Boolean statements.

We will start by defining (in Table 15-1) the specialized symbols of Boolean algebra, which have been introduced in Chapter IX. It is perhaps unfortunate that the symbols (•) and (+) have distinctly different meanings in the two algebras, but this convention is firmly established and we must conform. The dot symbol can be omitted if desired: AB is the same as $A \cdot B$.

TABLE 15-1

Symbols of Boolean Algebra

Operator	Symbol	Example
AND	•	$A \cdot B \cdot C$ = A **and** B **and** C
OR	+	$A + B + C$ = A **or** B **or** C
XOR	$\oplus$	$A \oplus B \oplus C$ = any *one* of A, B, **or** C
NOT	$\overline{}$	$\overline{A}$ = not-A = complement of A

The operations are similar to those of standard arithmetic, but there are major differences. The **AND** function, for example, leads to the expressions

$$A \cdot 1 = A \tag{15-1}$$

$$A \cdot 0 = 0 \tag{15-2}$$

$$A \cdot A = A \tag{15-3}$$

These can be demonstrated by considering the corresponding gate. If a two-input **AND** gate has a "1" at one input, the output will be "1" if the second input is "1" and "0" if it is "0." In other words, the output is "A" if the second input is "A," expressed by the statement $A \cdot 1 = A$.

Similarly, the **OR** function leads to the statements

$$A + 1 = 1 \tag{15-4}$$

$$A + 0 = A \tag{15-5}$$

$$A + A = A \tag{15-6}$$

These are verifiable by reference to the OR gate.

The **NOT** function tells us that

$$\overline{0} = 1 \tag{15-7}$$

$$\overline{1} = 0 \tag{15-8}$$

Combining this with the **AND** and **OR** functions we obtain the additional statements

$$A \cdot \overline{A} = 0 \tag{15-9}$$

$$A + \overline{A} = 1 \tag{15-10}$$

The *commutative* and *associative rules* offer no difficulties. We can write

$$A + B = B + A \tag{15-11}$$

$$A \cdot B = B \cdot A \tag{15-12}$$

$$A + (B + C) = (A + B) + C \tag{15-13}$$

$$A \cdot (B \cdot C) = (A \cdot B) \cdot C \tag{15-14}$$

The *distributive rules* that permit factoring of expressions are as follows:

$$A \cdot (B + C) = (A \cdot B) + (A \cdot C) \tag{15-15}$$

$$A + (B \cdot C) = (A + B) \cdot (A + C) \tag{15-16}$$

$$A + (\overline{A} \cdot B) = A + B \tag{15-17}$$

Each of these statements can be proved by combinations of prior statements. For example, Eq. 15-17 is derived as follows:

Truth Table

A	B	C	P	Q	X	R	S	Y	$\overline{Y}$
0	0	0	1	1	1	0	0	0	1
0	0	1	1	1	1	0	0	0	1
0	1	0	1	1	1	0	0	0	1
0	1	1	1	1	1	0	0	0	1
1	0	0	1	0	0	0	1	1	0
1	0	1	0	1	0	1	0	1	0
1	1	0	1	1	1	0	0	0	1
1	1	1	1	1	1	0	0	0	1

(*a*)

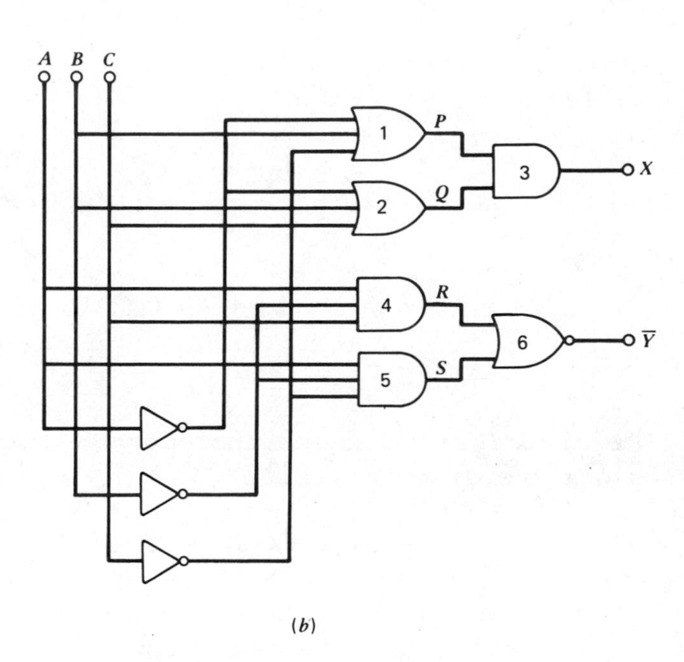

(*b*)

Figure 15-1. Two ways of implementing the same logic function. The truth table shows that X and $\overline{Y}$ are identical.

$$A + \bar{A}\cdot B = A(1 + B) + \bar{A}\cdot B$$

$$= A + A\cdot B + \bar{A}\cdot B$$

$$= A + B(A + \bar{A})$$

$$= A + B$$

The *De Morgan theorem* is stated as

$$\bar{A}\cdot\bar{B} = \overline{A + B} \qquad \text{or} \qquad \bar{A} + \bar{B} = \overline{A\cdot B} \qquad (15\text{-}18)$$

These relationships permit us to transform a sum-of-products into the equivalent product-of-sums or the reverse:

$$A\cdot\bar{B}\cdot C + A\cdot\bar{B}\cdot\bar{C} = \overline{(\bar{A} + B + \bar{C}) \cdot (\bar{A} + B + C)} \qquad (15\text{-}19)$$

$$\bar{A}(B + \bar{C}) = \overline{A + \bar{B}\cdot C} \qquad (15\text{-}20)$$

These statements can be proved by the use of truth tables for the corresponding gate structures. Take, for example, Eq. 15-19. This can be implemented by either two **OR** gates and an **AND** gate or two **AND** gates and a **NOR** gate, as shown in Figure 15-1, in which both assemblies are included. The truth table shows that output X, based on gates 1 through 3, is identical with Y, obtained using gates 4 through 6.

There is a set of rules that can be applied to making the De Morgan transformation:

1. Replace every (•) by (+) and every (+) by (•).
2. Place a bar over every unbarred variable and remove bars from all barred variables.
3. Place a bar over the entire resultant function.

As an example, we derive the statement given in Eq. 15-20:

$$\bar{A}(B + \bar{C}) \text{ becomes } \bar{A} + (B \bullet \bar{C}) \qquad \text{(by rule 1)}$$

$$\bar{A} + B \bullet \bar{C} \text{ becomes } A + \bar{B} \bullet C \qquad \text{(by rule 2)}$$

$$A + B \bullet \bar{C} \text{ becomes } \overline{A + \bar{B} \bullet C} \qquad \text{(by rule 3)}$$

The conversion of Eq. 15-19 can be carried out by similar steps. These rules are applicable to portions of a function as well as to the complete function. For example, in the expression $X = \bar{A} \bullet \bar{C} + \bar{C}(\bar{D} + E)$, the first term, $\bar{A} \bullet \bar{C}$, can be converted first, then combined with the rest of the statement:

$$X = \bar{A} \bullet \bar{C} + \bar{C}(\bar{D} + E)$$

$$= \bar{A} + \bar{C} + \bar{C}(\bar{D} + E) \qquad \text{(by Eq. 15-18)}$$

$$= \bar{A} + \bar{C}(1 + \bar{D} + E) \qquad \text{(by Eq. 15-15)}$$

$$= \bar{A} + \bar{C} \qquad \text{(by Eq. 15-4)}$$

$$= \overline{A \bullet C} \qquad \text{(by Eq. 15-18)}$$

Clearly, after some practice, the De Morgan conversion can be written down directly, without explicitly stating the intermediate steps.

The general statements in Eqs. 15-1 through 15-18 can be used to simplify logical expressions prior to implementing them in hardware. For example, consider the expression

$$X = A \bullet \bar{C} + A \bullet B \bullet C + A \bullet C$$

As it stands, this would call for three **AND** gates and

one three-input **OR** gate plus one inverter. However, it can be simplified by the following steps:

$$X = A \cdot \overline{C} + A \cdot B \cdot C + A \cdot C$$

$$= A(\overline{C} + C) + A \cdot B \cdot C \qquad \text{(by Eq. 15-15)}$$

$$= A(1) + A \cdot B \cdot C \qquad \text{(by Eq. 15-10)}$$

$$= A(1 + B \cdot C) \qquad \text{(by Eq. 15-15)}$$

$$= A(1) = A \qquad \text{(by Eq. 15-4)}$$

This shows that all five gates can be eliminated in favor of a direct wiring of X to A. It makes no difference whatever what values B and C may assume. The reader may want to verify this by constructing a truth table.

Whenever the exclusive-OR function appears, it can be replaced by its equivalent:

$$A + B = A \cdot \overline{B} + \overline{A} \cdot B \qquad (15\text{-}21)$$

as demonstrated by Figure 9-15.

MINTERMS AND KARNAUGH MAPS

The most important application of Boolean algebra in our context is the eventual simplification of systems of digital logic. The first step, after establishing a working expression, is to bring it to the *standard*, or *canonical* form. This is an equivalent expression consisting of the sum of terms, each of which contains all of the variables. The basic concept can be illustrated by conversion of the expression $A + B$. Each term in the canonical form must, in this case, contain both A and B. The conversion is performed as follows:

$$A + B = A(1) + B(1)$$

$$= A(B + \overline{B}) + B(A + \overline{A})$$

$$= A\bullet B + A\bullet\overline{B} + A\bullet B + \overline{A}\bullet B$$

$$= A\bullet B + A\bullet\overline{B} + \overline{A}\bullet B$$

For an expression containing the three variables A, B, and C, the term $A\bullet\overline{B}$, for example, would be expan-

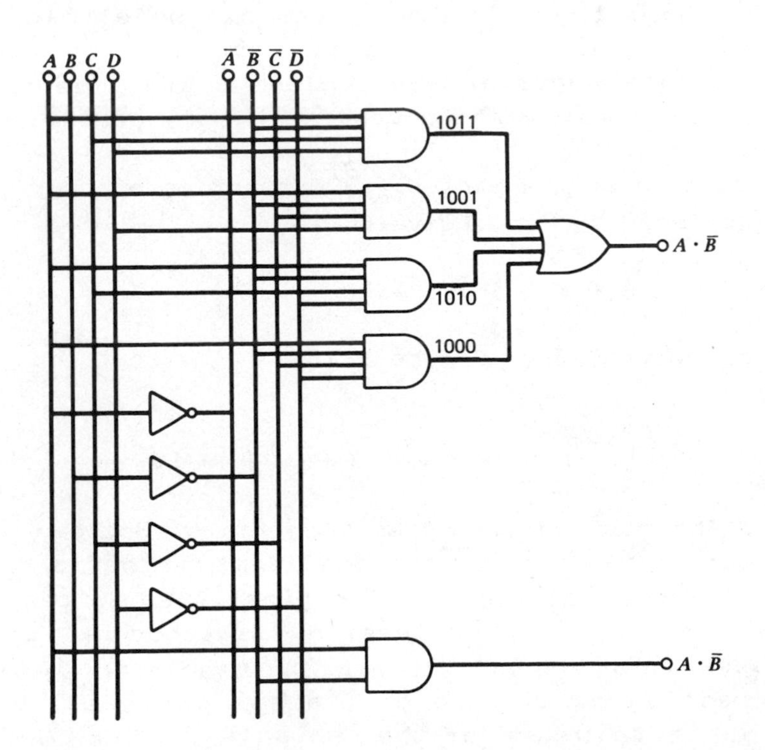

Figure 15-2. A direct implementation of Eq. 15-22. This is not a practical approach since the original function gives the same result with only one gate, as shown in the lower part of the figure.

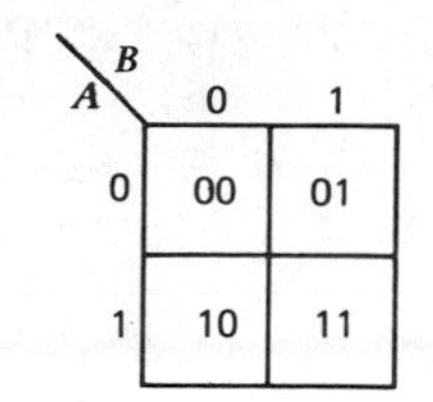

A \ BC	00	01	11	10
0	000	001	011	010
1	100	101	111	110

AB \ CD	00	01	11	10
00	0000	0001	0011	0010
01	0100	0101	0111	0110
11	1100	1101	1111	1110
10	1000	1001	1011	1010

Figure 15-3. Karnaugh maps for two, three, and four variables. Higher numbers of variables are difficult to handle.

ded to $A \cdot \overline{B} \cdot C + A \cdot \overline{B} \cdot \overline{C}$. If there were four variables in the expression, $A \cdot B$ would become $A \cdot B \cdot C \cdot D + A \cdot \overline{B} \cdot \overline{C} \cdot D + A \cdot \overline{B} \cdot C \cdot \overline{D} + A \cdot \overline{B} \cdot \overline{C} \cdot \overline{D}$. Each of these terms in the standard form is called a *minterm*. The advantage of using the canonical forms is that the minterms can be represented by unique binary numbers, by assigning digits "1" and "0" to the letters in sequence. An unbarred letter is given the digit "1," and a barred letter, the digit "0." For example, the four-variable minterms given above would be designated as

$$\begin{matrix} A\cdot\overline{B}\cdot C\cdot D & + & A\cdot\overline{B}\cdot\overline{C}\cdot D & + & A\cdot\overline{B}\cdot C\cdot\overline{D} & + & A\cdot\overline{B}\cdot\overline{C}\cdot\overline{D} \\ 1\ 0\ 1\ 1 & & 1\ 0\ 0\ 1 & & 1\ 0\ 1\ 0 & & 1\ 0\ 0\ 0 \end{matrix} \tag{15-22}$$

The electronic significance of the minterms is that they represent all possible input cases that give a HIGH output. A direct implementation with logic gates is shown in Figure 15-2. The total possible number of minterms is 2^n, where n is the number of variables. Of these Eq. 15-22 indicates those that give a true value at the output for the given function. Consequently it is desirable to form diagrams (called *Karnaugh maps* or *K-maps*) that provide numbered spaces for all possible minterms (Figure 15-3). Ones and zeros are to be inserted depending on whether or not a minterm appears in the expression.

The numbers down the left side are the binary digits corresponding to the first one or two characters of the minterms, whereas those across the top are the digits for the remaining characters. The sequences of binary numbers along both edges follow a Gray code; that is, only a single digit changes at a time.

The procedure for constructing a map is best explained with the aid of a few examples. Consider first the expression

$$X = \overline{A} + A\cdot\overline{B}$$

This can be expressed as the sequence of minterms:

$$\begin{matrix} X = & \overline{A}\cdot B & + & \overline{A}\cdot\overline{B} & + & A\cdot\overline{B} \\ & 0\ 1 & & 0\ 0 & & 1\ 0 \end{matrix}$$

The map is filled in with ones in the squares corresponding to minterms and zeros in the remainder, as in Figure 15-4*a*.

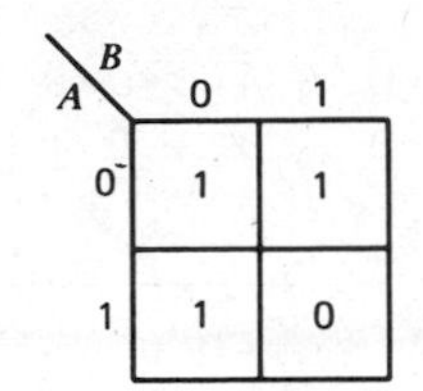

A \ BC	00	01	11	10
0	0	0	1	1
1	1	0	0	1

AB \ CD	00	01	11	10
00	0	0	0	1
01	1	1	0	0
11	1	1	1	0
10	0	0	1	1

Figure 15-4. Karnaugh maps of the statements given in the text.

A three-variable expression for which the minterms are

$$\overline{A}\cdot B\cdot C + \overline{A}\cdot B\cdot \overline{C} + A\cdot \overline{B}\cdot \overline{C} + A\cdot B\cdot \overline{C}$$
$$0\ 1\ 1 \quad 0\ 1\ 0 \quad 1\ 0\ 0 \quad 1\ 1\ 0$$

gives the map shown in Figure 15-4*b*.

An example requiring a four-variable map is the following Given the minterms

$$\overline{A}\cdot\overline{B}\cdot C\cdot\overline{D} + \overline{A}\cdot B\cdot\overline{C}\cdot\overline{D} + \overline{A}\cdot B\cdot\overline{C}\cdot D + A\cdot B\cdot\overline{C}\cdot D$$
$$0010 \quad 0100 \quad 0101 \quad 1101$$

$$+ A\cdot B\cdot\overline{C}\cdot D + A\cdot B\cdot C\cdot D + A\cdot\overline{B}\cdot C\cdot D + A\cdot\overline{B}\cdot C\cdot\overline{D}$$
$$1101 \quad 1111 \quad 1011 \quad 1010$$

The 16-place map is given in Figure 15-4*c*.

Interpretation of Karnaugh maps involves the generation of a new set of logical terms to express the original function in a simpler form. This is done by identifying adjacent pairs, quads, or octets of ones, forming "blocks." For this purpose, the term "adjacency" means juxtaposition in a vertical or horizontal sense, not diagonal, and permits "scrolling," which means that cells 0010 and 1010, for instance, are considered adjacent (Figure 15-5). Any two cells are adjacent if their binary designations differ only in a single digit. All cells occupied by ones must be included in at least one block.

Each block (pair, quad, or octet) contributes one term to a resultant expression. Its value is determined by observing which digits (*A*, *B*, *C*, or *D*) remain unchanged from one cell to another within the block;

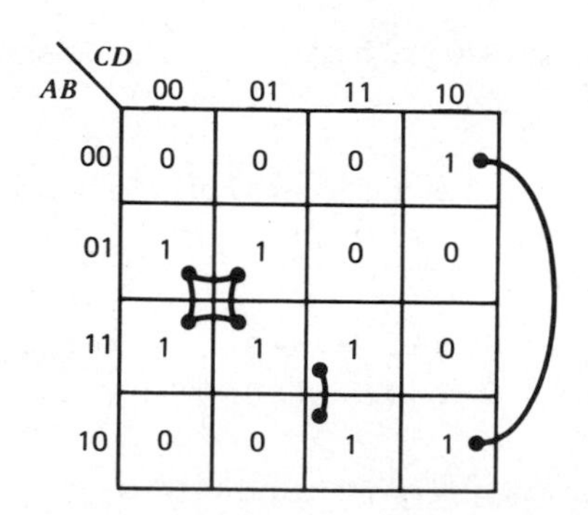

Figure 15-5. Selection of blocks in a Karnaugh map corresponding to the example presented in the text.

such digits are placed in the new term. For example the terms in the block mentioned in the preceding paragraph (0010, 1010) have in common the values of the bits B, C, and D; hence we can write a term corresponding to this block in the form $\overline{B} \cdot C \cdot \overline{D}$. There will thus be as many terms as there are blocks marked out on the map. As will be seen, each two-square block uses one less variable then the set, each four-square block uses two less, and an eight-square block three less.

To see how this works out, let us decipher the four-variable map given previously. It has been noted that square 0010 can be matched only with 1010, so we will make this the first block, with the value $\overline{B} \cdot C \cdot \overline{D}$. Cells 0100, 0101, 1100, and 1101 constitute a quad, making our second block. The cells in the quad share B (1s) and C (0s), to give a term $B \cdot \overline{C}$. Cells 1011 and 1111 then pair to form the third block. This gives a term $A \cdot C \cdot D$ since those digits are shared, all with ones. Thus the complete expression is

$$X = B \cdot C \cdot \overline{D} + B \cdot \overline{C} + A \cdot C \cdot D$$

It is often possible to assign blocks in different, equally correct, manners. This will give more than one resulting expression, all equally valid. This process, first reducing the given expression to its standard form (minterms), plotting a map, and then interpreting it, is useful as a means for simplification of a logic system.

PROBLEMS

15-1. Show that $A \cdot B + B \cdot C + \overline{A} \cdot C = A \cdot B + \overline{A} \cdot C$.

15-2. In the expression $Q = A \cdot \overline{B} + A \cdot \overline{C} + B \cdot C$, show whether Q is a "1" or a "0" under each of the following conditions:

(a) $A = 1, \quad B = 0, \quad C = 1$
(b) $A = 0, \quad B = 0, \quad C = 0$
(c) $A = 0, \quad B = 1, \quad C = 1$

15-3. In the expression $Q = \overline{A \cdot B + \overline{C}}$, determine the value of Q under each of the following conditions:

(a) $A = B = C = 0$
(b) $A = B = C = 1$
(c) $A = 0 \quad B = C = 1$

PART TWO

ANSWERS TO SELECTED PROBLEMS

2-2. (a) The transfer coefficient is 0.5 V/0.001 A = 500 Ω.

2-3. (e) $2 \angle 45^\circ$. The quadrant is given by the relative signs of x and y, indicating in this case the fourth quadrant.

2-4. (d) $(86.6 - j50)$ V.

2-5. (a) $400 \angle 60^\circ$ Ω. (b) 18 W.

2-8. (a) The square of the instantaneous voltage (E^2) is always 36; hence E_{RMS} is 6 V. (The frequency is irrelevant.) (b) In the triangle wave, the *average* amplitude is ±3 V; hence E_{RMS} = 3 V.

2-10. (a) $f = \omega/2\pi = 38.2$ Hz.

(b) $\phi = \phi_1 - \phi_2 = 90^\circ = \pi$ radians (current leading).

(c) $E_{pp} = 2(10) = 20$ V; $E_{RMS} = (10)(0.71) = 7.1$ V; $I_{pp} = 2(0.1) = 0.2$ A; $I_{RMS} = (0.1)(0.71) = 0.071$ A.

(d) $Z = 7.1/0.071 = 100\ \Omega$.

(e) $P = EI \cos(\phi_1 - \phi_2) = (7.1)(0.071) \cos 90^\circ = (0.5)(0) = 0$. The power factor is zero.

2-11. (a) $20 \log (0.01) = -40$ dB.

2-14. NEP = (noise output)/(gain) = $(10^{-3})^2/(1000)(100) = 10^{-11}$ W.

2-15. (b) $S/N = (0.1\ V)/(10^{-4} \sqrt{10}) \cong 320$.

CHAPTER 3

3-1. (a) Since $P = E^2/R$, we can write $E = \sqrt{RP}$. Hence $E_{max} = [(100)(2)]^{1/2} = 14.1$ V, or, with a reasonable safety factor, 10 V.

3-2. Each resistor can still dissipate 0.5 W, so the string can handle 5 W; $E_{max} = (5\times10^7)^{1/2} \cong 7000$ V.

3-3. $E_{max} = [(10^5)(5)]^{1/2} \cong 700$ V.

3-4. 18.5 kΩ.

3-5. (b) $10 + (5)(20)/25 = 14$ kΩ.
(c) $20 \log (15/4.3) = 10.8$ dB.

3-6. (a) 60 Ω.
(b) $Z_C = 1/(2\pi fC) \cong 9.64 \times 10^{4}$
$Z = RZ_C/(R + Z_C) \cong 9100\ \Omega$

3-7. $f = 1/(2\pi\sqrt{LC})$.

3-9. One possible choice is: $R_{in} = 15$ kΩ, $R_{out} = 15$ Ω.

3-10. $Z = R + 2\pi fH = 10{,}000 + 12.6 \cong 10{,}013\ \Omega$.

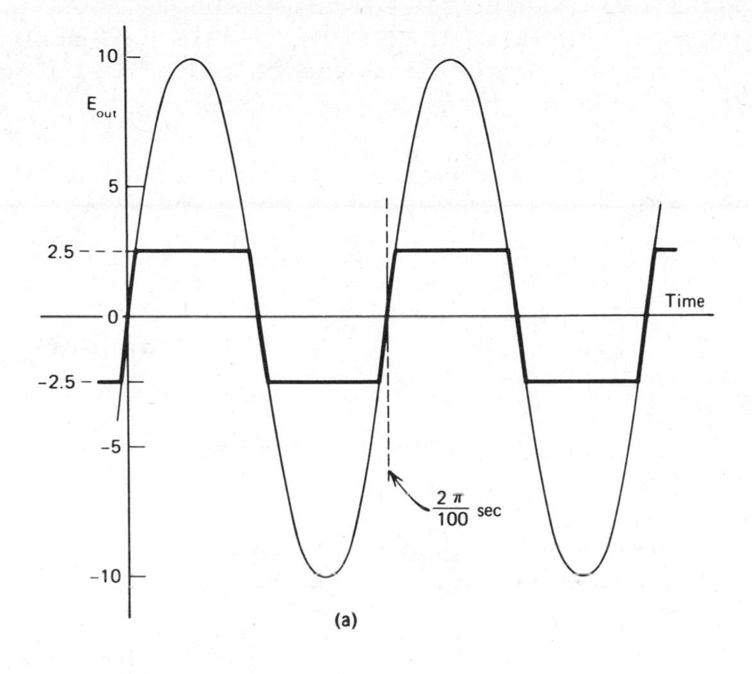

(a)

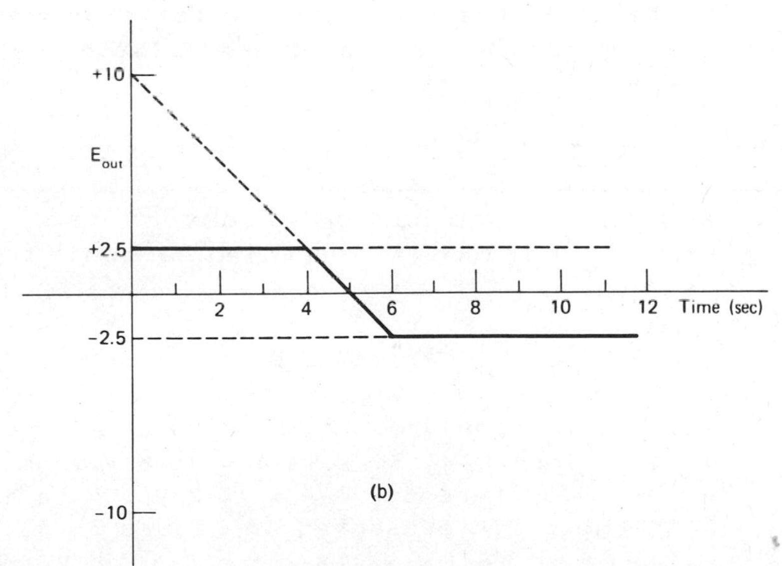

(b)

Figure P 3-12.

3-11. $f = 1/(2\pi\sqrt{LC}) \cong 6500$ Hz. The resistance value does not enter into the calculation but has some effect on the sharpness of tuning.

3-12. (a) Both diodes are reverse biased, so $E_{out} = E_{in} = 1$ V.

(b) Diode D_2 is forward biased, so the voltage across it is approximately −0.5 V, and $E_{out} \cong -2.0 - 0.5 = -2.5$ V. The drop across R is $10 - 2.5 = 7.5$ V.

(c) The sine wave is clipped at ± 2.5 V (Figure P3-12*a*).

(d) A ramp appears between +2.5 and −2.5 V (Figure P3-12*b*).

3-15. The slope of 40 dB/decade means that the ratio of voltages is the square of the ratio of frequencies. Hence the second harmonic (at 4 kHz) corresponds to

$$E_{out}/E_{ref} = (3000/4000)^2 = 0.56$$

and the second harmonic content is $(0.56) \times (2/10) = 0.11$ or 11%. The third harmonic content is 2.5%.

3-16. The S/N ratio is improved by a factor of 377.

3-19. (a) The voltage drop in R_C is 0.012 A × 1200 Ω = 14.4 V. Thus $V_C = 30 - 14.4 = 15.6$ V. And $V_b = 1.4$ V. By difference, $V_{CE} = 14.2$ V. This agrees with the graph presented in Figure 3-31*a*. The value of V_B is $1.4 + V_{BE} = 1.4 + 0.6 = 2.0$ V.

CHAPTER 4

4-1. (*c*) The circuit of Figure P 4-1, with $R_2/R_1 = 20$, will accomplish this task, provided that both inputs are positive.

4-6. For Figure 4-8*b*: $E_{out} = -E_{in} \cdot kR/R = -kE_{in}$.

4-7. $E_{out} = 1.0$ V.

4-8. It is convenient to establish an intermediate potential at point *E* in Figure 4-46. Then,

$$E = -\frac{1}{(10^6)(10^{-6})} \int 2dt = -2t; \qquad \frac{dE}{dt} = -2$$

$$E_{out} = -(2 \times 10^6)(2 \times 10^{-6}) \frac{dE}{dt} = -4 \frac{dE}{dt} = +8 \text{ V}$$

Note that the output is independent of time, until the amplifier saturates.

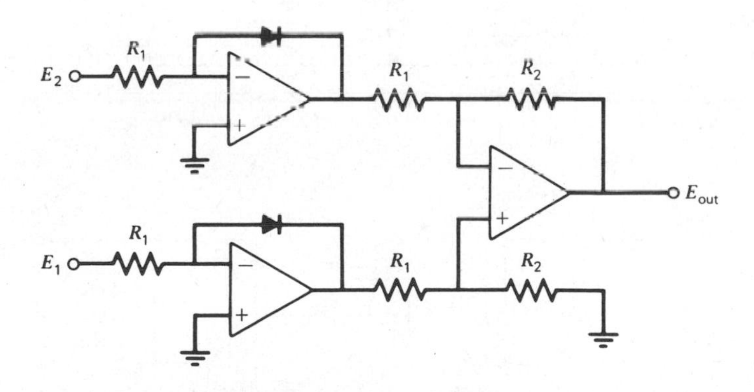

Figure P 4-1.

CHAPTER 5

5-15. (a) $dx/dt = -(1/RC)x$, *hence* $x = A \exp(-t/RC)$.

(b) $dx/dt = +(1/RC)x$, *hence* $x = A \exp(+t/RC)$.

5-16. See Figure P5-16. In *b*, note that since $E_2 = x$, it follows that $6 - E_2 \times (0.531) \times (10^6)/(5)(10^4) = 6 - (10.62)x$.

5-17. The general solution is $x = A \sin(\omega t + \phi)$, and $dx/dt = A\omega \cos(\omega t + \phi)$. For each case, substitute 0 for t (initial conditions refer to time 0). Note that the sine function is 0 for zero angle, where the cosine has its maximum amplitude.

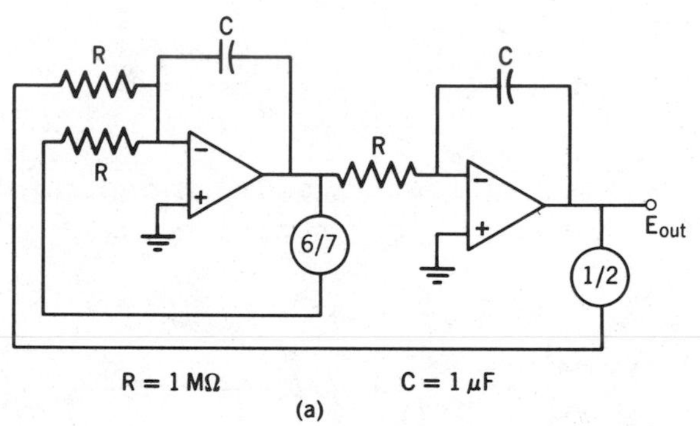

(a)

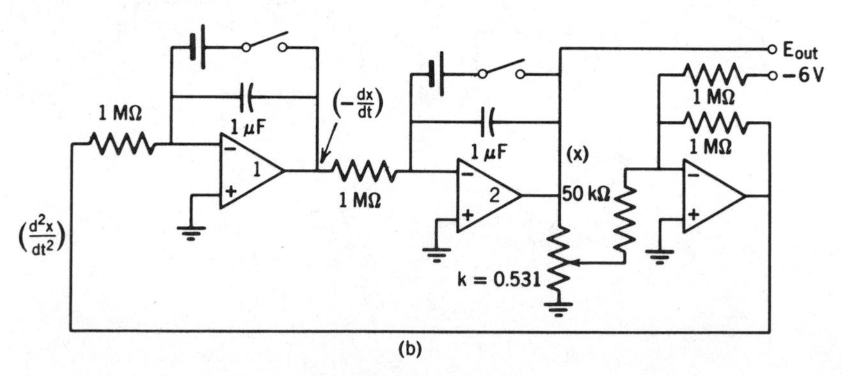

(b)

Figure P 5-16.

(a) $A\omega \cos (0 + \phi) = 0$ and $A \sin (0 + \phi) = 2$ V; consequently, $A = 2$ V, and $\phi = 90°$.

(d) In this case the two equations are compatible only for $A = 0$, and hence ϕ is not defined.

5-18. When the attenuation at R_Q is zero, the frequencies generated in the two loops are equal, and if the amplitudes and phases are identical, $E_4 = 0$. If either differs, a sine wave of the same frequency will appear. As the attenuation at R_Q is increased, the frequency of the lower loop decreases, and a mixed signal appears at the output:

$$-A \sin (\omega_1 t + \phi_1) + B \sin (\omega_2 t + \phi_2)$$

When the attenuation is complete, P' is grounded, and the B signal disappears, the output remaining at $-A \sin (\omega_1 t + \phi_1)$.

CHAPTER 6

6-1. (a) ($6\sqrt{2} \angle 45°$); (b) ($6\sqrt{2} \angle 135°$); (g) angle undefined.

6-2. (a) $(5.0 + j8.7)$; (d) $(10 + j0)$.

6-3. (a) $(27 + j0)$; (b) $(2 + j2)$; (c) $(-54 + j54)$; (d) $(3 - j4)$.

6-5. Parts (c) and (d) of problem 6-1 are the complex conjugates of (a) and (b), respectively.

6-6. (a) 1592 Hz.

6-9. (a) $E_{Th} = E_{in} \cdot R_2/(R_1 + R_2)$; $I_{sc} = E_{in}/R_1$, where I_{sc} is the short-circuit current; $Z_{Th} = E_{Th}/I_{sc} = R_1R_2/(R_1 + R_2)$, which is seen to be the parallel resistance of R_1 and R_2.

6-11. (a) The input resistance, given by the series combination of R_1 with the parallel system of R_2 with R_3 and R_L, is R_{in} = 600 Ω. The input current splits between R_2 and $(R_3 + R_L)$ in inverse ratio to the resistances. Since $I_2 = I_{in} - I_L$, we can write $I_L = 4.50\ I_2 = 4.50\ (I_{in} - I_L)$, from which $I_L/I_{in} = 0.818$. This corresponds to −1.7 dB. The voltage attenuation, given by E_{out}/E_{in}, is also 0.818, again equivalent to −1.7 dB. The power attenuation is −1.7 dB. (Note that the equality of decibel ratios is not general, but results from the equality of input and output impedances.)

(b) Suppose that R_L is replaced by another similar three-resistor network and a 600-Ω terminating load. Since the new unit has an impedance of 600 Ω, it can replace the former 600-Ω load without change. An observer at the input cannot determine by electrical measurements whether the load is connected directly or through any number of similar attenuators. An observer at the load sees voltage, power, and current attenuations of −1.7 dB per unit. Analogous networks with input impedance independent of length, especially those designed for high-frequency signals, are known as *transmission lines*.

CHAPTER 7

7-1. (a) 0.1 mA; (b) 20.0 mA; (c) 5.0 V; (d) 20.

7-2. (a) 23 V; (b) 0.2 MΩ; (c) 2.5 MΩ.

7-3.

$$\frac{E_{out}}{V_S} = \frac{\beta R_L}{R_S + R_{in}}$$

7-15. The frequency f = 2.55 kHz.

7-18. See Figure P7-18. In (b), the average output is zero if the zero-crossing of the signal coincides with a transition of the chopper signal. Any second harmonic present in the 50-Hz wave will not be averaged out and will appear at the output. The circuit is useful as a sensor of harmonic distortion.

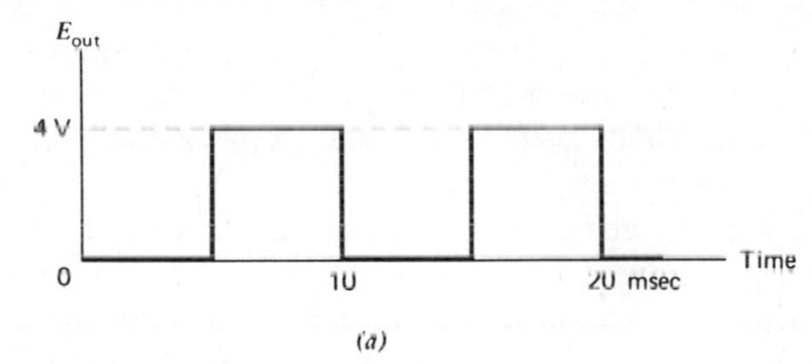

(a)

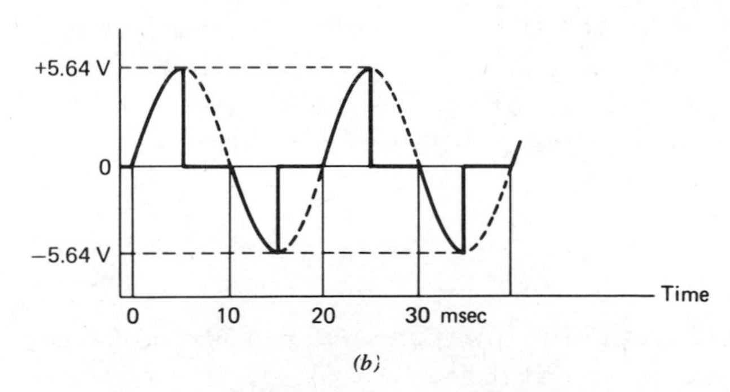

(b)

Figure P 7-18.

CHAPTER 8

8-2. When E_{in} is positive, the diode is turned ON, and $E_{out} = E_{in}(1000/1001)$, so $E_{out}/E_{in} = 0.999$. When the diode is OFF, $E_{out} = E_{in}(1000/1001000)$, and $E_{out}/E_{in} = 0.001$.

8-3. $I_R = (20 - 10.5)/100 = 0.095$ A, and $P_R = (0.095)(9.5) = 0.90$ W. The maximum transistor current occurs when the power supply is not loaded, and the power dissipation of the transistor and Zener diode is $P = (0.095)(10.5) = 1.0$ W. The dissipation is distributed between these two components in the ratio $P_T/P_Z \cong \beta$, a large number. Thus the power requirement of the Zener is negligible.

8-4. The drop across R is $15.0 - 12.0 = 3.0$ V. The maximum load current $I_{L,max}$ is $12.0/100 = 0.120$ A. The minimum, $I_{L,min}$, is $12.0/2000 = 6.0$ mA. At maximum load current, the Zener current is a minimum; let it be $I_{z,min} = 10$ mA. At minimum load current, the Zener current is a maximum, $I_{z,max} = I_{z,min} + I_{L,max} - I_{L,min} = 0.010 + 0.120 - 0.006 = 0.124$ A. The maximum Zener dissipation is $I_{z,max} \cdot E_Z = (0.124)(12) = 1.49$ W. (A 2-W resistor should be selected.) The value of R must be $3.0/(I_{z,min} + I_{L,max}) = 3.0/0.13 = 23\ \Omega$. The power rating of R should be $P = EI = (3)(0.13) = 0.39$ W. (Use a 0.5-watt or larger.)

CHAPTER 9

9-2. The output is low for $ABC = 100$, and high for all other combinations.

9-3. The circuit can be implemented with nine gates.

9-11. (a) Three equivalent forms are shown in Figure P9-11*a*.

(c) In all cases except the last (111), the **NAND** gate gives a "1," so that the output is equal to the value of *C* (Figure P 9-11*c*).

9-12. See Figure P9-12.

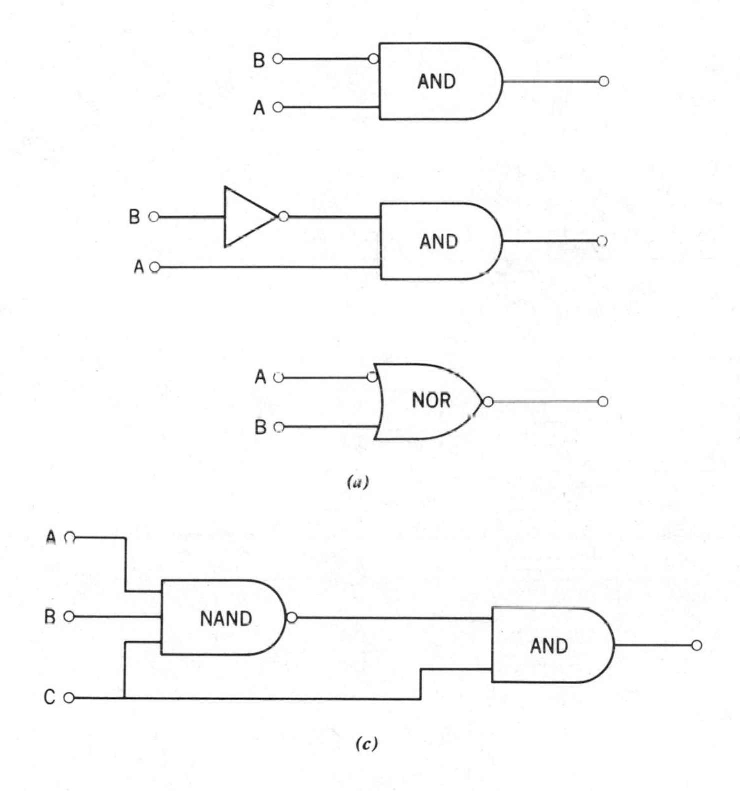

Figure P 9-11.

9-13. (a)

A	*B*	*E(out)*
0	0	1
0	1	0
1	0	1
1	1	0

(b)

A	*B*	*C*	*E(out)*
0	0	0	0
0	0	1	1
0	1	0	1
0	1	1	1
1	0	0	0
1	0	1	1
1	1	0	0
1	1	1	0

9-16. (a) 109; (b) 40.

9-17. (a) 100111; (c) 1110111.

9-18. (a) 0011 1001; (c) 0001 0001 1001.

9-19. See Figure P9-19.

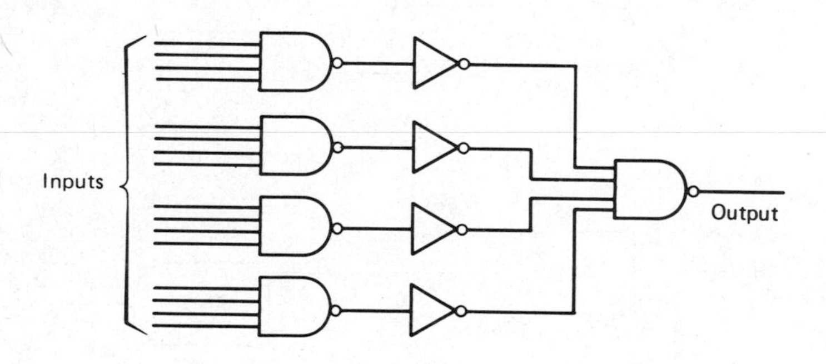

Figure P 9-12.

9-20.

A	B	C	D
0	0	0	0
0	0	1	1
0	1	0	1
0	1	1	0
1	0	0	1
1	0	1	0
1	1	0	0
1	1	1	1

Note that D is "1" if the number of ones at the inputs is odd and "0" if it is even. This circuit, called a *parity check*, is used to identify errors since most errors change only a single bit, hence causing a change in the number of ones.

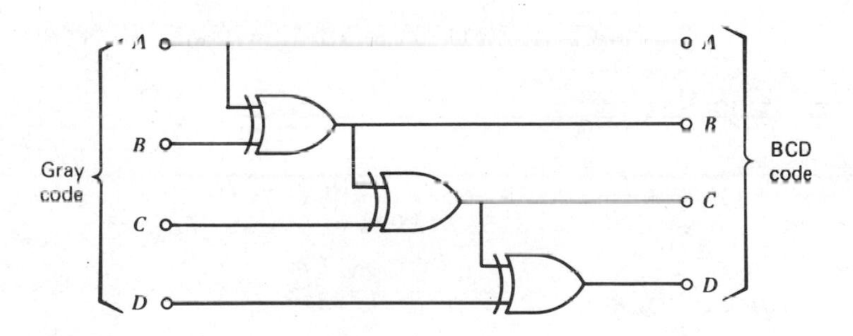

Figure P 9-19.

9-21.

A	B	Carry	Sum
0	0	0	0
0	1	0	1
1	0	0	1
1	1	1	0

This circuit is called a 2-bit "half-adder." It performs the addition operation with a "carry" output to permit multibit addition.

CHAPTER 10

10-2. (a) f = 320 Hz; duty cycle = 1/3.

(b) f = 160 Hz; duty cycle = 1/2.

CHAPTER 11

11-1. 200 Hz.

11-3. $E_{in} = E_{ref} \cdot (t_2 - t_1)/(t_1 - t_0)$
$= (10)(0.015)/(10) = 0.015$ V.

CHAPTER 13

13-1. The platinum would change by (0.003)(10,000) = 30 Ω, while for the nickel, (0.006)(1000) = 6 Ω.

CHAPTER 15

15-2. (a) $Q = 1$.

15-3. (b) $Q = 0$.

LABORATORY EXPERIMENTS

The following experiments are designed to accompany study of the text. The availability of a few standard items of test equipment is assumed. A digital multimeter (3 1/2-digit accuracy), a variable audiofrequency oscillator, and an oscilloscope are essential. A strip-chart recorder is highly desirable. By far the best way to construct experimental circuits is on a prototyping board with built-in power supplies to give +5, +15, and -15 V of regulated DC. These are available from E & L Instruments, Inc., 61 First St., Derby, CT 06418, Global Specialties, 70 Fulton Terrace, New Haven CT 06512, AP Products, Inc., Box 540, Mentor, OH 44060, and RSP Electronics Corp., P. O. Box 699, Branford, CT 06405.

EXPERIMENT 1
Capacitors

Objective: To study the laws governing combinations of capacitors.

Procedure

(1) Assemble the circuit shown in Figure Ex-1, in which *E* is a 9-V transistor battery or equivalent and *V*

is an electronic voltmeter with at least 10 MΩ input impedance.

(2) To start the experiment, open switch S_1, and close S_2 momentarily to discharge the capacitor. Then close S_1, simultaneously starting a timer. Record the voltage every 10 s until it becomes stable. (A strip-chart recorder can be used here to advantage.) Plot the readings as a function of time, and determine the RC time constant of the circuit by finding the time needed for the capacitor to attain 63% of its final voltage.

(3) Measure the value of R_1 (separated from the circuit) with an ohmmeter. Calculate the value of C and compare it with the labeled value. Comment on the closeness of agreement. Estimate the precision of the measurement in terms of possible errors in R_1, the time, and the measured voltages. What merits do you see in this method of measuring capacitance? Why is it important to use a high input-impedance voltmeter?

Additional Experiments

Electrolytic capacitors have "memory" effects. For example, when such a capacitor is discharged by

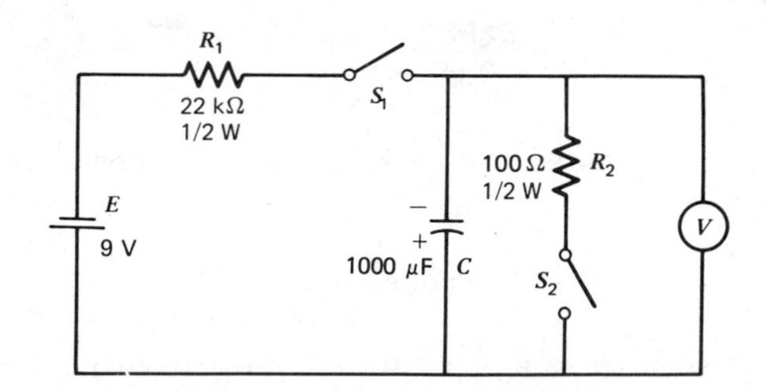

Figure Ex-1. Circuit for studying some properties of a capacitor.

momentarily short-circuiting its terminals, a lesser charge appears spontaneously. An enterprising student can devise a suitable procedure to measure this effect.

EXPERIMENT 2

Shunt and Series Resistors for a Meter.

Objective: To utilize the concepts of series and parallel combinations of resistors, and to match resistors to desired values.

Procedure

(1) Design measuring instruments similar to those

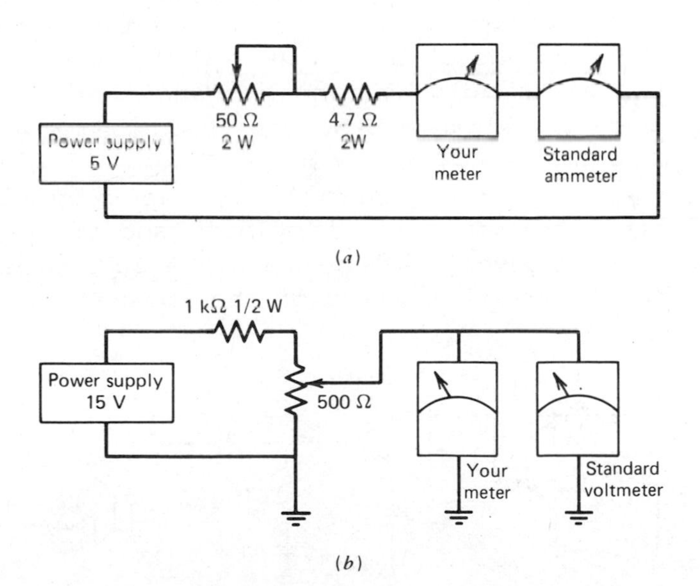

Figure Ex-2. Comparison circuits for calibrating meters by reference to standard meters: (*a*) for ammeters; (*b*) for voltmeters.

in Figure 3-7, using a 10-mV meter movement. Calculate the necessary resistors for a voltmeter with ranges of 0.01, 0.1, 1, and 10 V and an ammeter with 100-mA and 1-A ranges.

(2) Calibrate your two meters against commercial units, using the connections shown in Figure Ex-2.

EXPERIMENT 3

Amplification

<u>Objective</u>: To illustrate amplification of signals.

Procedure

(1) Construct the circuit diagrammed in Figure Ex-3. The pin numbers correspond to the type LM384, but other models can be used. A heat sink such as a piece of copper foil should be connected to pins 3, 4, 5, 7, 10, 11, and 12. The manufacturer recommends a "Staver V-7" heat sink for the purpose.

(2) Determine the gain in decibels by means of oscilloscope measurements at point *A* and at the output for various frequencies, and prepare a Bode plot. Use a 5-W, 8-Ω resistor in place of the speaker.

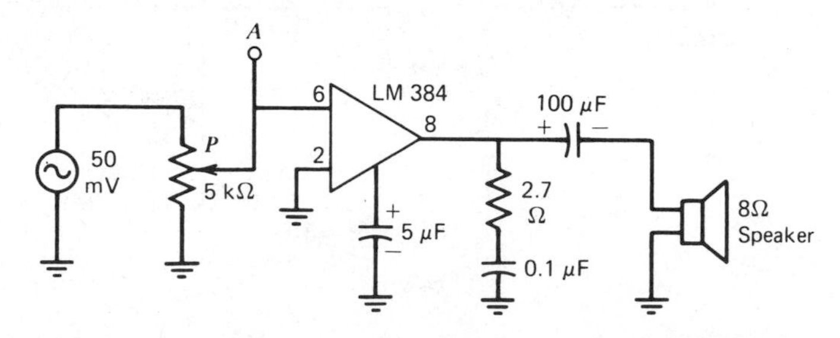

Figure Ex-3. Circuit for measuring some properties of an amplifier (pin 14: 15 to 22 V; pins 3, 4, 5, 7, 10, 11, and 12: grounded; pin 9: no connection).

(3) Connect the speaker, and determine the frequency range over which your ear is sensitive.

(4) Adjust the potentiometer *P* to give two gain settings such that the sound seems to you to be twice as loud with one as with the other. Express the ratio of the two in terms of decibels.

EXPERIMENT 4

AC Impedance

Objective: To illustrate the AC properties of capacitors and *RC* networks, including amplitude and phase relationships.

Procedure

(1) Study the instructions for the oscilloscope.

(2) Assemble the circuit shown in Figure Ex-4*a*. The resistors can be of low power (0.25-watt). Measure the amplitude of the AC wave at points *A* and *B*, both relative to ground, by estimating the peak-to-peak distance on the oscilloscope screen. The value at *B* should be just half as great as that at *A*. If a dual-trace scope is available, both signals can be displayed simultaneously, and both relative amplitudes and phase differences can be observed directly.

(3) Repeat the measurements obtained in (2), using an AC voltmeter in place of the oscilloscope. If the voltmeter is of the usual type, its measurements will be RMS. Show how this reading can be made compatible with the peak-to-peak readings from the scope.

(4) Now construct the circuit of Figure Ex-4*b*. Again measure the voltages at *A* and *B* with respect to ground, and then measure the voltage appearing across the resistor differentially, with scope or meter

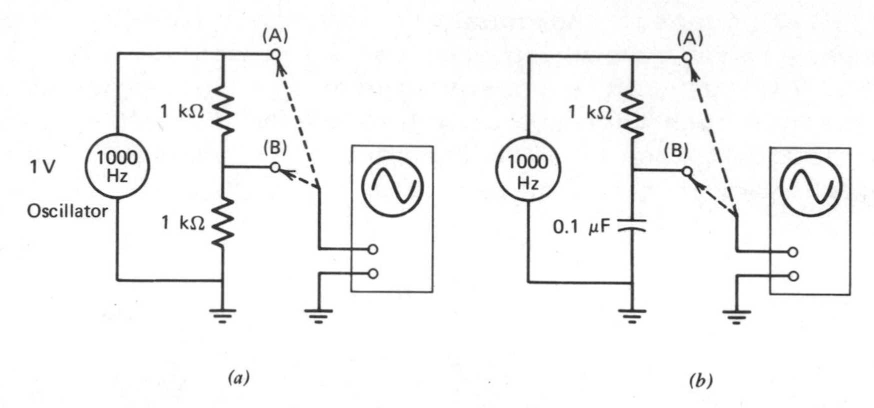

Figure Ex-4. Circuits for determining impedances at various frequencies.

connected directly to points *A* and *B*, with no ground contact. Do the observed voltages across the resistor and capacitor add to give the total applied voltage? If not, explain why not.

(5) With the dual-trace scope, determine the phase difference between the voltages appearing across *R* and *C*. Calculate the sum of these two voltages in complex notation, and discuss your results.

Additional Experiments

Perform similar studies of circuits containing inductors and resistors and of circuits with all three types of components included.

EXPERIMENT 5

RC Filters

Objective To illustrate the behavior of passive filters and to study component matching.

Procedure

(1) Assemble a low-pass filter with oscillator and oscilloscope as in Figure Ex-5*a*. Select components R and C such that f_0 = 1000 Hz, approximately. Keep in mind also that the impedances must be larger than about 1000 Ω to avoid loading the oscillator unduly, but less than about 100 kΩ to avoid making the circuit itself sensitive to loading. Note that at f_0, R and C have the same impedance. There are many permissible combinations of R and C that will give the same critical frequency. Because of the difficulty in finding capacitors of precisely known values, the following procedure is suggested. List several available resistors between 1000 Ω and 100 kΩ, and calculate for each the corresponding capacitance for the desired frequency. Select a capacitor from those available that is closest to one of your calculated values. Capacitors usually have large tolerances of ±10 or 20 percent, or even greater.

(2) Test the frequency response of your filter by determining the input and output amplitudes for various frequencies. Plot your results as a function of the logarithm of frequency (a Bode plot).

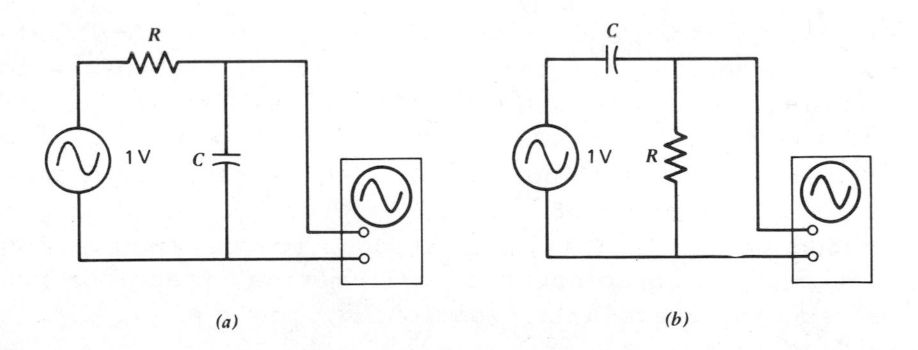

Figure Ex-5. Circuits for observing the response of passive filters.

(3) Repeat steps (1) and (2), using the high-pass circuit shown in Figure Ex-5*b*.

(4) With a dual-trace scope, determine the phase difference between input and output for each filter at various frequencies, and plot it against the frequency. Such plots are useful in predicting the dynamic behavior of filters.

Additional Experiments

Another area worth investigating is the effect of the filter on square and triangular waves. For this measurement, use a commercial generator of the appropriate waveforms. Note that for a low-pass filter beyond f_0, both square and triangular waves tend to become more like sine waves. Can you explain this?

EXPERIMENT 6

Thévenin Equivalent Circuits

Objective: To illustrate the formation of a Thévenin equivalent for purely resistive circuits.

Theory Consider a circuit containing one or more voltage sources, all of the same frequency, together with a variety of resistors, capacitors, and inductors. If outside connections are made at any two points, the circuit will appear indistinguishable externally from a series combination of a single source E_{Th} and a single impedance Z_{Th}. This is analogous to the observation that any sum or product of arithmetical fractions can be reduced to a single fraction.

Procedure

(1) In this experiment we use only a DC source and resistors. Construct the circuit shown in Figure Ex-6*a*. Use any resistors with values between 1000 Ω and 100 kΩ, and measure their exact values prior to assembly.

(2) Connect an ammeter across the output; its reading represents the short-circuit current I_{SC}, the maximum current obtainable from the network. Note from the Thévenin equivalent circuit shown in Figure Ex-6*b* that I_{SC} is given by the ratio E_{Th}/R_{Th}.

(3) With the help of an electronic voltmeter, measure the output voltage of the circuit. This "open-circuit" voltage is E_{Th}. From the values of I_{SC} and E_{Th}; calculate R_{Th}.

(4) The procedure for calculating the Thévenin quantities from known component values is described in the text. Make such a calculation for your circuit, and compare it with the measured values.

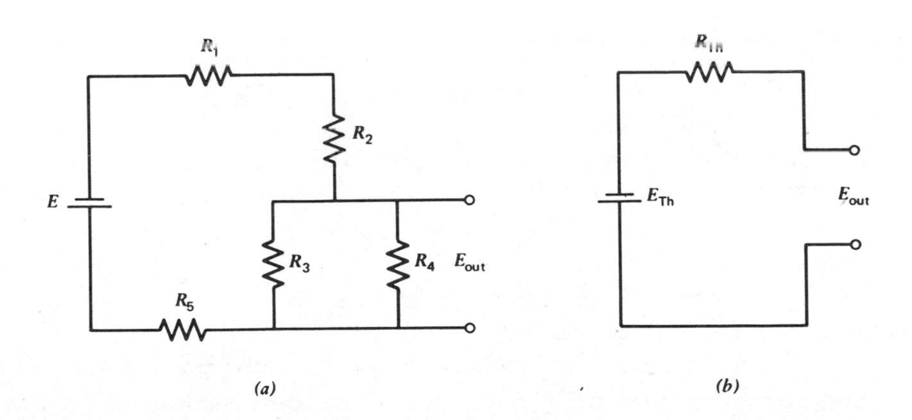

Figure Ex-6. Resistive circuits of which the Thévenin equivalents are to be determined.

Additional Experiments

Select any other two points in the circuit, and compare the theoretical and measured Thévenin quantities for them.

It is possible to measure R_{Th} by connecting a variable resistor at the output of the circuit and monitoring the voltage across it. When this voltage is equal to half of E_{Th}, the value of the resistance is equal to R_{Th}. Make this determination, and compare it with the value obtained by the short-circuit method. Which method seems to you most desirable, and why?

EXPERIMENT 7

Operational Amplifiers in the Open-Loop Mode

Objective: To examine the general behavior of open-loop operational amplifiers (op amps) and their limitations.

Procedure

Operational amplifiers are high-gain differential amplifiers of negligible input current. In this experiment, feedback will not be used, and hence the amplifiers will be at one of their voltage limits (i.e., saturated) most of the time, a direct result of the extremely high gain.

(1) A 741 op amp can be used for this experiment. A summary of some of its properties can be found in the text, and more details are available in the manufacturer's literature. The pin diagrams for the three common forms are given in Figure Ex-7*a*.

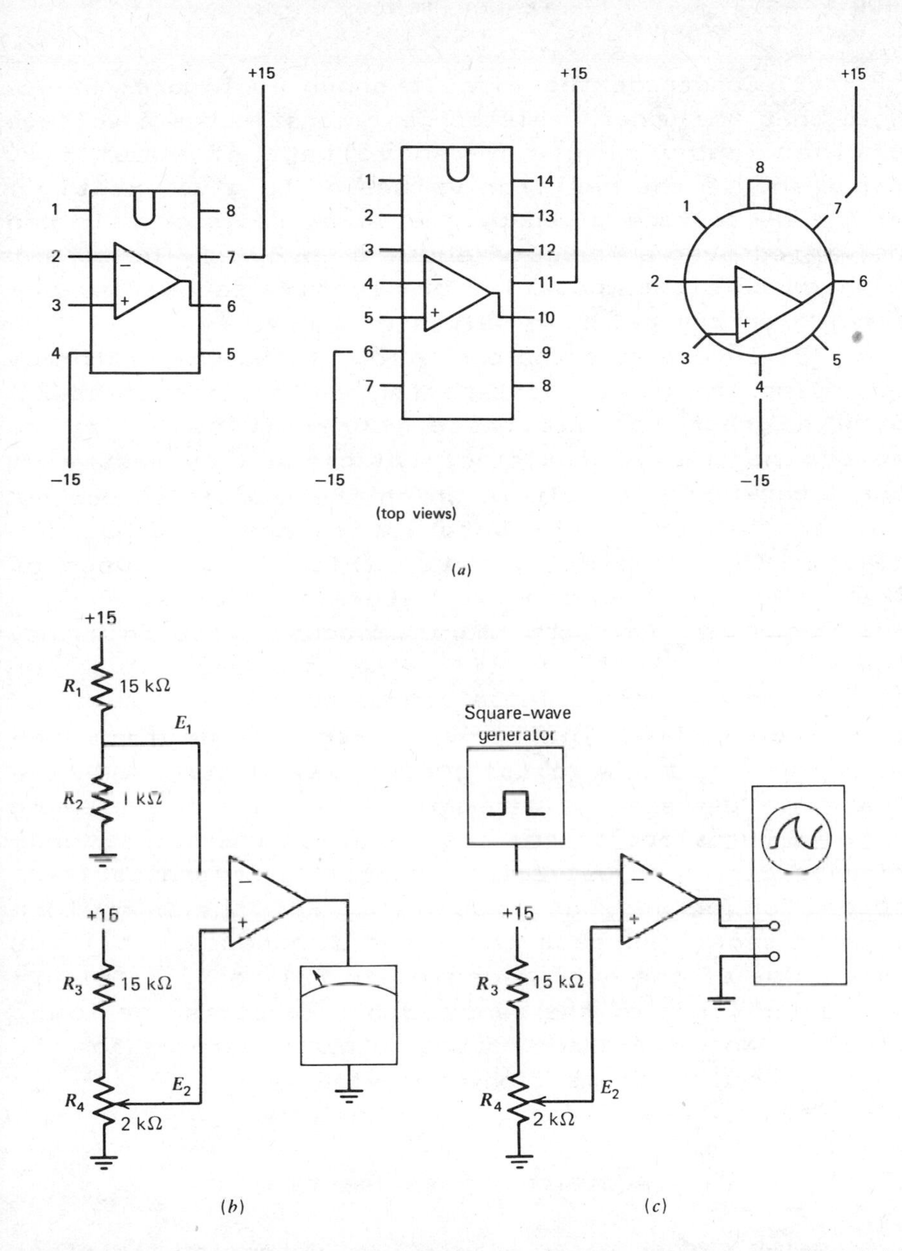

Figure Ex-7. (*a*) Pin diagrams of op amps of the 741 series. (*b* and *c*) Circuits for testing the open-loop characteristics of op amps.

(2) Construct the circuit shown in Figure Ex-7*b*. Note that the upper resistor pair constitutes a voltage divider, providing a fixed voltage of about 1 V. (What should the resistor values be to give exactly 1 V?) The voltage given by the lower resistor pair can be varied over a range of about 0 to 2 V. The op amp will go to either positive or negative saturation depending on the relative values of E_1 and E_2.

(3) Investigate the operation of the comparator by observing the output voltage for various inputs at E_2. Draw a graph to illustrate your results. Try to determine the resolution of the circuit by measuring the interval for E_2 within which the transition occurs.

(4) Two other characteristics are of great interest: (a) the slewing rate, which is the speed of transition between the two saturated states, and (b) the frequency limit, the maximum square-wave frequency for which the amplifier still attains output saturation at both polarities. These can be observed by means of the circuit shown in Figure Ex-7*c*. Using a square-wave input of a few volts, and with E_2 at zero, examine the scope patterns. Determine from them the slewing rate as the slope of the rise between the two saturation levels. Adjust both the oscillator and oscilloscope for convenient readout. For the bandwidth measurement, increase the input frequency until the amplitude of the output begins to fall off. This is the upper limit of the bandwidth. What is the lower limit? Note the shape of the output waveforms for the various conditions mentioned previously.

Additional Experiments

Using Figure 4-36 as a guide, determine the other dynamic parameters indicated: rise time, overshoot, and settling time. Consult the manufacturer's literature about specially optimized amplifiers for use as comparators. In what ways are they to be preferred over the 741 for this service?

EXPERIMENT 8

Operational Amplifiers with Feedback

Objective: To study two basic op amp circuits, the voltage follower and the integrator.

Procedure

(1) Construct the follower circuit shown in Figure Ex-8*a*, using a 741 op amp. Measure the voltages at points *A* and *B* with an electronic voltmeter. The results should be equal within a few millivolts. Now connect a 1000-Ω load from *B* to ground and read the two

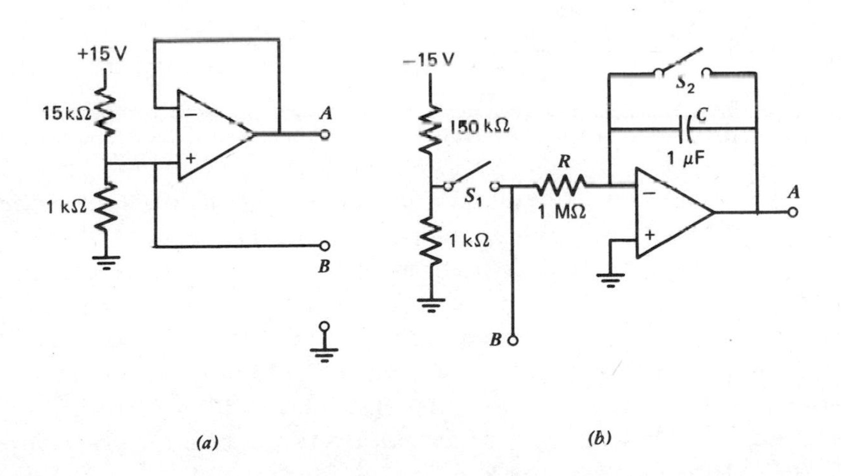

Figure Ex-8. Circuits for op amps with negative feedback (*a*) as a follower and (*b*) as an integrator.

voltages again. Note that the voltage at *B* is heavily affected, but not that at *A*. Explain this.

(2) Construct the integrator as in Figure Ex-8*b*. Connect a 10-V recorder or meter between the amplifier output and ground. Measure the voltage at point *B* with an electronic voltmeter. The output at *A* should be zero when switch S_2 is closed. With S_1 closed and S_2 open, the circuit will integrate the input as a function of time. Compare your observations with the theoretical slope of the output voltage against time.

(3) Open switch S_1. The circuit should retain its output voltage at a constant level. This is called the "hold" mode. The voltage will drop off slowly as a result of the leakage resistance of the capacitor. Measure this drift in your circuit by first integrating to approximately 1 V and then recording the output for a period of time with both switches open. Calculate the minimum voltage that can be held on the integrator without dropping more than 1% per minute. Also determine the longest time that a potential of 1 V can be held before the error becomes greater than 1%.

Additional Experiments

(1) Use the integrator as an instrument for measuring capacitance, and comment on its merits.

(2) Feed to the integrator a square wave of about 1 Hz. Examine the output waveform, and comment on the possible use of the integrator as a function generator.

(3) Determine the response of the integrator to sine waves of 0.1, 1, and 10 Hz. Could the integrator be used as a filter? What happens to the DC component of the signals, if any?

EXPERIMENT 9

An Active Band-Pass Filter

Objective: To design a filter to pass frequencies in the range 20 to 200 Hz.

Procedure

(1) Using the circuit shown in Figure 4-14, design and construct a band-pass filter with f_1 = 20 Hz and f_2 = 200 Hz.

(2) By means of a variable oscillator, plot the the attenutation in decibels as a function of the logarithm of frequency (the Bode diagram).

Additional Experiments

Insert your filter into a high-fidelity sound system, and note the effect on voice and music.

EXPERIMENT 10

A High-Precision Voltage Source

Objective: To construct a highly precise millivolt source.

Procedure

(1) Construct the circuit shown in Figure Ex-10. The OP-07 usually has an offset of less than 50 μV, so that the 20-kΩ trimming potentiometer may not be necessary.

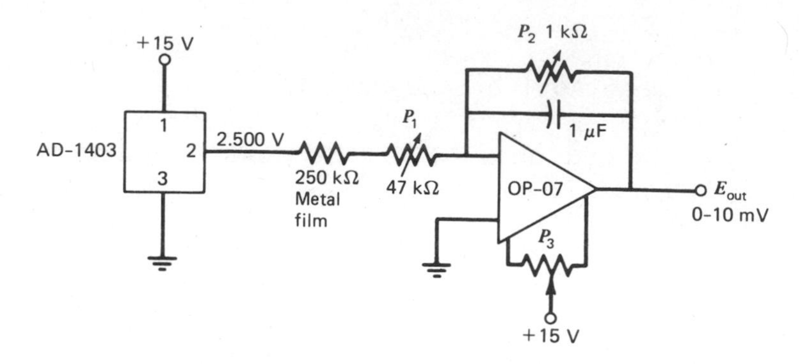

Figure Ex-10. Circuit for a precision adjustable voltage source. Potentiometers P_1 and P_3 should be multiturn trimmers; P_2 is a 10-turn precision type.

(2) Check your instrument against a digital voltmeter. If the resistors and potentiometer are of high quality, the output should be very precise.

(3) Determine the full-scale sensitivity of various recorders (1 or 10 mV). Express the result as percent deviation.

Additional Experiments

Determine the linearity of a strip-chart recorder by plotting the percent deviation betweeen applied and recorded voltages as a function of position on the scale.

EXPERIMENT 11

Low-Frequency, Average-AC Meter

Objective: To plot the average value of an AC signal as a function of time.

Procedure

(1) Design a precision rectifier system as in Figure 4-29, followed by a 1-Hz low-pass filter like that in Figure 4-17.

(2) Test the operation of your circuit with a sine-wave generator. The output should follow the amplitude of the AC, provided it does not change too rapidly.

(3) Determine the behavior at various frequencies, using an oscilloscope or AC voltmeter as a control.

(4) Plot the output with the aid of a recorder at various frequencies, such as 0.5, 1, 2, and 4 Hz. Interpret the results.

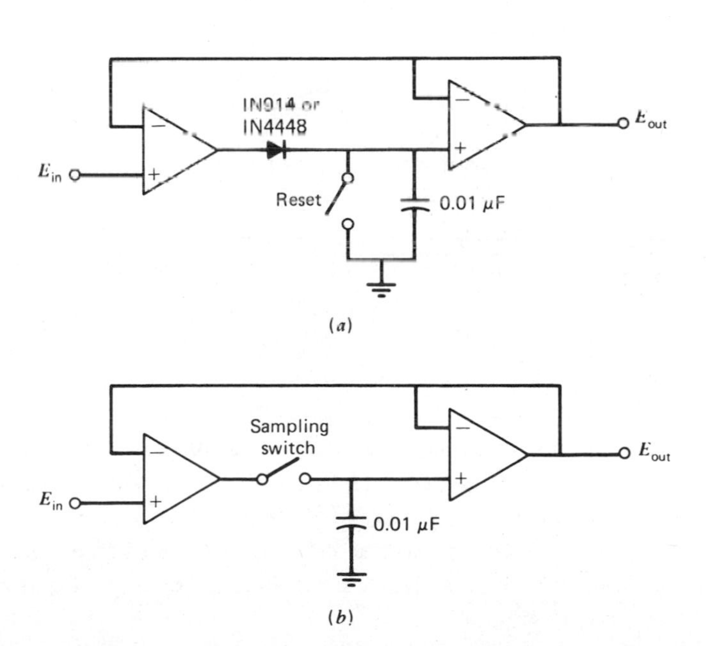

Figure Ex-12. (*a*) A peak detector. (*b*) A sample-and-hold amplifier. All op amps should be FET types.

EXPERIMENT 12

Peak Detection and Sample-and-Hold

Objective: To illustrate sampling techniques.

Procedure

(1) Construct the circuits shown in Figures Ex-12*a* and *b*.

(2) Check the operation with a variety of input signals (pulses, peaks, square waves, sine waves, etc.). A two-pen recorder can be used to advantage to show input and output waveforms simultaneously.

EXPERIMENT 13

Sine-Wave Oscillators

Objectives: To compare two types of oscillators and observe the interaction of two frequencies.

Procedure

(1) Design and construct a Wien bridge and a phase-shift oscillator (Figures 5-5 and 5-6), both tuned to give 1.0 kHz.

(2) Check the outputs of both oscillators with an oscilloscope. For each, determine the frequency and stability in time.

(3) Connect the two oscillators to the inputs of an analog multiplier (such as the AD533, shown in Figure 5-10), using high-pass filters as coupling

networks. Observe the output of the multiplier with the oscilloscope. Vary the setting of R_3 of the phase shift oscillator, and interpret your results.

EXPERIMENT 14

Analog Computation

Objective: To make a sine-wave oscillator by solving the differential equation of a sine wave.

Procedure

(1) Construct the circuit shown in Figure 5-15. The switch must be of the double-pole type to synchronize S_1 and S_2. For V_1 use a small (1.5-V) battery; replace V_2 by a short circuit. Select all resistors to be 1 MΩ and all capacitors 1 μF.

(2) Record the output as a function of time. If a multipen recorder or a multitrace oscilloscope is available, examine the outputs of all three amplifiers simultaneously and comment on the observed phase differences. Measure the amplitude stability if the oscillator is left in operation for a period of one hour.

EXPERIMENT 15

A Peak Integrator

Objective: To construct an instrument for integration of the area beneath a peak as seen in a graphical recorder.

Procedure

(1) Design the complete circuit shown in Figure 4-13, and connect it to a strip-chart recorder.

(2) Test this circuit with a pulse generator having a low repetition rate. Try using a low frequency sine or triangle wave (1 or 2 Hz), with and without a coupling capacitor. Interpret your results.

Additional Experiments

Connect your integrator to the output of a commercial gas chromatograph, and compare its operation with that of the built-in integrator.

EXPERIMENT 16

Fourier Expansions

Objective: To obtain physical evidence of the Fourier theorem.

Procedure

(1) With the circuit shown in Figure 4-16*c*, construct band-pass filters of 100.0, 200.0, 300.0, and 400.0 Hz. Match the components as carefully as you can. (Several students may pool their efforts to advantage here.)

(2) By means of the circuit in Figure Ex-16, determine the frequency at which the sine-wave oscillator is exactly in tune with the 100-Hz filter. Maintain this setting, and determine the harmonic components by switching S_2 so that the oscilloscope looks at 200,

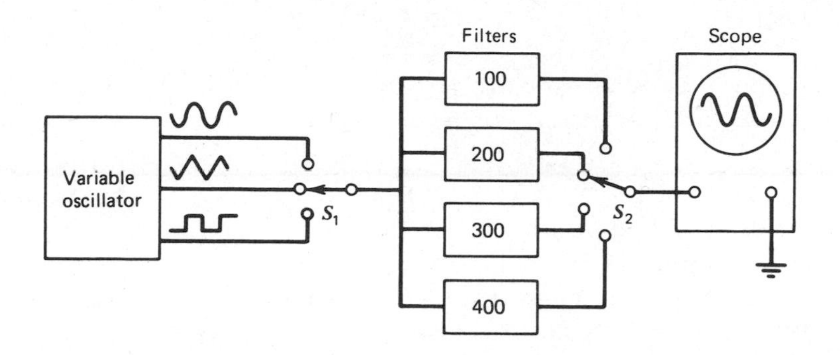

Figure Ex-16. Circuit for studying the Fourier components of alternating voltages of various waveforms. The oscillator can be made up from an 8038 IC, using the connections given in the manufacturer's literature.

300, and 400 Hz in turn. Determine the Fourier coefficients from the observed amplitudes (which are very small for the sine wave), and compare with theory.

(3) Repeat for triangle and square waves, without changing the oscillator setting.

Additional Experiments

If a pure sine wave from the oscillator is used as excitation, various devices, such as transistors, amplifiers, diodes, and filters, can be inserted in the circuit at point *S*, and the effect can be characterized in terms of Fourier series modifications.

EXPERIMENT 17

Power Supplies

Objectives To study a practical circuit employing transformer, diodes, and electrolytic capacitors.

Procedure

The circuit shown in Figure Ex-17 is to be constructed one section at a time.

(1) Assemble the power-line section, including the fuse, switch, and 25-V transformer. This should be approved by the instructor before connecting it to the line. After approval, plug the circuit into the line, and measure the transformer secondary voltage with an oscilloscope or AC voltmeter. It should be a sine wave with a peak-to-peak voltage of $2 \times 25 \times \sqrt{2} = 71$ V, as observed on the scope, or 25 V RMS on the meter. If the voltage differs appreciably from this, it may be due to a nonstandard line voltage or to the tolerance of the transformer. Check that the center tap (CT) does indeed divide the voltage into two equal parts.

(2) Unplug the transformer from the line, and add the two diodes and capacitor C_1. Measure the voltage across the capacitor with a DC voltmeter or with the oscilloscope. If a scope is used, you can observe the residual AC (the "ripple") by switching the scope to the AC mode and increasing the sensitivity. Note the wave shape and frequency of the ripple, and comment on your observations.

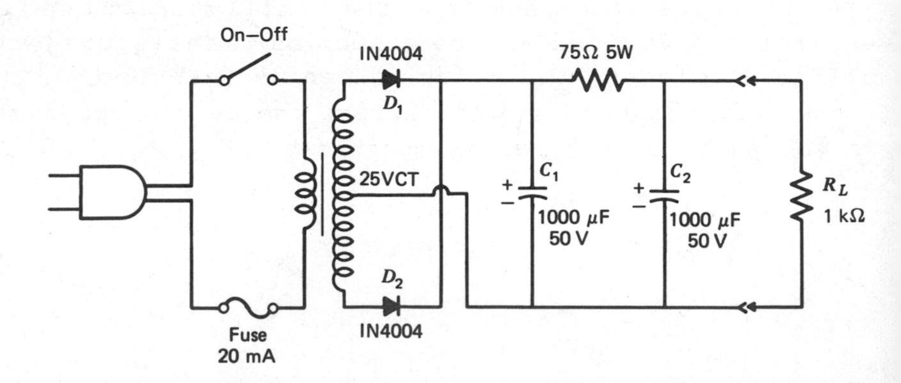

Figure Ex-17. A power supply circuit.

(3) Again disconnect the power line and complete the circuit. Measure the output voltage and ripple with and without the load R_L connected. The quality of the power supply can be described by the magnitude of the variation of the output on applying the load. How good is your power supply? Does the ripple increase or decrease on connecting the load?

(4) Remove one of the two diodes, and determine the effect on the ripple and regulation.

Additional Experiments

Determine the output impedance of your power supply by the Thévenin method. Is it desirable for the output impedance to be high or low?

EXPERIMENT 18

Constant-Current Sources

Objective: To construct current supplies for both low and high currents.

Procedure

(1) Construct the MOSFET unit illustrated in Figure Ex-18*a*. Using a milliammeter and resistor in series, determine the range of loads that can be accommodated. Calculate the voltage compliance.

(2) Measure the current range available by adjustment of the potentiometer *P*.

(3) Determine the regulation by plotting I_{out} against R_{load}. (The expected range is of the order of 1 to 6 mA.)

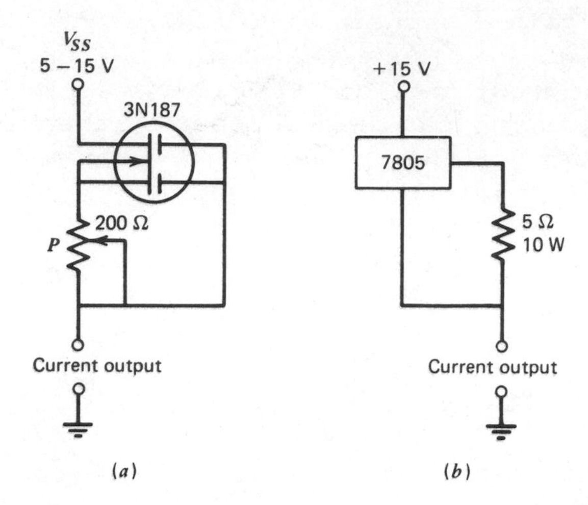

Figure Ex-18. Constant current sources using (*a*) a MOSFET transistor and (*b*) a 7805 IC.

(4) Construct the circuit at (b) and determine its sensitivity to load. Large power resistors (10 W or greater) should be used in this circuit.

EXPERIMENT 19

Frequency Doubler

Objective: To convert a square wave to twice the frequency, with synchronization.

Procedure

Construct the circuit shown in Figure Ex-19. The duration of the output pulses is dependent on the values of resistors R_1 and R_2. Determine the form of this functional dependence.

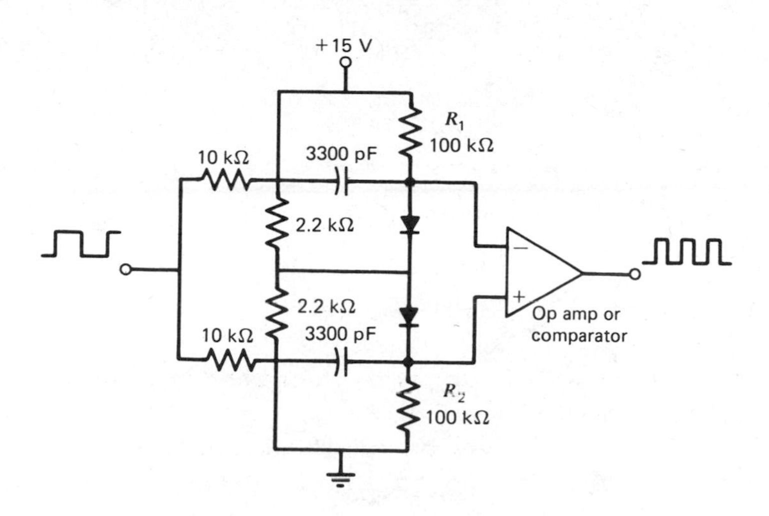

Figure Ex-19. Use of a comparator as a frequency doubler for square waves. The square-wave source can be a 555 or 8038. The diodes are 1N914, 1N4448, or equivalent.

EXPERIMENT 20

Relaxation Oscillator

Objective: To study the operation of a unijunction transistor.

Procedure

(1) Construct the circuit shown in Figure 8-24, employing a UJT of type 2N4870 or equivalent. The power supply can be +15 V, and the component values R_1 = 47 Ω, R_2 = 220 Ω, R_T = 22 kΩ, and C_T = 1 μF. Determine the frequency and output waveform.

EXPERIMENT 21

The Schmitt Trigger

Objective: To illustrate the digital and analog behavior of a Schmitt trigger.

Procedure

(1) Using a 741 op amp, construct the Schmitt trigger shown in Figure Ex-12.

(2) Study the behavior of the circuit when a slowly rising voltage is fed to E_{in}. Determine the curve of E_{out} as a function of E_{in}, with its accompanying hysteresis loop, which becomes apparent on lowering the voltage again.

(3) Study the response of the trigger with sine waves as input. Vary E_{ref}, and determine how to obtain a symmetrical output.

(4) Describe the changes that take place when frequencies approach 1 MHz.

Additional Experiments

Compare the high-frequency performance of your Schmitt trigger with a type 7413 TTL unit.

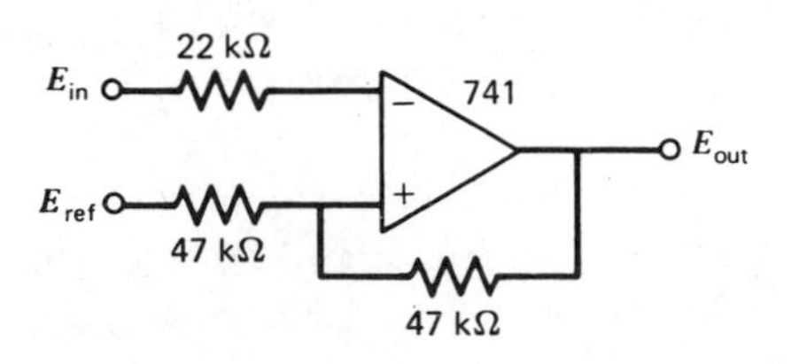

Figure Ex-21. A 741 op amp connected as a Schmitt trigger.

EXPERIMENT 22

Astable Multivibrator

Objective: To illustrate the internal operation of a multivibrator flip-flop.

Procedure

Construct the circuit shown in Figure 7-15. Start with all resistors of 47 kΩ and both capacitors 1μF. Study the effect of changing the values of the various components. The transistors can be type 2N2222.

EXPERIMENT 23

Logic Gates

Objective: To study the properties of some logic gates and establish their truth tables.

Procedure

Figure 9-8 gives the pin diagrams of several common TTL gates. Note that pin 7 is ground and pin 14 is +5 V in all diagrams.

(1) Start with a 7400 **NAND** gate, using only one of the four segments. Connect pins 7 and 14 appropriately. Connect a voltmeter between pin 3 and ground. Ground both pins 1 and 2, and record the voltmeter reading. Now connect each input pin (1 and 2) first separately and then together to the +5 V supply, and note the results. Assemble your four observations into the form of a truth table.

(2) Repeat step (1) with each of the other basic types of gates (7402, 7408, and 7432).

(3) Interconnect segments of 7408 and 7404 in the arrangement shown in Figure 9-15, and demonstrate that the resulting truth table is identical with that obtained from the 7486.

(4) The text shows several ways in which flip-flops can be synthesized from unit gates. Make up flip-flops of increasing sophistication following the diagrams in Figures 10-1, 10-2, and 10-3, and prepare truth tables for each. Can you discern any advantages of one over another?

(5) The 7451 is an example of a more complex logic element on a single chip. (Nothing should be connected to pins 11 and 12 in this IC). The two segments can be interconnected to form a particularly useful form of flip-flop known as a "data latch." Refer to the literature for the description of the 7475 latch, and reproduce its circuitry with the 7451. Establish its truth table.

EXPERIMENT 24

Flip-Flops and Counters

Objective: To illustrate *RS* flip-flops and their use in counters and frequency division.

Procedure

Light-emitting diodes (LEDs) can be used to indicate the state of a logic element in place of the voltmeter specified in the previous experiment. To avoid burnout, a resistor must be connected in series with each LED; 470 Ω is a reasonable value.

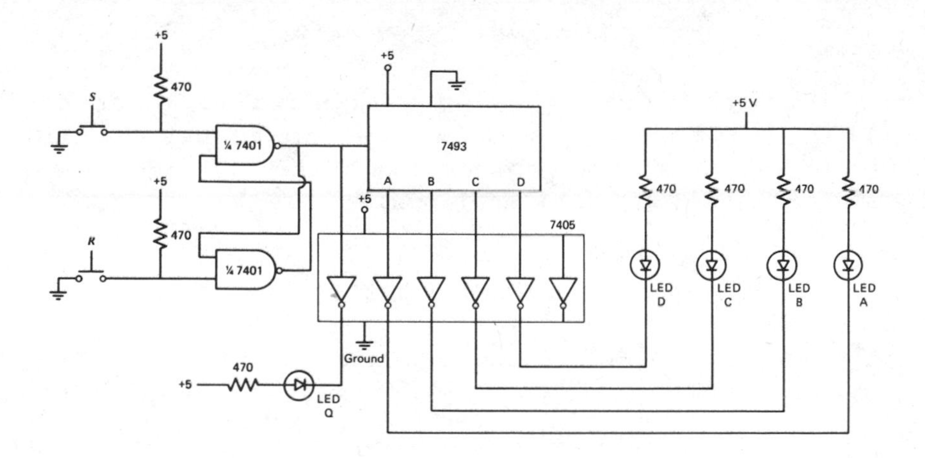

Figure Ex-24. A binary counter.

(1) Construct the circuit shown in Figure Ex-24, using the manufacturer's literature to determine the pin connections. The two **NAND** gates form an *RS* flip-flop that can be made to change state by pressing alternately the *R* and *S* momentary contact switches. When LED *Q* is on, it indicates a high state at *Q*, whereas LEDs *A*, *B*, *C*, and *D* indicate highs at consecutive outputs from the 7493 binary counter. The inverters (7405) serve simply as high current lamp drivers. (Type 7404 inverters will not give sufficient current for this; neither will the outputs of the 7493.)

(2) Test the operation of this circuit, and prepare a table showing the overall system response, comparable to that in Figure 10-3.

(3) Connect a low-frequency square wave (about 1 Hz) in place of the *RS* flip-flop, and note the sequence of lamp signals. Can you suggest some uses for such a system?

Additional Experiments

Design a system using a monostable (half of a 74121) in place of the *RS* flip-flop to respond to a single switch contact rather than two. Design a decimal counter along similar lines.

EXPERIMENT 25

A Logic Latch

Objective: Construct a latching register with manual entry.

Procedure

(1) Assemble the circuit shown in Figure 10-9, using the 7475 *D* latches as in Figure Ex-25. Both clock (enable) terminals must be actuated at the same time.

(2) Connect inputs *A*, *B*, *C*, and *D* to switches to permit selecting +5 V or ground. Note that a byte can

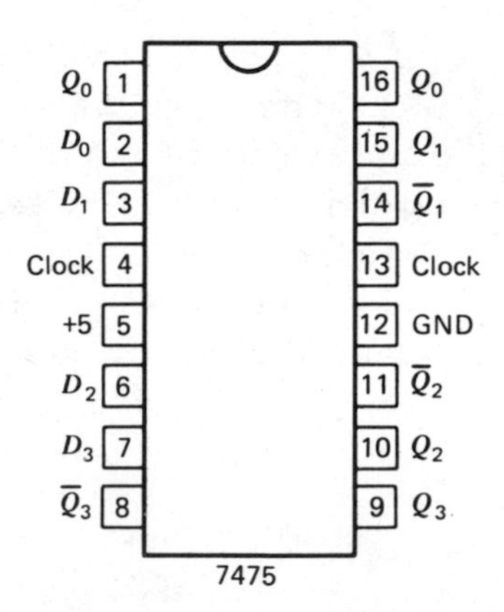

Figure Ex-25. Pin-out diagram of a 7475 data latch.

be entered by the switches without altering the lamps and then transferred instantly when the enable button is pressed.

EXPERIMENT 26

The 555 Timer

Objective: To construct a square-wave generator.

Procedure

(1) Use Figure 10-15 as a guide, and construct an oscillator to give f = 1 kHz, approximately. Use a variable resistor for R_B, and explore its effect on the waveform and frequency.

(2) Follow the oscillator by a divide-by-2 counter (7493; see Figure Ex-26). Note the symmetry of the output.

(3) Connect together pins 1 and 12 of the 7493 and

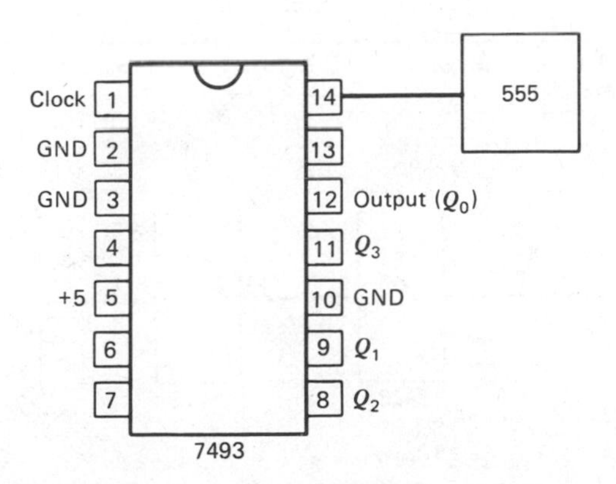

Figure Ex-26. Connections of a 555 square-wave generator with a binary counter.

examine on the oscilloscope the relationship between the outputs at Q_0, Q_1, Q_2, and Q_3.

(4) Observe the effect of connecting the pair of pins 2 and 3 to +5 V rather than ground.

EXPERIMENT 27

Counters

Objective: To generate low-frequency square waves synchronized with the line frequency.

Procedure

(1) Assemble the circuit shown in Figure Ex-27, and determine the frequency at the output (pin 11 of the 7493).

(2) Observe the waveforms at pins 8, 9, 11, and 12 of the 7492 and pins 8 and 9 of the 7493. Comment on your observations.

(3) Note the effect of disconnecting pins 2 and 3 of the 7493 and forcing them high.

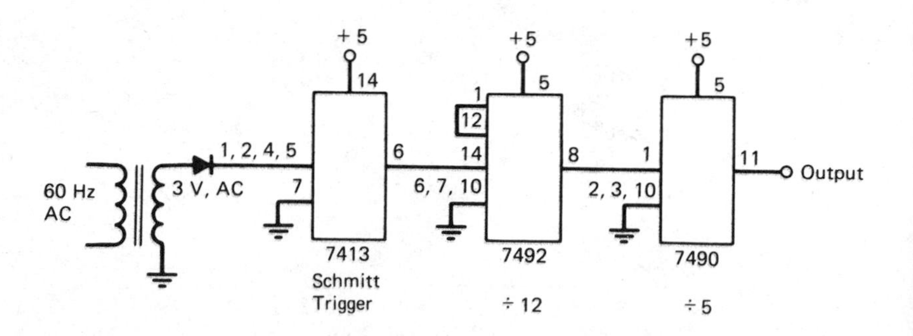

Figure Ex-27. The use of counters to generate low-frequency square waves.

EXPERIMENT 28

Analog-to-Digital Conversion

Objectives: To explore some relationships between analog and digital signals.

Procedure

The 8-bit DAC for this experiment can be the MC1408L8 (Motorola), AD7523 (Analog Devices), DAC-08 (Datel-Intersil), or the equivalent. In place of the 555, an external square-wave or pulse generator can be used if desired.

(1) Assemble the ramp generator shown in Figure Ex-28*a*, and observe its output on the oscilloscope. By adjusting the scope controls and the frequency of the 555, you should be able to see the ramp either as a series of steps or as an apparently continuous straight line.

(2) Plot your results at low frequency on a strip-chart recorder. If any nonlinearity appears, do you consider it a fault of your generator or the recorder?

Additional Experiments

Connect the circuit shown in Figure Ex-28*b*, and make similar observations. Generate the output with and without inserting the MM522 between the counter and the DAC. (The MM522 is a ROM that converts the up-down staircase into a sine wave.) What limits the frequency, at both high and low ends, of the sine or triangle wave that can be generated? How much harmonic distortion would you expect in the sine wave?

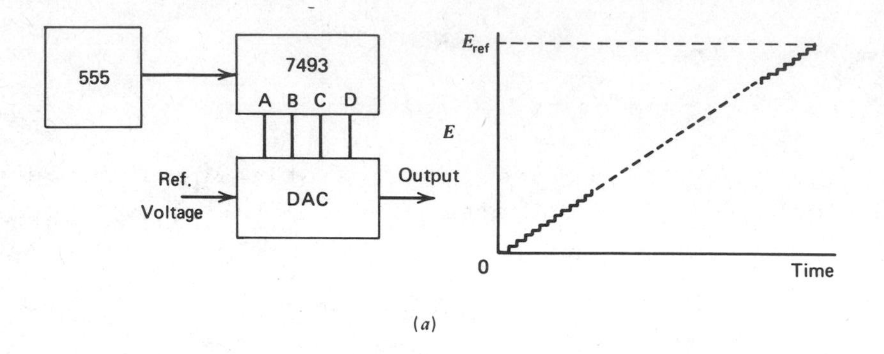

(a)

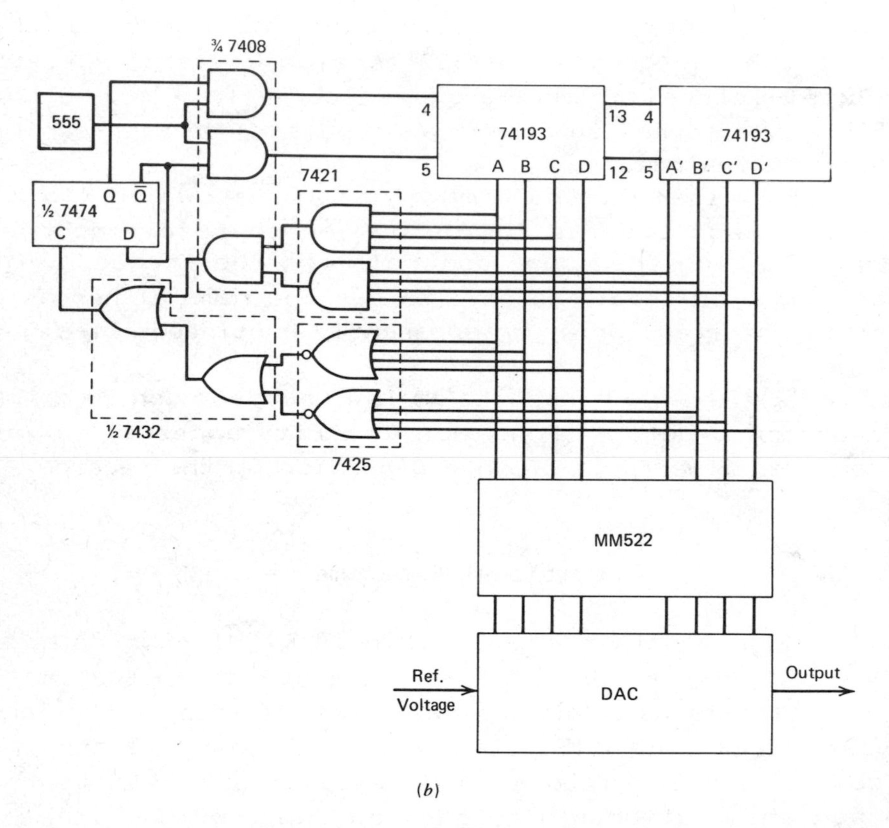

(b)

Figure Ex-28. (*a*) A DAC used as a generator of a staircase ramp. (*b*) A DAC generator of an up-down staircase or sine wave.

APPENDIX I

DECIBEL TABLE

dB	Voltage Gain	Power Gain	Voltage Loss	Power Loss
0.0	1.00	1.00	1.00	1.00
0.1	1.01	1.02	0.99	0.98
0.2	1.02	1.05	0.98	0.96
0.3	1.04	1.07	0.97	0.93
0.4	1.05	1.10	0.96	0.91
0.5	1.06	1.12	0.94	0.89
0.6	1.07	1.15	0.93	0.87
0.7	1.08	1.17	0.92	0.85
0.8	1.10	1.20	0.91	0.83
0.9	1.11	1.23	0.90	0.81
1.0	1.12	1.26	0.89	0.79
2.0	1.26	1.58	0.79	0.63
3.0	1.41	2.00	0.71	0.501
4.0	1.58	2.51	0.63	0.398
5.0	1.78	3.16	0.56	0.316
6.0	2.00	3.98	0.501	0.251
7.0	2.24	5.01	0.447	0.200
8.0	2.51	6.31	0.398	0.158
9.0	2.82	7.94	0.355	0.126
10.0	3.16	10.00	0.316	0.100
20.0	10.00	100.00	0.100	0.010

Note that the voltage ratio for 3 dB is $\sqrt{2}$, whereas that for 6 dB is 2. The corresponding power ratios are 2 and 4. If a value not listed in the table is required, one should break down the decibel number into a sum of known values, take them from the table, and multiply together the results. Thus 43.5 dB as a voltage gain can be written as 20 + 20 + 3 + 0.5, corresponding to gains of 10, 10, 1.41, and 1.06. The net gain is given by the product of these, 149.5.

APPENDIX II
TRIGONOMETRIC TABLE

Angle in				
Degrees	Radians	Sine	Cosine	Tangent
0	0	0	1	0
30	$\pi/6$	$1/2$	$\sqrt{3}/2$	$1/\sqrt{3}$
45	$\pi/4$	$\sqrt{2}/2$	$\sqrt{2}/2$	1
60	$\pi/3$	$\sqrt{3}/2$	$1/2$	$\sqrt{3}$
90	$\pi/2$	1	0	∞
120	$2\pi/3$	$\sqrt{3}/2$	$-1/2$	$-\sqrt{3}$
135	$3\pi/4$	$\sqrt{2}/2$	$-\sqrt{2}/2$	-1
150	$5\pi/6$	$1/2$	$-\sqrt{3}/2$	$-1/\sqrt{3}$
180	π	0	-1	0
$-a$	--	$-\sin a$	$\cos a$	$-\tan a$
$90 + a$	--	$\cos a$	$-\sin a$	$-\text{ctn}\ a$
$90 - a$	--	$\cos a$	$\sin a$	$\text{ctn}\ a$

APPENDIX III
CAPACITIVE IMPEDANCE

The following table gives the impedance (the capacitive reactance) of selected values of capacitance calculated from the formula $Z_C = X_C = 1/(2\pi f C)$. Values less than 0.1 Ω are not likely to be useful and thus are not listed. Actual capacitors will obey this relationship only over limited frequency ranges, because of inductive and dissipative effects.

(μF)	50 Hz	100 Hz	1 kHz	10 kHz	100 kHz	1 MHz
0.001	3.2 M	1.6 M	160. k	16. k	1.6 k	160.
0.005	640. k	320. k	32. k	3.2 k	320.	32.
0.01	320. k	160. k	16. k	1.6 k	160.	16.
0.05	64. k	32. k	3.2 k	320.	32.	3.2
0.1	32. k	16. k	1.6 k	160.	16.	1.6
0.5	6.4 k	3.2 k	320.	32.	3.2	0.32
1.0	3.2 k	1.6 k	160.	16.	1.6	0.16
5.0	640.	320.	32.	3.2	0.32	
10.0	320.	160.	16.	1.6	0.16	
50.0	64.	32.	3.2	0.32		
00.0	32.	16.	1.6	0.16		
00.0	6.4	3.2	0.32			

ll values given in ohms, except k = kilohms, M = megohms.

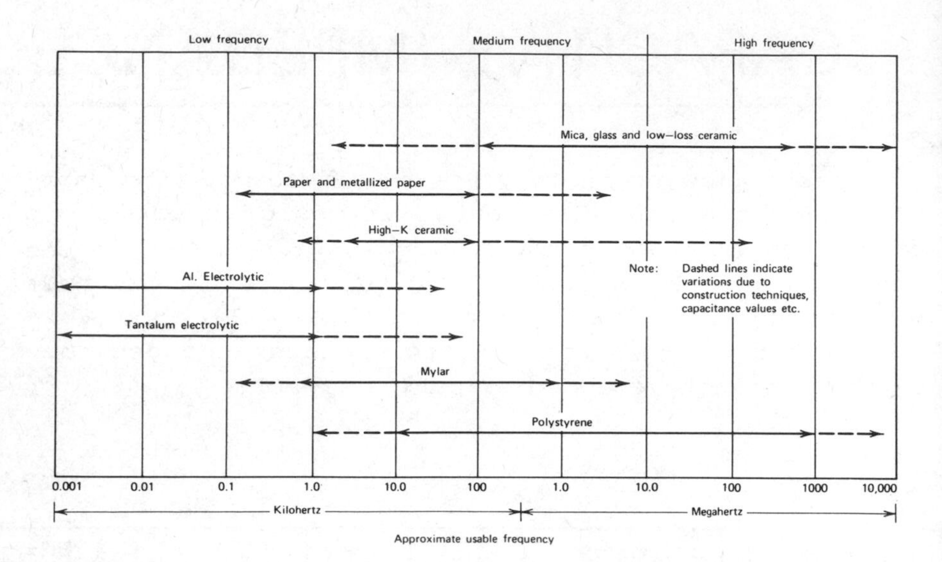

The accompanying figure* shows the approximate ranges over which various types of capacitors are useful.

* Reproduced, with permission, from H. W. Ott, *Noise reduction Techniques in Electronic Systems*, Wiley, New York, 1976; Copyright Bell Telephone Laboratories, Inc.

APPENDIX IV
LAPLACE TRANSFORMS

The Laplace transform $\pounds$ of a function $F(t)$ is defined by the equation

$$\pounds\{F(t)\} = \overline{F}(s) = \int_0^{\infty} F(t) \exp(-st)\, dt$$

The inverse transform is given by

$$F(t) = \frac{1}{2\pi j} \int_{c-j\infty}^{c+j\infty} \overline{F}(s) \exp(st)\, ds$$

The following table gives pairs of equivalent expressions for selected functions. More extensive tables are available in many reference works.

$\overline{F}(s)$	$F(t)$
$1/s$	Unit Step
$1/s^2$	t
$1/s^n$	$t^{n-1}/(n-1)!$
$1/\sqrt{s}$	$1/\sqrt{\pi t}$
$\dfrac{1}{s+a}$	$\exp(-at)$
$\dfrac{1}{s(s+a)}$	$(1/a)[1 - \exp(-at)]$
$\dfrac{1}{s(s+a)(s+b)}$	$\dfrac{1}{ab}\left(1 + \dfrac{1}{a-b}[b \exp(-at) - a \exp(-bt)]\right)$
$\dfrac{1}{(s+a)^2}$	$t \exp(-at)$
$\dfrac{1}{(s+a)^n}$	$\dfrac{t^{n-1} \exp(-at)}{(n-1)!}$

Laplace Transforms (*Continued*)

$\bar{F}(s)$	$F(t)$
$\dfrac{1}{s(s + a)^2}$	$\dfrac{1}{a^2}[1 - \exp(-at) - at \exp(-at)]$
$\dfrac{1}{s^2 + a^2}$	$\dfrac{1}{a} \sin at$
$\dfrac{1}{s(s^2 + a^2)}$	$\dfrac{1}{a^2}(1 - \cos at)$
$\dfrac{s}{s^2 + a^2}$	$\cos at$
$\dfrac{1}{(s^2 + a^2)^2}$	$\dfrac{1}{2a^3}(\sin at - at \cos at)$
$\dfrac{1}{(s + a)^2 + b^2}$	$\dfrac{1}{b} \exp(-at) \sin bt$
$\dfrac{1}{s^2 - a^2}$	$\dfrac{1}{a} \sinh at$
$\dfrac{s}{s^2 - a^2}$	$\cosh at$
$\dfrac{s}{(s + a)^2}$	$(1 - at) \exp(-at)$
$\dfrac{s}{(s + a)(s + b)}$	$\dfrac{1}{a - b}[a \exp(-at) - b \exp(-bt)]$
$\dfrac{s}{(s^2 + a^2)^2}$	$\dfrac{1}{2a} t \sin at$
$\dfrac{s + a}{s(s + b)}$	$\dfrac{a}{b} - \dfrac{a - b}{b} \exp(-bt)$
$\dfrac{s + a}{(s + b)^2}$	$[1 + (a - b)t] \exp(-bt)$
$\dfrac{s^2 - a^2}{(s^2 + a^2)^2}$	$t \cos at$

APPENDIX V
OCTAL AND HEXADECIMAL NUMBERS

In addition to the decimal and binary systems of numeration, it is often advantageous to use number systems with bases 8 and 16.

The *octal system*, based on eight, uses the following sequence:

Decimal	0	1	2	3	4	5	6	7	8	9	10	•••	16	17
Octal	0	1	2	3	4	5	6	7	10	11	12	•••	20	21

One of the principal uses of octal numbers is to organize binary numbers in more manageable form. For example, the binary number 110100111010 is difficult to identify and memorize, but arrangement in groups of 3 bits each, such as 110 100 111 010 is more acceptable. Even easier is to deal with the octal representation of the groups:

$$\begin{array}{cccc} 110 & 100 & 111 & 010 \\ 6 & 4 & 7 & 2 \end{array} \quad = 6472_8$$

where the subscripted 8 indicates that the number is in octal.

A second way in which binary numbers can be organized is in groups of four, representing values between $0000 = 0_{10}$ and $1111 = 15_{10}$. This classification parallels the division of computer memory in "nibbles" (groups of 4 bits) and bytes (groups of 8 bits). Thus a computer word consisting of two bytes might be grouped as

0000 0010 1110 1111

The 16 possible values of each group can be represented by a single digit in the hexadecimal system. Evidently this system must use digits beyond 9, namely, *A*, *B*, *C*, *D*, *E*, and *F*, to represent decimal values 10 through 15, respectively. In the hexadecimal notation, the word given in the preceding example becomes $02EF_{16}$. Numbers in hexadecimal are sometimes designated by the letters "hex" or the letter "h." Thus $02EF_{16}$ is equivalent to 02EFhex or 02EFh. The correspondance between decimal, binary, octal, and hexadecimal is shown in the following table. This should be compared with Table 9-2, which includes also the binary-coded decimal (BCD) system.

Decimal	Binary	Octal	Hexadecimal
0	0	0	0
1	1	1	1
2	10	2	2
3	11	3	3
4	100	4	4
5	101	5	5
6	110	6	6
7	111	7	7
8	1000	10	8
9	1001	11	9
10	1010	12	A
11	1011	13	B
12	1100	14	C
13	1101	15	D
14	1110	16	E
15	1111	17	F
16	10000	20	10
17	10001	21	11
••	•••	••	••
255	---	377	FF

The conversion between different number system can be done by the following sequence of divisions. Consider, for example, the conversion of 44_{10} into binary:

$$
\begin{aligned}
44/2 &= 22 + 0 \\
22/2 &= 11 + 0 \\
11/2 &= 5 + 1 \\
5/2 &= 2 + 1 \\
2/2 &= 1 + 0 \\
1/2 &= 0 + 1
\end{aligned}
$$

The binary number can be read upward in the remainder column, as 101100_2. Let us verify the result. By the positional values of binary numbers, there are $(1 \times 32) + (0 \times 16) + (1 \times 8) + (1 \times 4) + (0 \times 2) + (0 \times 1) = 44_{10}$.

APPENDIX VI
ASCII CODE

There are a number of *codes* in use for representing letters and digits in the form of binary words. The most common of these, called the *ASCII code* (American Standard Code for Information Interchange), uses 7 bits to represent the desired character, plus an eighth for internal consistency checking. For example, the letter "A" is represented by the word "01000001," in which the leading "0" is the check bit. The standard code consists of 94 characters, numbered from 32 to 126. As used in many computers, the numbers from 1 to 31 are reserved for various nonprintable control characters and those from 127 to 254, for specialized graphic symbols.

Decimal	Hex	ASCII	Decimal	Hex	ASCII
33	21h	!	54	36h	6
34	22h	"	55	37h	7
35	23h	#	56	38h	8
36	24h	$	57	39h	9
37	25h	%	58	3Ah	:
38	26h	&	59	3Bh	;
39	27h	'	60	3Ch	<
40	28h	(	61	3Dh	=
41	29h	)	62	3Eh	>
42	2Ah	*	63	3Fh	?
43	2Bh	+	64	40h	@
44	2Ch	,	65	41h	A
45	2Dh	–	66	42h	B
46	2Eh	.	67	43h	C
47	2Fh	/	68	44h	D
48	30h	0	69	45h	E
49	31h	1	70	46h	F
50	32h	2	71	47h	G
51	33h	3	72	48h	H
52	34h	4	73	49h	I
53	35h	5	74	4Ah	J

ASCII CODE (*Continued*)

Decimal	Hex	ASCII	Decimal	Hex	ASCII
75	4Bh	K	101	65h	e
76	4Ch	L	102	66h	f
77	4Dh	M	103	67h	g
78	4Eh	N	104	68h	h
79	4Fh	O	105	69h	i
80	50h	P	106	6Ah	j
81	51h	Q	107	6Bh	k
82	52h	R	108	6Ch	l
83	53h	S	109	6Dh	m
84	54h	T	110	6Eh	n
85	55h	U	111	6Fh	o
86	56h	V	112	70h	p
87	57h	W	113	71h	q
88	58h	X	114	72h	r
89	59h	Y	115	73h	s
90	5Ah	Z	116	74h	t
91	5Bh	[	117	75h	u
92	5Ch	\	118	76h	v
93	5Dh	]	119	77h	w
94	5Eh	^	120	78h	x
95	5Fh	_	121	79h	y
96	60h	`	122	7Ah	z
97	61h	a	123	7Bh	{
98	62h	b	124	7Ch	¦
99	63h	c	125	7Dh	}
100	64h	d	126	7Eh	~

APPENDIX VII
COLOR CODE FOR RESISTORS

The values of low-wattage resistors are coded by a series of colored bands near one end of the resistor body. The code is given in the following table.

Color	As Digit	As Number of Zeros
Black	0	None
Brown	1	1
Red	2	2
Orange	3	3
Yellow	4	4
Green	5	5
Blue	6	6
Violet	7	7
Gray	8	8
White	9	9

The first two bands from the end represent digits, the next indicates the number of zeros following the two digits, and the fourth (if present) is a tolerance indicator: gold signifies ±5%, silver 10%, and no band 20%. Thus a resistor banded brown-black-green-silver has a value of 1-0-00000 or 1,000,000 Ω ± 10%.

APPENDIX VIII
BIBLIOGRAPHY

Artwick, B. A., *Microcomputer Interfacing*, Prentice-Hall, Englewood Cliffs, NJ, 1980.

Barnaal, D., *Digital and Microprocessor Electronics for Scientific Application*, Breton Publishers, North Scituate, MA, 1982.

Brophy, J. J., *Basic Electronics for Scientists*, 4th ed., McGraw-Hill, New York, 1983.

Connelly, J. A. ed., *Analog Integrated Circuits*, Wiley, New York, 1975.

Evans, C. H., *Electronic Amplifiers*, Van Nostrand-Reinhold, New York, 1979.

Gothmann, W. H., *Digital Electronics--An Introduction to Theory and Practice*, Prentice-Hall, Englewood Cliffs, NJ, 1977.

Graeme, J. G., *Applications of Operational Amplifiers*, McGraw Hill, New York, 1973.

Greenfield, J. D., *Practical Digital Design Using ICs*, Wiley, New York, 1977.

Hnatek, E. R., *Applications of Linear Integrated Circuits*, Wiley-Interscience, New York, 1975.

Hnatek, E. R., *A User's Handbook of Semiconductor Memories*, Wiley-Interscience, New York, 1977.

Millman, J., *Microelectronics*, McGraw-Hill, New York, 1979.

Ott, H. W., *Noise Reduction Techniques in Electronic Systems*, Wiley, New York, 1976.

Shaklette, L. W., and Ashworth, H. A., *Using Digital and Analog Integrated Circuits*, Prentice-Hall, Englewood Cliffs, NJ, 1978.

Subbarao, W. V., *Microprocessors--Hardware, Software, and Design Applications*, Reston Publishing Company, Reston, VA, 1984.

INDEX

Devices specifically mentioned or described: